Generalist Case Management

Generalist Case Management

A Method of Human Service Delivery

Marianne Woodside
Tricia McClam
University of Tennessee, Knoxville

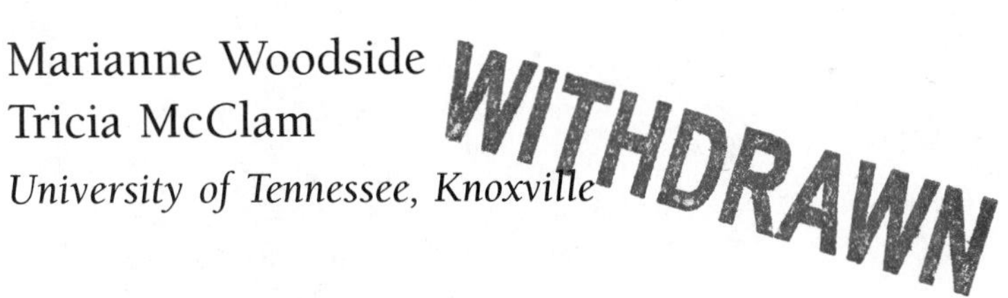

Brooks/Cole Publishing Company

I(T)P® An International Thomson Publishing Company

Pacific Grove • Albany • Belmont • Bonn • Boston • Cincinnati • Detroit • Johannesburg • London
Madrid • Melbourne • Mexico City • New York • Paris • Singapore • Tokyo • Toronto • Washington

Sponsoring Editor: *Lisa Gebo*
Marketing Team: *Jean Vevers Thompson, Deanne Brown, and Margaret Parks*
Editorial Assistant: *Susan Carlson*
Production Editor: *Laurel Jackson*
Manuscript Editor: *David Hoyt*
Permissions Editor: *Cathleen Collins Morrison*
Interior and Cover Design: *Terri Wright*
Cover Illustration: *Diana Ong/SuperStock*
Art Editor: *Lisa Torri*
Photo Editor: *Kathleen Olson*
Typesetting: *Carlisle Communications*
Cover Printing: *Phoenix Color Corporation*
Printing and Binding: *The Maple-Vail Book Mfg. Group*

For more information, contact:

BROOKS/COLE PUBLISHING COMPANY
511 Forest Lodge Road
Pacific Grove, CA 93950
USA

International Thomson Publishing Europe
Berkshire House 168–173
High Holborn
London WC1V 7AA
England

Thomas Nelson Australia
102 Dodds Street
South Melbourne, 3205
Victoria, Australia

Nelson Canada
1120 Birchmount Road
Scarborough, Ontario
Canada, M1K 5G4

International Thomson Editores
Seneca 53
Col. Polanco
11560 México, D.F., México

International Thomson Publishing GmbH
Königswinterer Strasse 418
53227 Bonn
Germany

International Thomson Publishing Asia
221 Henderson Road
#05–10 Henderson Building
Singapore 0315

International Thomson Publishing Japan
Hirakawacho Kyowa Building, 3F
2-2-1 Hirakawacho
Chiyoda-ku, Tokyo 102
Japan

Printed in the United States of America

10 9 8 7 6 5 4 3

Library of Congress Cataloging-in-Publication Data

Woodside, Marianne.
Generalist case management: a method of human service delivery / Marianne Woodside, Tricia McClam.
p. cm.
Includes bibliographical references and index.
ISBN 0-534-34897-1 (alk. paper)
1. Social case work—United States. 2. Human services—United States. I. McClam, Tricia. II. Title.
HV43.W643 1998
361.3'2—dc21 97-8770
CIP

THIS BOOK IS PRINTED ON ACID-FREE RECYCLED PAPER

About the Authors

Together, Tricia McClam and Marianne Woodside have more than 40 years of experience in human service education, as well as many years working as practitioners in education and vocational rehabilitation. Currently they are professors in the Human Service Education Program in the College of Education at the University of Tennessee, Knoxville. They are committed to research in teaching and learning in the human services and have written two other texts, *An Introduction to Human Services* and *Problem Solving in the Helping Professions.*

Contents

1 Introduction to Case Management 1

Case Management Today 3
The Process of Case Management 7
Assessment 9
Planning 18
Implementation 21
Three Components of Case Management 23
Principles and Goals of Case Management 25
Integration of Services 25
Continuity of Care 25
Equal Access to Services 26
Quality Care 26
Advocacy 27
The Whole Person 28
Client Empowerment 28
Evaluation 29
Summary 29
Chapter Review 30

2 Historical Perspectives on Case Management 33

Perspectives on Case Management 35
Case Management as a Process 35
Client Involvement 38
The Rule of the Case Manager 39
Utilization Review and Cost–Benefit Analysis 40
The History of Case Management 41
Early Organizations 42
Early Pioneers 43

The Impact of World Wars I and II and the American Red Cross 46
The Impact of Federal Legislation 48
The Impact of Managed Care 51
History of Managed Care 51
Defining Managed Care 53
Models of Managed Care 54
Expanding the Responsibilities of Case Management 57
Summary 57
Chapter Review 58

3 Models of Case Management 61

Roles in Case Management 63
Advocate 63
Coordinator 63
Broker 64
Colleague and Collaborator 64
Community Organizer 65
Consultant 65
Counselor/Therapist 65
Evaluator 66
Expediter 66
Planner 67
Problem Solver 67
Recordkeeper 68
Service Monitor and System Modifier 68
Models of Case Management 69
Role-Based Case Management 70
Organization-Based Case Management 73
Responsibility-Based Case Management 77
Case Management and the Problem-Solving Process 82
Summary 83
Chapter Review 83

The Assessment Phase of Case Management 87

Application for Services 89
The Interview 92
Evaluating the Application for Services 98

Case Assignment 101
Documentation and Report Writing 101
Process Recording and Summary Recording 102
Intake Summaries 104
Staff Notes 105
Summary 108
Chapter Review 108

5 **Effective Intake Interviewing Skills 111**

Attitudes and Characteristics of Interviewers 113
Essential Communication Skills 118
Interviewing Skills 120
Interviewing Pitfalls 135
Summary 136
Chapter Review 137

6 **Service Delivery Planning 139**

Revisiting the Assessment Phase 141
Developing a Plan for Services 143
Identifying Services 150
Information and Referral Systems 150
Setting Up a System 152
Gathering Additional Information 153
Data Collection Methods for the Case Manager 153
Interviewing 153
Testing 156
Summary 165
Chapter Review 166

7 **Building a Case File 169**

Medical Information 171
Medical Exams 172
Medical Terminology 175

Psychological Evaluation 181
Referral 182
The Process of Psychological Evaluation 184
Social History 191
Other Types of Information 197
Summary 207
Chapter Review 215

8 Service Coordination 217

Coordinating Services 219
Resource Selection 220
Making the Referral 222
Monitoring Services 226
Working with Other Professionals 229
Advocacy 231
How to Be a Good Advocate 234
Teamwork 236
Types of Teams 237
Leading a Team 238
Effective Teamwork: Barriers and Solutions 240
Supporting Others 242
Being a Role Model 242
Managing Conflict 243
Stages of Conflict and Strategies for Resolving It 244
Summary 246
Chapter Review 246

9 Working within the Organizational Context 249

Understanding the Organizational Structure 251
The Organization Plan 251
Structure of the Organization 253
The Informal Structure 258
The Organizational Climate 260
Managing Resources 262
What Exactly Is a Budget? 262
Features of a Budget 264
Sources of Revenue 265

Improving Services 267
What Is Quality? *268*
Conducting a Utilization Review *269*
Planning Quality Assurance Programs *269*
Summary 275
Chapter Review 275

10 Ethical and Legal Issues 279

Family Disagreements 281
Working with Potentially Violent Clients 283
Confidentiality 285
Duty to Warn 288
Autonomy 291
Client Preferences: One Component of Autonomy *293*
Breaking the Rules 295
Summary 298
Chapter Review 298

11 Professional Development 301

Understanding Burnout 303
Burnout and Case Management *304*
Recognizing and Preventing Burnout *306*
Developing a Personal Philosophy 306
Step One: The Principles of Case Management *307*
Step Two: The Ethics of Case Management *308*
Step Three: Personal Values of Case Management *308*
Managing Time 311
Personal Characteristics *311*
How Time Is Spent *313*
Time Management Techniques *313*
When Time Is Managed Poorly *317*
Acting Assertively 319
Defining Assertiveness *319*
How Case Managers Use Assertiveness *321*
Summary 325
Chapter Review 326

12 Case Management Today 329

Case Managers Speak 331
Knowing and Learning 331
Handling the Bureaucracy 332
Gathering Information 333
Client Problems and Client Goals 333
Documentation and Report Writing 334
The Importance of Documentation 334
Effective Writing Skills 336
Using Supervision 338
Responsibilities of the Supervisor 338
Performance Review 343
Technology and Case Management 348
Summary 356
Chapter Review 356

Glossary 360
Index 367

Preface

For us, the purpose of writing textbooks is to share with our colleagues and students what we have learned about the profession of human services during our years of teaching and working in the field. Our first text, *An Introduction to Human Services,* emerged from many years of teaching the introductory course in the Human Service Education Program at the University of Tennessee, Knoxville. The warm reception given to this work encouraged us to continue our writing, and *Generalist Case Management: A Method of Human Service Delivery* is the result.

The concept for this text began with a methods/research course that we teach together. This course has several goals: to identify necessary information about clients; to learn how to gather information; to teach quality in report writing and other types of documentation; to learn how to review and interpret reports from other professionals; to use information to assess client status; and to work with clients and colleagues to meet the goals and objectives of the helping process.

Although the course was difficult, feedback from our students was overwhelmingly positive. Many students had their first field experience immediately after, and they told us semester after semester, "I wouldn't have known how to do the work requested of me in my field placement had I not taken this class." From former students working at their first jobs, the feedback was even more encouraging. They had been able to identify what skills were necessary to work effectively in many of the jobs they were accepting: intake interviewing, writing a social history, developing and writing a case study, and documenting the progress of clients, among others.

No course, however successful, retains its usefulness if it remains static. The course in information assessment and interpretation was no exception. We have continued to fine-tune—and in some instances, change dramatically—the topics and the focus of this course. As we received feedback from our students and engaged in dialogue with practitioners, it became clear to us that case management was emerging as an effective way of working with clients to deliver services. Case management was not a new concept for us; we had both studied and practiced it in our respective fields of education and rehabilitation. But we were surprised by its prominence in the fields of education, corrections, geriatrics, health care, nursing, youth and family services, and many others.

Our interest was heightened, and we were curious about how professionals in the field were defining and using case management. We wanted to understand why the use of this method had become so widespread. To answer these questions, we studied case management in three ways. We interviewed human service case managers nationwide about the nature of their jobs, the skills they need, the challenges they face, and the clients they serve. We also studied the research literature that reported the many ways in which case management is used in the field today. Finally, we asked students, who were studying human services and working in the field, to give us examples of how they were using the concepts of case management. The result of this intensive study is *Generalist Case Management: A Method of Human Service Delivery.*

WHAT IS CASE MANAGEMENT?

Case management is a method of human service delivery that has been used in this country for many years. Early on, the actual term *case management* was not used, but many of the elements of the method were present. In the movement to deinstitutionalize people with mental illness, case management was used to support community-based service delivery. From then on, the use of case management has continued to expand, especially with populations (such as people with disabilities and the elderly) that required long-term, comprehensive services. Today, case management is an idea whose time has come. At present, there are many definitions of the term, represented by various models. The case management process has been adapted for different settings and populations, varying with client needs, the availability of professionals, and agency goals and organizational structure.

Whatever model is used for case management, it includes the dual role of coordinating and providing direct service. Professionals who perform the case management responsibility work with the client, family members, and other professionals to provide services in a creative way. Case managers spend their days gathering information about clients, conducting assessments, planning treatments, linking clients to services, monitoring treatment plans, and providing for aftercare. For these human service professionals, the ultimate goal of the case management process is to teach clients to manage their own lives within the scope of their resources and abilities.

The concept of case management is dynamic. Just as the process has changed during the last decade, it will continue to evolve during the 21st century. Many factors are likely to influence service delivery: the managed care environment, the scarcity of resources, demands for accountability, and the conservative political climate. We have defined and described case management as it is practiced today, but with an eye to the future.

OUR GOALS AND OBJECTIVES

We have written this book to serve as an introduction to the concept of case management and explain how it is used to provide human services. In addition, we expect that this text will also serve as a reference as you begin your fieldwork and then your professional career. Our in-depth approach to the study of the process will make it valuable for you to review relevant chapters when confronted with a difficulty or a dilemma in your work.

We begin by defining case management, discussing its history, and describing the models that are used. The case management process is traced from the intake interview to termination. In addition, professional issues and skills such as time management, stress management, and supervision are explored. Finally, the most up-to-date aspects of case management are discussed. In short, our goals for this text are fourfold: to define case management; to describe many of the responsibilities that case managers assume; to discuss and illustrate the many skills that case managers need; and to describe the context in which case management occurs. More critical than these goals, however, are the human service values and principles that guide them.

Human service professionals, and specifically case managers, have a commitment to working with the whole person. This concept has permeated the helping professions and guides the entire case management process. Individual clients are recognized as multifaceted human beings who have various physical, psychological, social, financial, vocational, educational, and spiritual needs.

A second guiding principle is that case managers work with clients in the context of their environments. Clients exist in a world that can support or deter their progress toward self-sufficiency, and effective case managers can help clients manage their environments.

A third value that guides the work of the case manager is helping clients build on their strengths and the strengths that they find in their environments. Ultimately, the case manager hopes that clients can learn to serve as their own case managers, needing professional support only in times of crisis or transition. This emphasis on empowering clients is a focus of the case management process from the intake interview until termination.

Finally, professionals who perform case management assume certain legal and ethical responsibilities in working with clients. Many of the traditional ethical principles of human service professionals, such as confidentiality and self-determination, are relevant to the case management process. Other principles, such as competence and family responsibilities, are also important for case managers to understand.

ORGANIZATION OF THE TEXT

The first three chapters of *Generalist Case Management: A Method of Human Service Delivery* focus on defining case management. Chapter 1 introduces the concept of case management and describes the context in which human service delivery occurs today. This chapter begins the exploration of the differences between traditional case management and what is practiced today. Vignettes illustrate the many ways in which the process is used in service delivery. Roy Roger Johnson's case illustrates the three phases of case management: assessment, planning, and implementation. We also introduce the guiding principles that undergird the work of the professional case manager.

Chapter 2 expands the definition of case management by reviewing its history. We use the history of Sam, who was institutionalized early in childhood, to illustrate how the changing definition of case management has been reflected in the care of clients. Among the main historical roots of case management are the contributions of Mary Richmond, Jane Addams, and Lillian Wald, as well as the impact of federal legislation. First-person accounts of clients in the early days, as well as excerpts from relevant legislation, enliven the history. Managed care, which has a strong influence on human service delivery today, is defined and discussed in terms of its effects on the case management process.

Chapter 3 covers the roles and responsibilities assumed by the case manager and the models used in service delivery. Vignettes and examples illustrate the many roles of the case manager and the various models of the process.

Chapters 4 through 8 describe in detail the phases of the case management process. Chapter 4 introduces the assessment phase, emphasizing the application for services, case assignment, and documentation. Discussion of this phase covers conducting an intake interview, evaluating the application for services, and reviewing the information gathered. Guidelines for documentation conclude the chapter.

Interviewing is a critical skill for the case manager. Chapter 5, Effective Intake Interviewing Skills, tells how to conduct a good interview. The focus is on attitudes and characteristics of interviewers, the skills that make them effective, the application of skills in structured interviews, and pitfalls to avoid while interviewing.

Chapter 6 introduces the second phase of case management: planning. It describes the activities and skills needed to work with clients and colleagues. It is action-oriented, covering such useful topics as developing a plan of services, identifying service providers, and gathering additional information. Tests and their appropriate uses are discussed.

Building the case file is the focus of Chapter 7. Medical, psychological, social, vocational, and educational information complete a picture of the client. This chapter describes such reports, identifies sources of information, provides examples, and illustrates how to make sense of the case file as an integrated whole.

Chapter 8 focuses on the case manager's interaction with other colleagues during the process. A discussion of service coordination explores the process, including referrals and effective communication with other professionals. Advocacy, a major responsibility of the case manager, is discussed in depth. We give several examples of the ways in which effective advocacy can help meet client needs better. This chapter also examines how to work effectively as a team member and as the team leader, as well as ways for a case manager to become a valued colleague.

Chapters 9, 10, and 11 address the context in which the case management process occurs, and introduce the professional issues case managers encounter. Chapter 9 covers the organizational structure of human service agencies. We believe that it is critical for case managers to understand the organizational context in which they work, in order to manage resources well and to provide quality services.

Chapter 10 presents legal and ethical issues confronting the case manager, including confidentiality and autonomy. Other relevant issues are working with violent clients, the duty to warn, and the question of when to break the rules. Case managers also deal constantly with issues of professional development. This chapter explores four important areas: understanding burnout, developing a personal philosophy, managing time, and acting assertively.

The final chapter describes case management today. Professionals who have undertaken case management responsibilities now are able to identify many of the challenges the future holds. As the case managers we interviewed discussed their jobs, with an eye to the future, they identified four key themes: writing reports, documenting services, using supervision, and mastering technology. This final chapter makes liberal use of the words of today's practitioners to illuminate important concepts on which case managers will focus in the future. Chapter 12, like those that precede it, is full of real-life examples and quotations from the professionals who are practicing and shaping case management today.

ACKNOWLEDGMENTS

Many people contribute to an undertaking such as this text, and we would be remiss if we failed to acknowledge them. Our colleagues in the National Organization for Human Service and the Council for Standards in Human Service Education have encouraged and supported our efforts to investigate case management—offering suggestions, reviewing materials, and sending information. Howard Harris, our colleague at Bronx Community College, deserves special thanks for inviting us to come to the Bronx and visit a number of human service agencies there. Our students continue to be a major source of experiences, assistance, and feedback—particularly Kara Fowler, Jody Butler, and Melinda Johnson, who worked with us for a semester. Ray Vaughn also has our gratitude for sharing his perspectives on case management.

We are indebted to the College of Education and the Chancellor's Office at the University of Tennessee, Knoxville, for providing financial support for the interviews that we conducted across the nation and for other costs encountered in writing this text. For computer support, word processing, reference assistance, and form reproduction, we are deeply grateful to Rhonda Green and Betsy Johnson in those offices. Their patience and expertise have been invaluable.

The individuals that we interviewed during the past three years made many contributions to this book. They shared their time, experiences, successes, and failures to enlighten us about the complexities of case management. It is their words that give this text a firm grounding in reality. Among their contributions are definitions of case management, perspectives on the components of the process, and evidence of the trends and challenges that the future holds. Most of all, we thank them for helping us understand the dynamics of the rich and varied process of case management.

Throughout our careers we have valued the review process. The comments and suggestions of Mary Jo Blazek, University of Maine at Augusta; Ann Bonner, Mt. Hood Community College; Elin J. Cormican, Mohawk Valley Community College; William Lynn McKinney, University of Rhode Island; Deborah L. Phelps, Fontbonne College; and Daniel L. Yazak, Montana State University–Billings were critical to the development of this text. As they read the printed version, all of them will be able to see how their unique contributions have helped us.

Of course, our friends at Brooks/Cole deserve our renewed thanks. Their expertise and assistance have been central to the project. Lisa Gebo, our editor, provided enthusiastic support and direction along the way. Others who deserve special mention are Deanne Brown, Susan Carlson, Carline Haga, Laurie Jackson, Cat Collins Morrison, Kathleen Olson, Margaret Parks, Kelly Shoemaker, Jean Thompson, and Lisa Torri. We would also like to thank copy editor David Hoyt and designer Terri Wright.

Last, but not least, we thank our families for their support during this effort. We have spouses who encourage our writing and support us in times of frustration as well as celebration. Their steadfastness has been superhuman.

As human services continues to grow and develop, we look forward to hearing from you. We hope you will share with us your observations and experiences with case management in the field, as well as your reactions to this text. Please send us your comments.

—Marianne Woodside
Tricia McClam

Chapter One

Introduction to Case Management

There is only one mission here and that is to see to it that persons with very complex medical needs go home and stay home safely. This agency delivers services to individuals with over 200 different diagnoses; these medical problems coupled with social, educational, financial, and other family concerns are difficult to address, and there is no single service delivery system to meet the needs of every person. We work with the individuals and their families as our clients. We coordinate and integrate an array of community services. Some clients need the services of both mental health and mental retardation. Professionals are not provided the practical experience to enable them to cross service delivery systems.

—Margaret Mikol, Sick Kids Need Involved People of New York, personal communication, May 4, 1994

People come into day care and, if it is the first point of entry for service, through our assessment we can see what services are needed. If other services are needed such as home care or long-term home health care, we will connect the client to the agency providing such service. The person may need a referral to a mental health program or clinic because their problems are more severe and he/she will need a lot more intervention or a psychiatric evaluation. When a person comes to our center, we see what their various problems are. Our mandate is to connect them to the right agencies and to make the correct or appropriate referrals. This is a medical day center . . . people coming to our center have significant medical problems that need supervision, monitoring, and coordination of care.

—Roz Jaffe, Jewish Home and Hospital for the Aged, Bronx, New York, personal communication, May 3, 1994

Intensive Case Management Program is one that is set up to service long-term clients, people who have needs that you just can't meet in two or three months. . . . We do about everything there is to do to try to aid and help with our clients. . . . Normalization is real important in mainstreaming them into the community . . . and trying to erase the stigma of mental illness . . . providing the necessary services that we feel that will assist them from daily living skills to transportation, health needs, and medications.

—Jana Berry Morgan and Paula Hudson, Helen Ross McNabb Center, Inc., Knoxville, Tennessee, personal communication, April 27, 1995

The preceding quotations are the words of case managers who are involved in the delivery of human services. This chapter introduces you to the subject and presents a model of case management that guides many helping professionals who work in human service delivery. Within each listed section of the chapter, focus your reading and study on the following objectives.

CASE MANAGEMENT TODAY

- Describe the context in which human service delivery occurs today.
- Differentiate between traditional case management and case management today.

THE PROCESS OF CASE MANAGEMENT

- List the three phases of case management.
- Identify the two activities of the assessment phase.
- Illustrate the role of data gathering in assessment and planning.
- Describe the helper's role in service coordination.

THREE COMPONENTS OF CASE MANAGEMENT

- Define case review and list its benefits.
- Support the need for documentation and report writing in case management.
- Trace the client's participation in the three phases of case management.

PRINCIPLES AND GOALS OF CASE MANAGEMENT

- List the principles and goals that guide the case management process.
- Describe how each principle influences the delivery of services.

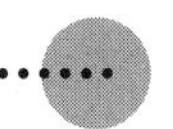

Case Management Today

The world in which case managers function is changing rapidly. Because of client tracking systems, the electronic transfer of records, dual-diagnosis clients, limited resources, and rapid communication capabilities, service delivery is vastly different from that of just a few years ago. One result is that the time between policy development and implementation is much shorter. Also, many human service agencies and organizations have chosen to limit the services they provide. More and more, case managers need skills in teamwork, networking, referral, and coordination in order to obtain the services clients need. All of this takes place in a constellation of service providers that continues to grow and change.

In addition, service delivery is affected by a political climate in which the role of government in human services comes under close scrutiny. How involved should government be in meeting human needs? What is its role? What is the proper relationship between state and federal governments? As these questions are examined and debated, case managers sometimes find themselves working under a cloud of uncertainty that influences the work they do, their professional identity, and their professional development.

The quotations that introduced this chapter share a common theme: All three situations require providing and coordinating services for the individuals served. Margaret Mikol directs an agency that provides intensive case management to children and families with complex medical problems. In this agency, the case management process begins as early as the diagnosis of a medical problem and can be terminated once clients are back home and able to manage their own care. Clients are supported by an assessment, planning, and coordination process.

The services provided by the agency where Roz Jaffee works are different. Her primary responsibility is to provide assessment and then appropriate referral for elderly clients. The contact she has with a client is usually brief, unless that client is referred to the agency's day care program. Day care services focus on a daily plan for individuals, based on their needs, strengths, and interests.

Paula Hudson and Jana Berry Morgan work in an agency that provides long-term managed care for persons with mental illness. Rarely do they close a case. People with severe mental illness who reside in the community require service coordination that is long-term, closely monitored, and supportive. The agency's commitment to these clients is to assess their needs periodically and adjust plans and provide services accordingly. Often this agency is the only lifeline for these adults.

These diverse examples illustrate service delivery today. As you can see, it varies from agency to agency, from helper to helper, and from client to client. Many of us think about service delivery in terms of **case management,** a term that includes all the activities involved in moving an individual through the service delivery process from intake to closure. Often confused with case management is the broader term **caseload management**, which includes case management. Caseload management encompasses the knowledge, skills, and activities involved in managing an entire caseload. This text focuses on providing case management services to an individual or a group of individuals rather than the problems of juggling a number of cases.

To define case management, it is helpful to look at the ways in which case management was traditionally regarded. According to Lourie (1978), case management is a necessary component of service delivery because clients need assistance with access, especially when treatment involves multiple services. He suggests that the case management process of linking clients to services begins with assessment and continues through intervention. In the 1980s, there was a shift in the focus of case management. Many professionals and clients objected to the use of the word *manage* because its definition is "to control and direct; to handle; to treat" (*Webster's New Collegiate Dictionary,* 1973, p. 697). This language did not seem to reflect a commitment to client involvement or empowerment. Terms such as *service coordination* and *care coordination* were considered to indicate better these new goals of case management. According to Betz (1991), the term *service coordination* more accurately represents the primary work of the case management process—linking the client to services and monitoring progress. Jackson, Finkler, and Robinson (1992) describe the development of the term *care coordination* during their work with Project Continuity, which facilitated care for infants and toddlers who required repeated hospitalization and who qualified for intervention under Public Law 94-457, the Individuals with Disabilities Act of 1986.

> Over the course of this project, the term *care coordination* evolved from what is popularly described as case management. Staff expressed dissatisfaction with the case management term because they did not feel families should be viewed as cases needing to be managed. Therefore, the project changed the description to care coordinator, which reflects the role as coordinator of care services for the child and family. (p. 224)

Another common term is *human development liaison specialist.* Wheeley (1981) proposed this title to reflect the responsibility of the professional to work with the service system in meeting the needs of the client. In her definition, the human development liaison specialist (case manager) helps other human service professionals assess needs, identify resources, and assist those participating in the problem-solving process.

We have conducted numerous interviews with service providers who are doing case management, and some indicate a preference for terms other than *case management* and *case manager* in describing their jobs and job titles (McClam & Woodside, 1994). Three primary objections to these terms surfaced. One is that the practitioners find it objectionable to think of clients as "cases." A second relates to the resentment clients may feel at "being managed."

Third, these helpers believe that they do more than case management. Also, many of the helpers interviewed did refer to themselves as case managers, but not necessarily in the traditional sense of the term.

What has emerged is a broader perspective on service delivery, one that encompasses traditional case management. In some situations, it includes case management with a new focus. The human service literature also supports this view. According to Austin (1993), "case management is an idea whose time has come" (p. 451); it is omnipresent in the provision of human services. Case management is defined and mandated through federal legislation, has become part of the services offered by insurance companies, and is now accepted by helping professionals as a way to serve long-term clients who have multiple problems. Clearly, case management includes the dual role of coordinating and providing direct service. It is "a creative and collaborative process, involving skill in assessment, consulting, teaching, modeling and advocacy that aim to enhance the optimum social functioning of the client served" (Sullivan, Wolk, & Hartmann, 1992, p. 198). The goal of **case managers** is to teach those who need assistance to manage their own lives but to support them when expertise is needed or a crisis occurs. These professionals gather information, make assessments, and monitor services. They find themselves working with other professionals, arranging for services from other agencies, serving as advocates for their clients, and monitoring resource allocation and quality assurance. They also provide direct services, describing their responsibilities broadly as doing whatever is necessary to help the client.

The evidence is clear that service delivery is changing. The following examples illustrate situations in which new concepts of case management have emerged.

Myron Blackman has noticed that his wife, June, has been behaving strangely this past month. She rarely goes out anymore and doesn't appear to be in touch with her friends. He became truly alarmed yesterday when he arrived home to find all the drapes and blinds closed, the house dark, and June cowering in the hallway. She explained that someone had been watching her through the windows all day. Myron calmed her down and stayed by her side that evening. The next morning, he called their family physician, who referred him to a psychiatrist. Myron called his managed care organization's 800 number to get authorization for a visit. He spoke with Judy Blum, a case manager. After explaining the problem, Myron received a short list of approved providers for a single consultation. He was instructed that if more sessions were needed, the therapist would have to get prior approval. Ms. Blum suggested that a prescription medication for anxiety would probably take care of the situation.

Three youths are accused of the satanic killing of a fourth, an 18-year-old female who was attending a training program with them. The three who are charged are all 16 or 17 years of age. Media coverage of the murder has been extensive. The judge, who has experience with high-profile cases, wants everything done "by the book" so that there will be no grounds for appeal of his eventual ruling. Before the first defendant's preliminary hearing, the judge has ordered both physical and psychiatric examinations. In addition, the judge wants a social history and any available education records for his review.

Mr. Harris is a case manager for a new "wraparound" agency that provides services to small children. He works as a liaison between managed care organizations and families to plan treatment, locate services, negotiate costs, and maintain the provision of services. Mr. Harris has a unique position in his organization, in that he also provides counseling services to families and to the children for whom he coordinates services.

In these examples, the professionals are managing cases, but in very different ways. The first case manager described is an employee of a managed care organization. Her primary responsibilities are member needs, available services, and cost containment. For her, case management entails listening to a problem, determining whether services are needed, and (if so) authorizing those services. All of this is usually done by phone. She has never met a client face to face. In the second case, the judge is likewise not an actual provider of services. Rather, he identifies what information is needed to complete the court's investigation and then orders the appropriate services. Others will coordinate and provide the services. In the third case, Mr. Harris of the "wraparound" agency both coordinates services and provides direct service himself. His responsibilities as case manager are comprehensive and continue until families are able to function as their own case managers.

These are only three examples of the changes that are occurring in the delivery of case management services. The diversity of professionals performing case management responsibilities is reflected in the many job titles involved: case manager, service coordinator, counselor, social worker, service provider, care coordinator, caseworker, and liaison worker. In some cases, these professionals provide services themselves; in others, they coordinate services or manage them. Increasingly, they are assuming new responsibilities such as cost containment and budget management. There is little agreement about what to call those they serve, but most frequently they talk about "clients," "individuals," or "participants."

The diversity of job titles, the range of individuals and groups served, and the variety of job responsibilities are all indications that service delivery is changing. This text explores case management as a complex, evolving, and diverse process. You will review traditional case management, learn about the new ways in which case management is being applied, and explore the new responsibilities given to helpers.

One of the important ways of learning about case management is through the voices of helping professionals themselves, as in the many concrete examples in this book. As you read, note their different job titles, responsibilities, service delivery methods, and terminology. The examples that illustrate concepts and principles generally use the terminology of the particular setting involved. When a case or example does not define the terminology, the term *case management* will be used to mean the responsibilities of both service provision (e.g., counseling) and service coordination (e.g., arranging for services from others). The term also refers to the management skills needed to move a case from intake to closure. In referring to the service provider, the term *case manager* will mean the professional who performs the responsibilities of case management.

The section that follows describes the process of case management and its three phases. The case of Roy Roger Johnson illustrates each phase.

The Process of Case Management

The three phases of case management are assessment, planning, and implementation. (See Figure 1.1.) Human service delivery has become increasingly complex in terms of the number of organizations involved, government regulations, policy guidelines, accountability, and clients with multiple problems. Therefore, the case manager needs an extensive repertoire of knowledge, skills, techniques, and strategies.

Let's see how these phases occur in three different settings. Thom Prassa is a case manager at the East Tennessee Community Health Agency, which has initial responsibility for all children who come through the juvenile court system. He spends much of his time making assessments of young people who are transferred to correctional facilities in the state and the community. For him, assessment is complex and multifaceted. He describes it this way:

> I will gather every bit of information there is about a child: school records, medical records, prior psychological evaluations. I might arrange for a psychological evaluation or interview a guidance counselor and parents. Do home visits. If the child is referred to us after coming into custody, then one of my main responsibilities is to get that child a physical. (Personal communication, July 10, 1994)

Yolanda Vega, CSW, Director of Agency Services at Casita Maria Settlement House in the Bronx, describes the process of planning how her staff will provide services to clients.

> It is very important to take one step at a time. Narcotics Anonymous and Alcoholics Anonymous talk about steps. Our clients want to do so

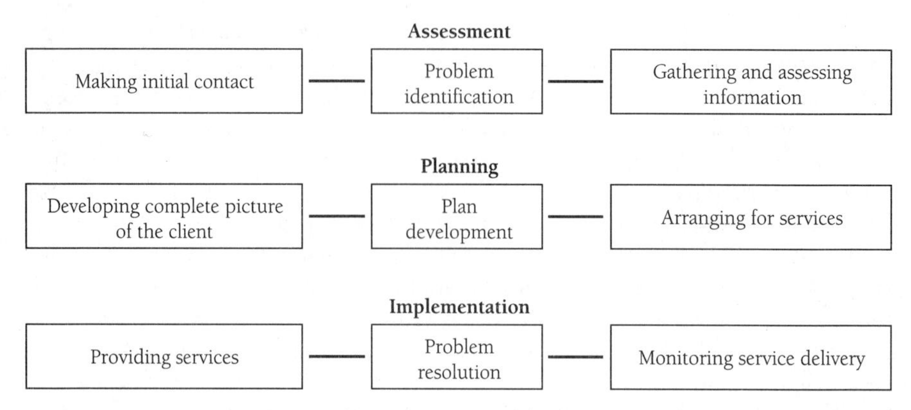

Figure 1.1 The process of case management

> much all at once. You have to talk with them and convince them that the right way to go is one day at a time. You do not try to battle all these demons at once. Sometimes the worker might have high expectations and the client is really in a rush to get there. But oftentimes that does not work. (Personal communication, May 4, 1994)

Finally, Leslie Badaines, a behavior specialist at a state school for the students who are hearing-impaired, describes the third phase of service delivery—implementation. Since students enter at an early age and remain until they graduate from high school, the responsibilities of implementation and monitoring extend over years of involvement with an individual.

> We've got our kids all the time. They do not leave here until they graduate. So you are always going to be doing some ongoing kind of monitoring one way or the other. This is one of the biggest factors in a setting like ours. (Personal communication, October 27, 1993)

As you can see, the responsibilities at each phase vary, depending on the setting and the case manager's job description. The important thing to understand is that the three phases represent the *flow* of case management rather than rigidly defined steps to successful case closure. An activity that occurs in the first phase (for example, the information gathering that Thom Prassa does) may also appear in the second or third phases of Yolanda Vega's planning and Leslie Badaines's monitoring. Other key components in effective case management appear throughout the process, including case review, report writing and documentation, and client participation. The case of Roy Roger Johnson will illustrate how this happens.

Roy Roger Johnson is a real person, but his name and other identifying information have been changed. The case as presented here is an accurate account of Roy's experience with the human service delivery system and the case management process. The case exemplifies the three phases of case management. The agency that served Roy uses the terms *counselor* and *client*. The following background information will help you follow his case through assessment, planning, and implementation.

Roy referred himself for services after suffering a back injury at work. He was 29 years old and had been employed for five years as a plumber's assistant; he hurt his back lifting plumbing materials. After back surgery, he wanted help finding work. Although he had received a settlement, he knew that the money would not last long, especially since he had contracted to have a house built. He heard about the agency from a friend who knew someone who had received services there and was now working. The agency helps people with disabilities that limit the kind of work they can do. An important consideration in accepting a person for services at the agency is determining whether services will enable that person to return to work. Roy's case was opened in 1990; we will follow it to closure.

Assessment

The **assessment** phase of case management involves initial contact with an applicant as well as gathering and assessing information. These two activities focus on evaluating the need or request for services and determining eligibility for services. Until eligibility is established, the individual is considered an applicant. When eligibility criteria have been met and the individual accepted for services, he or she becomes a client.

THE INITIAL CONTACT

The initial contact is the starting point for gathering and assessing information about the applicant, so as to establish eligibility and evaluate the need for services. In most organizations, the data gathered during the initial contact is basic and demographic: age, marital status, educational level, employment information, and the like. Other information may be obtained to provide detail about aspects of the client's life—for example, medical evaluations, social histories, educational reports, and references from employers.

Roy was self-referred to the agency. He initiated contact by telephoning for an appointment. Fortunately, a counselor was able to see him that week, so he made an appointment for May 24 at 10:30. He was sent a brochure about the agency and a confirmation of his appointment. When he arrived at the agency, Roy completed an application for services. (See Figure 1.2.) He was able to complete it without too much trouble, although he wasn't sure how to answer the question about where he had heard about the agency. He didn't know the name of his friend's friend. The receptionist

APPLICATION FOR SERVICES

Part 1

A. Office No. ☐☐ B. Counselor No. ☐☐☐ C. Client SSN 123-45-6789

D. Review Date ☐☐-☐☐-☐☐ E. Referral Date ☐☐-☐☐-☐☐

F. Name Johnson, (Last) Roy (First) Roger (Middle)

G. Address Rt. #51 (Street) Centerville (City) ______ (County) ☐☐ TN (State) 99999 (Zip)

H. Phone No. 111-589-7104 Directions to home: ______

I. Birthdate 07-16-60 Age 29 Place of birth: Michigan East Lansing

J. Referral Source ☐☐ self

K. Disability ☐☐☐ bad back L. Sex M M-Male F-Female

L. Cause of Disability: accident Age at beginning of disability: 28

M. How does disability limit activities? walking, standing, pushing, pulling, lifting

N. Other physical or mental problems: none

O. Have you previously received Services? Yes____ No ✓ State______ Date______

Part 2

A. Race 1 1-White 2-Black 3-American Indian or Alaskan Native 4-Asian or Pacific Islander

B. Highest grade of school completed 12 (MR=99) Year____ Name & Address: ______

Other Training: ______

C. Martial status 5 1-M,2-W,3-Div,4-Sep,5-NM D. No. of Dep ☐☐ E. Hisp. Origin: Yes ☐ No ☑ F. Vet: Yes ☐ No ☑

G. Wk status ☐ H. Wkly Earn ☐☐☐ I. Hrs Worked ☐☐ J. Primary source of support ☐ K. Institution ☐☐

L. Public Assistance/Support:

SSI-Aged	Yes ☐	No ☐	Amount____	SSDI	Yes ☐	No ☑	Amount____
SSI-Blind	Yes ☐	No ☐	Amount____	VA Disability	Yes ☐	No ☑	Amount____
SSI-Disabled	Yes ☐	No ☐	Amount____	Other Disability	Yes ☐	No ☑	Amount____
AFDC	Yes ☐	No ☐	Amount____	Other PA/PS	Yes ☐	No ☑	Amount____
Gen. Ass't	Yes ☐	No ☐	Amount____	Total monthly amount (Nearest $) ☐☐☐ (none 000)			

M. Medical Insurance Coverage: Yes ☐ No ☐ Name & Type Coverage: ______

______ Coverage ID No. ______

N. Availability of Medical Insurance through Client's Employment for a Salary or Wages at Application: ☐
0 - Insurance not available 1 - Insurance is available 2 - Client not working at application

Figure 1.2 Application for services

helpfully told him to write in "self-referral." She suggested that he leave any questions blank if he wasn't sure about the response. She also asked him not to sign the application until he had met with a counselor. She stated that each counselor liked to explain the paragraph at the end of the application in order to make sure that applicants understood the implications of applying for services and the conditions that apply to the release of any client information.

Name & Address of Family Physician: Dr. Alderman

Name & Address of Physicians and Dates Seen for Disability: Dr Alderman 06/16/84

Date Last Hospitalized: May 89 Hospital: Centerville Reason: Surgery

Do you wear any type artificial appliance? Yes_____ No ✓ Type:_____

Family Members Living in Home:

NAME	AGE	RELATION	EDUCATION	JOB	MONTHLY WAGE
Separate	family	entity			

Past Work Record: [LIST LAST JOB FIRST]

EMPLOYER	ADDRESS	JOB TITLE	DATES	WEEKLY WAGE	WHY DID YOU LEAVE THIS JOB?
				$	
Memorial Hosp.	Centerville	Plumber	2 mos. '87	$ 7.50/hr.	hurt on job
Rock City Mechanical				$	
				$	
	Construction work all			life	
				$	

Type or Area of Vocational Interest: undecided

List 2 persons (Other than listed in home) who would always know your address:

Name: Terry Jones Address:_____ Phone: 987-6543

Name: Mae Johnson Address:_____ Phone: 123-4567

APPLICATION FOR REHABILITATION SERVICES*

I hereby make application to receive services I may be eligible to receive. I understand that the information contained in my record will not be disclosed, other than in the administration of the program unless written consent is obtained from me or my parent/guardian. Also, I hereby request and authorize any person(s), agency or institution to release to the agency any medical, social, psychological, vocational and/or financial information they may have or may receive, pertaining to me.

Signature of Client Roy R. Johnson Date ☐☐-☐☐-☐☐

Signature of Parent/Guardian (if required)_____ Date_____

Signature of Counselor Tom Chapman Date 5/24/90

*To be explained at the time of initial interview

Review Date ☐☐-☐☐-☐☐

Figure 1.2 *(continued)*

Roy had brought a copy of a letter prepared by his orthopedic surgeon, Dr. Alderman, for his attorney a year earlier. (See Figure 1.3.) Dr. Alderman had written that, in his opinion, Roy would be left with a disability of 10% as a result of the injury. Dr. Alderman was also careful to clarify that Roy's condition did not reflect a preexisting disability even though he had suffered back problems previously. Tom Chapman, the counselor who saw Roy, made a copy of the letter and returned Roy's copy to him.

ORTHOPEDIC ASSOCIATES

200 W. MAIN STREET
DOUGLAS, TN 37210-2915

June 16, 1989

Mr. Jefferson Maupin
Attorney at Law
215 Fourth Street
Douglas, TN 37210

Re: Roy Roger Johnson

Dear Mr. Maupin:

HISTORY: Mr. Roy Johnson presented to my office on April 12, 1989, with a history of an injury to the lumbar spine which occurred at work on March 15, 1989. He reported that he was lifting an object weighing approximately 60 lb. at the time of onset. Prior to that event, he had not experienced significant low back or lower extremity pain for the preceding four years. The patient does have a significant history, including previous lumbar spine surgery. The surgical procedure was performed in July, 1985, and included a bilateral L4 laminectomy and diskectomy with a left L5 foraminotomy. When the patient presented to my office on this occasion, he complained of right lower extremity pain rather than left lower extremity pain. My initial disposition was to refer the patient to Physical Therapy and ask him not to work for two weeks.

The patient returned to my office on April 25, 1989, at which time he continued to complain of severe low back and right lower extremity pain. On that occasion, an MRI study was ordered which revealed evidence of a herniated disc at the L4 level primarily on the right consistent with the patient's right lower extremity pain.

Figure 1.3 Dr. Alderman's letter

During the initial contact, the case manager determines who the applicant is, establishes a relationship, and takes care of such routine matters as filling out the initial intake form. An important part of getting to know the applicant is learning about the individual's previous experiences with helping, his or her perception of the presenting problem, the referral source, and the applicant's expectations. As these matters are discussed, the case manager uses appropriate verbal and nonverbal communication skills to establish rapport with the applicant. (These skills

Mr. Johnson was admitted to the hospital for definitive surgery and, on May 4, 1989, he underwent a right L4 diskectomy with the operating microscope. The procedure was more difficult because of previous surgical scar tissue. At surgery, the patient had several free extruded disc fragments consistent with acute lumbar disc herniation.

Since surgery, the patient has returned to the office on several occasions for routine follow-up. He has also been attending Physical Therapy for routine postoperative physical therapy program.

DISCUSSION: This patient sustained an acute lumbar disc herniation on March 15, 1989, as evidenced by acute onset of back and leg pain documented by MRI study and by positive surgical findings. Because the patient has a history of significant previous lumbar disc disease, it is my opinion that he will undergo a gradual recovery and that he will be left with a disability to the body as a whole as a result of the injury described above of 10 percent to the body as a whole. This disability rating does not reflect his preexisting disability. The disability rating is based on anatomical findings and is in accordance with the AMA guidelines.

Sincerely,

MF Alderman

Marvin F. Alderman, M.D.

MFA:bj

Figure 1.3 *(continued)*

will be discussed in Chapter 5.) Skillful use of interviewing techniques facilitates the gathering of information and puts the applicant at ease. By providing information about routine matters, the case manager demystifies the process for the applicant and makes him or her more comfortable in the agency setting. Some of the routine matters addressed during the initial meeting are completing forms, gathering insurance information, outlining the purpose and services of the agency, giving assurances of confidentiality, and obtaining information releases.

Counselor's Page

NAME: Roy R. Johnson DATE: 5/25/90

Mr. Johnson is a 29-year-old referral who has an orthopedic back problem and brought with him a doctor's report from the first accident as well as the second. The client has limitations in most of all his daily living activities. The client has just settled with unemployment for $40,000 and is presently building a home which will deplete this very soon. The client has a 12th-grade education and stated he had been through 2 college quarters. The client was pleasant and well-mannered and answered the questions without any problems.

The rules, regulations and time limits were explained and understood by the client as well as the order of selection. The client is seeking possible training, but really is undecided about what he can and cannot do. He would meet the economic guidelines for certain services at this time and, if the statement is correct, all services once his settlement is depleted.

The client was given a functional limitation sheet for Dr. Alderman which he is to return to this office. We will sponsor a general medical and psychological with Barbara Hillman. Possibly we will put him in a vocational evaluation at the TVTC. We will place this case in status 02 as of this date.

TC:bj

Figure 1.4 Counselor's Page

Documentation records the initial contact. In the agency Roy went to, case managers fill out a Counselor's Page (Figure 1.4.), which describes the initial meeting, and a Client Master Record (Figure 1.5.). The Client Master Record provides basic information about the client, his or her sources of support, and his or her employment. Its format was designed so that data could easily be entered into the computer, thus simplifying the agency's recordkeeping. At this point, Roy was still considered an applicant for services in accordance with agency guidelines.

CLIENT MASTER RECORD

APPLICANT STATUS 02

A. Office No. ☐☐

B. Counselor No. 0 5 0

C. Client No. 1 2 3 ■ 4 5 ■ 6 7 8 9

D. Client Name Johnson, Roy R.
LAST FIRST MIDDLE

E. Review Date: ☐☐■☐☐■☐☐

F. Date of: Status 02 0 2 ■ 1 3 ■ 9 0

G. Race
(1) White
2. Black
3. American Indian or Alaskan Native
4. Asian or Pacific Islander

H. Highest Grade Of School Completed 1 2
(M.R. = 99)

I. Marital Status
1. Married
2. Widowed
3. Divorced
4. Separated
(5) Never married

J. Number of Dependents 0 0

K. Hispanic Origin Yes ☐ No ☒

L. Veteran Yes ☐ No ☒

M. Occupation (Title) none

Occupation Code 0 0 0 0 0 0

N. Weekly Earnings (Nearest $) $☐☐☐ None ☒ 000

O. Primary Source of Support
A. Current earnings, interest, dividends, rent
B. Family and friends
C. Private relief agency
D. Public assistance, at least partly with Federal funds
E. Public assistance, without Federal funds (General Assistance only)
F. Public institution-tax supported
G. Workmen's compensation
H. Social Security Disability Insurance benefits
J. All other public sources
(K.) Annuity or other non-disability insurance benefits (private insurance)
L. All other sources of support

P. Work Status
1. Wage or salaried worker-competitive labor market
2. Wage or salaried worker-sheltered workshop
3. Self-employed-except state agency managed business
4. State agency-managed business enterprise
5. Homemaker
6. Unpaid family worker
7. Not working-student
(8.) Not working-other
9. Trainee or worker (non-competitive labor marker)

Q. Hours Worked 0 0

R. Institution
(00) Not in institution
01 - Public Mental Hospital
02 - Private Mental Hospital
03 - Psychiatric inpatient unit of General Hospital
04 - Community Mental Health Center-inpatient
05 - Public institution for the mentally retarded
06 - Private institution for the mentally retarded
07 - Alcoholism treatment center
08 - Drug abuse treatment center
09 - School and other institution for the blind
10 - School and other institution for the deaf
11 - General Hospital
12 - Hospital or specialized facility for chronic illness
13 - Institution for aged
14 - Halfway House
15 - Correctional Institution - adult
16 - Correctional Institution - juvenile
17 - Other institutions and special living arrangements including group home

S. Public Assistance/Public Support
(0) Not on public assistance
1 - SSI - aged
2 - SSI - blind
3 - SSI-disabled
4 - AFDC
5 - GA only
6 - SSDI
7 - Veterans' disability benefits
8 - Other disability benefits
9 - All other PA/PS payments

T. Public Assistance Monthly Amount
(Nearest $) $☐☐☐ None ☒ 000

U. Weeks Unemployed 2 7

Tom Chapman 6/25/90
COUNSELOR'S SIGNATURE DATE

Figure 1.5 Client Master Record

Although Dr. Alderman's letter provided helpful information about Roy's presenting problem, agency guidelines stated that all applicants must have a physical examination by a physician on the agency's approved list. Mr. Chapman also felt that a psychological evaluation would provide important information about Roy's mental capabilities. He discussed both of these with Roy, who was eager to get started. As Roy prepared to leave, Mr. Chapman explained that it would take time to process the forms and review his

July 20, 1990
Roy R. Johnson
Rt. 51
Centerville, TN 15971

Dear Mr. Johnson:

We have scheduled the following appointment(s) for you:

We have authorized a general physical for you with Dr. Jones, Suite 201, Physicians' Office Building, 172 Lake Road. Please call 589-2111 to schedule the appointment. We have also authorized a psychological evaluation with Barbara Hillman. She will call you to schedule an appointment.

Please make every effort to keep the appointment(s) which we have scheduled. If, however, you will be unable to keep the appointment(s) please contact this office *prior* to the date of the appointment. Our phone number is 596-5120.

Your cooperation is appreciated. If you have any questions, please feel free to contact us.

Sincerely,

Tom Chapman

Rehabilitation Counselor

TC:bj

Figure 1.6 Tom Chapman's memo

application for services. He would be in touch with Roy very soon. On July 20, he received a memo from Tom Chapman, explaining the next steps. (See Figure 1.6.)

GATHERING AND ASSESSING INFORMATION

If the applicant is accepted for services, the client and the case manager will become partners in reaching the goals that are established. Therefore, as they work through the initial information gathering and routine agency matters, it is important that they identify and clarify their respective roles, as well as their expectations for each other and the agency. Client participation and service coordination are critical components in the success of the process, beginning during the initial contact. The case manager must make clear that the client is to be involved in all phases of the process. A skillful case manager makes sure that client involvement begins during the initial meeting.

In Roy's case, the counselor reviewed the application with him. There were some blanks on the application, and they completed them together. Roy had not been sure how to respond to the questions about primary source of support and members of his household. As Roy elaborated on his family situation, the counselor completed these items. Roy felt positive about his interactions with Tom Chapman because Tom listened to what he said, accepted his explanations, and showed insight, empathy, and good humor.

In gathering data, the case manager must determine what types of information are needed to establish eligibility and to evaluate the need for services. Once the types of information are identified, the case manager decides on appropriate sources of information and data collection methods. His or her next task is making sense of the information that has been gathered. In these tasks, assessment is involved: The case manager addresses the relevance and validity of data and pieces together information about problem identification, eligibility for services, plan development, service provision, and evaluation.

Client participation continues to play an important role throughout the information-gathering and assessment activities. In many cases, the client is the primary source of information, giving historical data, perceptions about the presenting problem, and desired outcomes. The client also participates as an evaluator of information, agreeing with or challenging information from other sources.

The counselor needed other information before a certification of eligibility could be written. In addition to Dr. Alderman's letter, a general medical examination, and a psychological evaluation, the counselor requested a period of vocational evaluation at a regional center that assesses people's vocational capabilities, interests, and aptitudes. Tom Chapman had worked with all these professionals before, so he followed up the written reports he received with further conversations and consultations. Following a two-week period at the vocational center, the evaluators met with Roy and Mr. Chapman to discuss his performance and make recommendations for vocational objectives. A report was completed in early November. At that time, Mr. Chapman and Roy met several times to review information, identify possibilities, and discuss the choices available to Roy. Mr. Chapman's knowledge of career counseling served him well as he and Roy discussed the future. Unfortunately, an unforeseen complication occurred, delaying the delivery of services. Tom Chapman changed districts, and another counselor, Susan Fields, assumed his caseload. Also, Roy moved to another town to attend college. Although he was still in the same state, Roy was now about 200 miles from his counselor. While Roy was attending school in January, Ms. Fields completed a certificate of eligibility for him. (See Figure 1.7.) This meant that he was accepted as a client of the agency and could now receive services. In May, his case was transferred to another counselor (his third) in the town where he lived and attended college.

Planning

The second phase of case management is **planning.** At this point, the agency has usually accepted the applicant for services. The individual has met the eligibility criteria and is now a client of the agency. During this planning process, the counselor and the client turn their attention to developing a service plan and arranging for service delivery. Client participation continues to be important as desired outcomes are identified, services suggested, and the need for additional information determined. The actual plan addresses what services will be provided and how they will be arranged.

A plan for services may call for additional information to be gathered so as to round out the agency's knowledge of the client. Some case managers suggest that the service delivery process is like a jigsaw puzzle, with each piece of information providing another clue to the big picture. During this stage, the case manager may realize that a social history, a psychological evaluation, a medical evaluation, or educational information might provide the missing pieces. The plan identifies what services are needed, who will provide them, and when they will be given. The case manager must then make the appropriate arrangements for the services.

During the assessment phase, Tom Chapman did a comprehensive job of gathering information about Roy. When Roy was accepted for services, the task facing him and his new counselor was to develop a plan of services. The agency's term for this is an Individualized Written Rehabilitation Plan (IWRP). (See Figure 1.8.) Clarity and succinctness characterize an IWRP, which the counselor and the client complete together. The plan lists each objective, the services needed to reach that objective, and the method(s) of checking progress.

Suppose that Tom Chapman had believed a psychological evaluation to be unnecessary and had been able to establish eligibility solely on the basis of the medical and vocational evaluations. Susan Fields, the new counselor, might still find that a psychological evaluation would be beneficial, especially since the agency was contemplating providing tuition and support for a college education. One objective of the plan would then be to provide a psychological evaluation of the client. This is an example of continuing to gather data during the planning phase.

Read Roy's IWRP carefully. First, it indicates that he is eligible for services and meets agency criteria. His program objective, Business Communications,

Contact Report and Case Memorandum

Name Roy R. Johnson Date of contact: 1/5/91

District 31 Counselor: S. B. Fields

Status: Last report 02 This report 10 Date of change 1/5/91

Certification of Eligibility

Disability: The client's primary disability is a back condition. The enclosed general medical exam and orthopedic report from Dr. Porter substantiate this. At present, there are no secondary disabling conditions identified.

Functional limitations/vocational handicap: The client has had long-term treatment for his back condition. He is able to lift only 20 lb; is restricted to walking, standing, sitting, and occasionally bending; and has restrictions on his lower extremities. The client's vocational handicap restricts him from numerous job areas, such as construction work and truck driving.

Reasonable expectations: I feel that the client can benefit from an educational program before being placed in selective job placement.

Severely disabled: The client presently meets our guidelines for severe disability, as he will require multiple services over an extended period of time. He clearly has long-term impairment.

Priority category: The client is being assigned to Priority Category I.

Susan Fields 1/5/91

Counselor's signature Date

bj

Figure 1.7 Certificate of eligibility

was established as a result of evaluation services, counseling sessions with Mr. Chapman, and Roy's stated vocational interests. The three intermediate objectives that are stated will help Roy achieve the program objective. There is also a place to identify the respective responsibilities of Roy and of the agency in carrying out the plan. Note that this agency takes very seriously the participation of the client in the development of the IWRP, even asking that the client as well as the counselor sign the plan. There is one additional part of the plan to

INDIVIDUALIZED WRITTEN REHABILITATION PROGRAM

1. NAME JOHNSON, ROY PROGRAM TYPE ☒ INITIAL ☐ AMENDMENT

2. YOU ARE ELIGIBLE FOR: ☒ VOCATIONAL REHABILITATION SERVICES ☐ EXTENDED EVALUATION SERVICES ☐ PAST EMPLOYMENT SERVICES

BECAUSE: ☒ A. YOU HAVE A PHYSICAL OR MENTAL DISABILITY WHICH CONSTITUTES A SUBSTANTIAL HANDICAP TO EMPLOYMENT AND:
☒ B. YOU CAN REASONABLY BE EXPECTED TO BENEFIT IN TERMS OF EMPLOYABILITY FROM SERVICES.
☐ C. IT CANNOT BE DETERMINED WHETHER OR NOT YOU CAN BENEFIT IN TERMS OF EMPLOYABILITY FROM REHABILITATION SERVICES.
☐ D. POST EMPLOYMENT SERVICES ARE NEEDED FOR YOU TO MAINTAIN EMPLOYMENT.

3. PROGRAM OBJECTIVE: Business Communications ANTICIPATED DATE OF ACHIEVEMENT: MONTH 1 YEAR 92

ESTIMATED DATES TO REACH OBJECTIVE & RECEIVE SERVICES

4. INTERMEDIATE OBJECTIVE, SERVICES METHODS OF CHECKING PROGRESS.

	RESPONSIBILITY	FROM	TO
OBJECTIVE To correct physical impairment so that client might			
SERVICES reach vocational objective	Client	1/91	1/92
Possible office visit with the doctor			

METHOD OF CHECKING PROGRESS Medical information

	RESPONSIBILITY	FROM	TO
OBJECTIVE To provide background information and educational skills			
SERVICES so that client might reach vocational objective	VR	1/91	1/92
A. Tuition/UT Knoxville		1/91	1/92
B. Miscellaneous Educational Expenditures		1/91	1/92

METHOD OF CHECKING PROGRESS R-11, Grade Reports

	RESPONSIBILITY	FROM	TO
OBJECTIVE To follow client's progress and develop plan amendment if needed so that client might			
SERVICES reach objective	Client	1/91	1/92
A. Possible RP-B		1/91	1/92
B. Client/Counselor Contacts		1/91	1/92

METHOD OF CHECKING PROGRESS R-11

5. CLIENT OR FAMILY AND AGENCY RESPONSIBILITIES AND CONDITIONS: I. Client is responsible to maintain contact with counselor twice each semester by mail, phone or in person. II. Client is responsible to furnish VR Counselor with a copy of grades at the end of each term. III. Client is responsible to maintain an average load of classes and average grades throughout his program. IV. Client is responsible to file for any similar benefits which might help him pay for his program. V. Client is responsible to furnish VR counselor with a resume and a list of potential employers to interview with during the first part of his senior year. VI. Client is responsible to notify counselor of any significant change of address, health, phones number of financial status.

6. CLIENT'S VIEW OF PROGRAM The client and I have discussed the services necessary to help him reach his vocational objective and we are in mutual agreement with his plan.

I HAVE PARTICIPATED IN THE DEVELOPMENT OF THIS PROGRAM AND I UNDERSTAND IT.
I UNDERSTAND AND ACCEPT THE STATEMENT OF UNDERSTANDING WHICH HAS BEEN EXPLAINED TO ME.

Roy Johnson	5-6-91	Susan Fields	5/6/91
Client's Signature	Date	Supervisor Signature	Date

Figure 1.8 Individualized Written Rehabilitation Plan (IWRP)

be signed by the client. Because one of Roy's objectives is to attend college, which involves a significant expenditure on the agency's part, Roy has also signed a Student Letter of Understanding that further describes his responsibilities. In a sense, this letter is a contract between the student and the agency; the counselor signs it as the agency's representative.

Once the plan is completed, the counselor begins to arrange for the provision of services. He or she must review the network of service providers

Case managers must work closely with other professionals during the case management process.

that has been established over time. Case managers who have been in the business for a while know who provides what services, and who does the best work. Nonetheless, they should continue to develop their networks. For beginning helpers, the challenge is to develop their own networks: identifying their own resources and building their own files of contacts, agencies and organizations, and services. Chapter 6 will provide information about developing, maintaining, and evaluating a network of community resources.

Implementation

The third phase of case management is the **implementation** of the service plan. This is the point at which service delivery begins, and the case manager's task becomes either providing services or overseeing service delivery. He or she addresses the questions of who provides each service, how to monitor implementation, and how to work with other professionals.

In general, the approval of a supervisor is needed before services can be delivered, particularly when funds will be expended. Many agencies, in fact, have a cap (a fee limit) for particular services. In addition, a written rationale is often required, to justify the service and the funds. As resources become increasingly limited, agencies redouble their efforts to contain the costs of service delivery. In Roy's case, the agency's commitment to pay his college tuition represented a significant expenditure. Susan Fields submitted the IWRP and a written rationale to the agency's statewide central office for approval.

Who provides services to clients? The answer to this question often depends on the nature of the agency. Some are full-service operations that offer a client whatever services are needed in-house, as described in Chapter 3. As a rule, however, the client does not receive all services from a single worker or agency. It is usually necessary for him or her to go to other agencies or organizations for needed services. This makes it essential for the case manager to possess referral skills, knowledge of the client's capabilities, and information about community resources.

No doubt you remember that Roy's first counselor, Tom Chapman, arranged for a psychological evaluation. Many agencies like Tom's have so many clients needing psychological evaluations that they hire a staff psychologist to do in-house evaluations of applicants and clients. Other agencies simply contract with individuals—in this case, licensed psychological examiners or licensed psychologists—or with other agencies to provide the service. Whatever the situation, the counselor's skills in referral and in framing the evaluation request help determine the quality of the resulting evaluation.

Another task of the case manager at this stage is to monitor services as they are delivered. This is important in several respects: for client satisfaction, for the effectiveness of service delivery, and for the development of the case manager's network. Monitoring is doubly important because of the changes in personnel that constantly occur in human service agencies. Moreover, there may be a need to revise the plan as problems arise and situations change.

The implementation phase also involves working closely with other professionals, whether they are employees of the same agency or at another organization. A case manager who knows how to work successfully with other professionals is in a better position to make referrals that are beneficial to the client. These skills also contribute to effective communication among professionals about policy limitations and procedures that govern service delivery, the development of new services, and expansion of the service delivery network.

Perhaps there is no other point in service delivery at which the need for flexibility is so pronounced. For example, during the implementation stage it often becomes necessary to revise the service plan, which must be regarded as a dynamic document to be changed as necessary to improve service delivery to the client. Changes in the presenting problem or in the client's life circumstances, or the development or discovery of other problems, may make plan modification necessary. Such developments may also call for additional data gathering.

In his second semester at the local college, Roy heard about a course of study that prepared individuals as interpreters for the deaf. This intrigued him, because he was already proficient in sign language. His mother was severely hearing impaired, and as a child, Roy signed before he talked. He also thought back to the evaluation staff meeting, at which the team discussed the possibility of making interpreter certification a vocational objective for him. Roy liked the interpreting program and the faculty, so he applied for acceptance into the program. This change in vocational objective made it necessary to modify his IWRP. His counselor (by now, his fourth) revised the plan at the next annual review, so as to include his new vocational objective of educational interpreting.

Three Components of Case Management

Case review, report writing and documentation, and client participation appear in all three phases of case management; they are discussed in detail in later chapters. Here we will examine how each applies to Roy's case.

Case review is the periodic examination of a client's case. It may occur in meetings between the case manager and the client or the case manager and a supervisor, or in a group meeting of helpers called a *staffing* or *case conference*. A case review may occur at any point in the case management process, but it is most common whenever an assessment of the case takes place. Case review is an integral part of the accountability structure of an organization; its objective is to ensure effective service delivery to the client.

Roy's case was reviewed in several ways. Each time a new counselor assumed the case (unfortunately, this was often), a periodic review occurred. There were also reviews on the occasion of the two contacts each semester that Roy had with his counselor. At the end of each semester, his grades were checked—also part of the case review. The staffing regarding Roy's vocational evaluation is an example of case review by a team. In this case, the client was an active participant in the case review. Roy also participated in developing the IWRP, which involved a review of the information gathered, the eligibility criteria, and the setting of objectives. The agency serving Roy implemented the important component of case review in various ways at different times throughout the process.

The particular form of **case documentation** used depends on the nature of the agency, the services offered, the length of the program, and the providers. A **record** is any information relating to a client's case, including history, observations, examinations, diagnoses, consultations, and financial and social information. Also important are "all reports pertaining to a client's care by the provider, reports originating from orders written within the facility for tests completed elsewhere, client instruction sheets, and forms documenting emergency treatment, stabilization, and transfer" (Mitchell, 1991, p. 17). The case manager's professional expertise must include documenting appropriately and in a timely manner and preparing reports and summaries concisely but comprehensively.

Roy's file includes many different types of **documentation.** The figures in this chapter have included computer forms, applications for services, counselors' notes, medical evaluations, reports, and letters. Other documentation (not shown here) in Roy's file are a psychological evaluation, a vocational evaluation, specialized medical reports, and medical updates. In Roy's case, all this documentation turned out to be indispensable, because he worked with five different counselors. For continuity of service, good case documentation is essential.

Client participation is important to the partnership that makes service delivery accountable and effective. A primary goal of client participation is client empowerment. One of the many factors involved in engaging the client as an active participant is clarity. The case manager must make clear to the client the respective goals, purposes, roles, and positions of the case manager, the agency, the service providers, and the client. It is helpful for the case manager to have knowledge of subcultures, deviant groups, reference groups, and ethnic minorities so as to communicate effectively with the client about roles and responsibilities. Other factors can affect client involvement, including the timing, the setting, and the structure of the helping process. Minimizing interruptions, inconveniences, and distractions will always enhance client participation.

Encouraging client participation has identifiable components. The first is the initial contact between the client and the case manager. It is easier to involve those clients who initiate the contact for help, as Roy did, because they usually have a clearer idea what the problem is and are motivated to do something about it. In Roy's case, the clarification of roles and responsibilities occurred at two points in the assessment phase. When Roy completed his application, the counselor went over it with him, especially the statement at the bottom of the second page. Upon signing the statement, the client has voluntarily placed himself or herself in the care of the agency. With this agreement come roles and responsibilities, which the counselor reviews at that point. A second opportunity to clarify roles and responsibilities comes with the completion of the IWRP. Both the client and the counselor sign the IWRP, which designates both the responsibility for each task and the time frame for completion of each service.

The middle phase of case management is devoted to identifying problem areas and developing and implementing a plan of services. It was during this phase that Roy decided to change his major from Business Communications to Interpreting, necessitating an amendment to the plan. Roy also needed help paying for a tutor in a science course that was particularly difficult for him. Client participation during this phase ensures that the client's perceptions of the problem and its potential resolution are taken into account.

The final phase of client participation comes at the termination of the case. At this time, the client and the case manager together review the problem, the goals, the service plan, the delivery of services, and the outcomes. They may also discuss their roles in the process. Thus, in terms of client participation, *termination* means more than just closing the case.

Now that you have some sense of the flow of the process and its component parts, we will review the responsibilities of managing cases in all three phases.

This section introduces the principles and goals that guide service delivery and discusses how they influence the work of case management.

Principles and Goals of Case Management

The principles and goals that guide case management have emerged through the work of early pioneers in helping, through federal legislation, and through current practice. They are integration of services, continuity of care, equal access to services, quality care, advocacy, working with the whole person, client empowerment, and evaluation. The subsections that follow discuss these principles and their relevance to case managers.

Integration of Services

Integration of services is effective in meeting client needs. It is also a primary focus of human service activities and a guiding principle of case management. Many people enter the system with multiple needs or have needs that change over a long period of time. To address multiple needs, case managers integrate the work of many agencies and professionals. With service integration, there is less chance of fragmentation and duplication. Integrated services also facilitate effective priority setting and encourage positive interaction among the services provided.

> We create a plan of care for each child who comes into state custody. It is a court document that covers health and medical factors, educational records, vocational interests, social skills, emotional behaviors. It also says who is responsible for providing what. Once the plan of care is created, then everyone involved meets to approve it. It then goes to the court for ratification. (Thom Prassa, East Tennessee Community Health Agency, Knoxville, Tennessee, personal communication, July 10, 1994)

Continuity of Care

Continuity of care has two meanings. First, continuity means that an individual's needs are addressed, if possible, even before he or she makes the first visit to the agency. For example, the case manager may phone before the initial visit or the intake interview, asking if there are any questions; or he or she might mail a welcoming letter with information about the agency. Continuity of care carries beyond termination, through a transition period when services are no longer needed.

Second, the term *continuity of care* refers to the comprehensiveness of the care provided. This aspect of continuity involves therapeutic intervention along with support in the environment, maintaining a relationship with the client's family and significant others, crisis intervention, and social networking beyond mere linking of services (Kanter, 1991).

Roz Jaffee at the Jewish Home and Hospital for the Aged in the Bronx provides the following illustration of continuity of care.

> We have one case right now where the person died in the hospital. . . . She became more and more frail, wheelchair bound, incontinent, working through problems as to the support you need and working with the family. . . . We worked to get the family to cooperate with us and to cooperate in getting things that the mother needed. The woman had her wish. She never wanted to go into a home. And she came here [day care for the aged]. No matter how much pain she had, she came every single day, until the end. (Personal communication, May 3, 1994)

In the case described, case managers were able to provide a continuity of care that extended beyond the end of the client's life, as they continued to work with the family. They also provided a social network as well as links between the client, the family, and the needed services.

Equal Access to Services

There must be **equal access to services** for clients in need of assistance. The commitment to equal access is reflected in the case manager's assumption of the role of advocate. Attention is also given to developing ways to extend access to services, such as fee waivers, transportation, and outreach efforts. Affordability is directly linked to the issues of eligibility and access. To ensure access to services, eligibility must be defined so as to include those who lack traditional economic, social, and political access. There must also be a plan for individuals who are on the margins of eligibility. Paula Hudson discusses the importance of access to services.

> The community aspect of case management is the wave of the future. Someone will go out to a person's home and do a home visit, and be able to go in and find out what is happening with that client. It is also being able to decrease their hospitalizations. The hospitalization of our clients has diminished greatly after providing case management services to them. Our intensive case management program has a consumer choice component. There is a lot of consumer choice involved with the goal of making them more independent versus dependent. (Paula Hudson, Helen Ross McNabb Center, Inc., Knoxville, Tennessee, personal communication, April 27, 1995)

Quality Care

Quality care implies a commitment that respects the rights of the client and demands accountability on the part of human service professionals. The watchwords are *effective* and *efficient* care; both are key considerations in service delivery. Quality assurance is a component of case management; the term signifies professional excellence, high standards of care, and continuous improvement (Sullivan, Wolk, & Hartmann, 1992).

Janelle Stueck, a program manager at the Private Industry Council in Knoxville, Tennessee, made these observations on the subject:

> The difference between the traditional case manager approach, which has to do with coordinating resources, and what we try to do here . . . is . . . guiding, mentoring, leading, big-brothering or -sistering, as the case may be, the kind of role that goes significantly beyond the relationship of the case manager and client. The ability to simply know what the rules are with regard to what is out there and how to assemble things so that you put the best package together. For the client I think it is that extra dimension that makes it worth doing. (Personal communication, August 1993)

Effectiveness means getting results and producing outcomes. In this age of scarce resources, it is important that the available resources are used wisely—that is, efficiently. In service delivery, the productive use of resources involves making a determination of the outcomes desired and developing a plan to reach those outcomes. Efficiency is measured in terms of the resources required, the time expended, the costs of services, and the outcomes achieved. The plan is constantly monitored, necessary adjustments are made, and appropriate justification is presented when additional resources are needed.

The case manager must maintain efficiency because of the complexity of many of the problems he or she faces. It is essential that professionals work together to deliver quality services in an efficient manner.

Advocacy

The purpose of **advocacy** is to represent the interests of the client. This may involve speaking or writing in defense of a person or cause. In acting as **advocates** for their clients, case managers often assume a dual role, representing both the institution and the client. Advocacy is essential because of imperfections in the delivery system: duplication, fragmentation, and often resources that are insufficient to provide the services needed. Sometimes the needed services do not exist. According to Austin (1990), "The myth that case managers can effectively manage scarce resources on behalf of their clients assumes that there are resources to manage" (p. 402).

Linda Washington and her colleagues at the Third Avenue Family Service Center in the Bronx describe their advocacy work:

> When we go into the home we also advocate services. . . . A homemaker may need something that has not been done. Appointments have not been kept, so we need to reschedule new appointments. . . . Transportation is needed. . . . Services have not been provided at a speedy rate; so we again intervene. (Personal communication, May 5, 1994)

The Whole Person

Case managers must be committed to a holistic view of the individual receiving help. Recognizing that there are many human dimensions to be considered in service delivery, the case manager and the client address social, psychological, medical, financial, and vocational problems. Most likely, the client has problems in more than one of these areas.

One family counseling agency uses a 12-page standardized form during the intake interview. The Pima Client Assessment Plan is designed to uncover a range of problems that clients may be experiencing.

> It is a review of their medical functioning, social support system, needs for medical equipment. . . . This form is an outline. . . . It is the ticket to services within our system. . . . We also look for more than just what is on the form. . . . We are looking for a picture of their lives. (Suzy Bourque, Family Counseling Center, Tucson, Arizona, personal communication, October 7, 1994)

Client Empowerment

Respect for the client stems from the long-standing belief that all individuals, regardless of their needs or disabilities, have integrity and worth. This belief guides the case manager to place the client in a central role in the helping process. Ensuring the client's full participation during the service delivery process means involving him or her in every step, including identifying the problem, gathering information, establishing goals, planning, implementing the plan, and evaluating the plan and outcomes.

Client empowerment within the process expands this respect for the client to include teaching the client basic coordination skills and, as time and skill development allow, encouraging the client to manage his or her own case. The goal is to develop self-sufficiency so that the client can manage his or her own life without dependence on the human service delivery system.

In many agencies, self-sufficiency is a primary goal. Janelle Stueck says:

> [We are advocates] for the participant . . . trying to help that person become self-sufficient. We try to help them make plans so they will know from step 1 where they need to go, how they need to go about it, and what the end result will be. . . . The person may come in very dependent . . . but hopefully, as things progress, they become more sure of themselves . . . and they are able to do more and more on their own. And so hopefully by the end they are quite ready to leave the nest, as it were. (Personal communication, August 30, 1993)

Another way of showing respect for the client is to treat him or her as a customer. This means asking clients about their needs, providing services that match those needs, offering clients an opportunity to evaluate the services they receive, and making changes to improve services based on their feedback. SKIP of New York is an agency that has worked hard to treat its clients as customers. It was a struggle for the agency to make this transition.

> To empower the families, a lot of emphasis is placed upon parent partnerships. Parents or caregivers should be provided with information and choices so, as knowledgeable participants, they can make informed choices and decisions on behalf of their children. Professional input is often riddled with highfalutin terminology and statements of "best care" for the child. Parents need to understand professional input; parents deserve the respect of all parties involved with their child. As [customers], they seek the best and often require guidance, information, and training to realize the goals they have designed for the child. This agency was founded by the parents of a child with complex medical needs. Based upon this genesis, the organization has evolved and succeeded based upon parent [customer] feedback, involvement, and partnership. (Margaret Mikol, Sick Kids Need Involved People of New York, personal communication, May 4, 1994)

Evaluation

Evaluation is a critical part of case management; it takes place throughout the process. Case managers evaluate the effectiveness of the process itself as well as that of the many services provided. The focus should be on relevance for the client, client progress and satisfaction, integration of services, and quality of services. Both professionals and clients are involved in the evaluation process.

> They review the program every 6 months unless we see a change in behavior and we think that something needs changing . . . or there is no need for this program . . . or someone is told this medication is not working. . . . (Kim Ehlers, Mitchell Area Adjustment Training Center, Parkston, SD, personal communication, October 14, 1995)

Chapter Summary

Managing client services is an exciting and challenging responsibility for helping professionals who work as case managers. To assist clients with multiple problems, case managers must know the process of case management and be able to use it. This process can be adapted to many different settings, for work with a variety of populations. The coordination and the provision of direct services are guided by common principles and goals, committing the agency to give the client quality services and ensuring that he or she participates in the case management process.

Chapter Review

Key Terms

Case management
Case manager
Caseload management
Assessment
Planning
Implementation
Case review
Case documentation
Record
Integration of services
Continuity of care
Equal access to services
Quality care
Advocacy
Client empowerment

Reviewing the Chapter

1. Describe the context in which case management services are delivered today.
2. How has the definition of case management changed over the years?
3. What is the expanded definition of case management?
4. Distinguish between the terms *applicant* and *client.*
5. What should be accomplished during the assessment phase?
6. What occurs during the initial contact between the case manager and the individual seeking services?
7. Describe the routine matters that are discussed during the initial contact.
8. Identify the types of information that are gathered during the initial interview.
9. What factors determine what information has been obtained by the end of the initial interview?
10. Using the case presented in this chapter, discuss the advantages of a partnership between the case manager and the client.
11. Describe the case manager's activities during the planning phase.
12. What questions guide implementation?
13. Why is flexibility so important during the implementation phase?
14. Define *case review.*
15. List the three keys to successful case review.
16. Why is documentation important in service coordination?
17. How can the case manager promote client participation?
18. How will a client's resistance affect his or her participation in the service coordination process?

Questions for Discussion

1. Why do you think that case management as a method of service delivery is changing?

2. From your own work and study of human services, what evidence do you have of the importance of assessment and planning?
3. If you were a case manager, what three principles would guide your work? Provide a rationale for your choices.
4. What do you think Roy Roger Johnson would say about his experience with the case management process?

References

Austin, C. (1990). Case management: Myths and realities. *Families in Society: The Journal of Contemporary Human Services, 71*(7), 398–405.

Austin, C. (1993). Case management: A systems perspective. *Families in Society: The Journal of Contemporary Human Services, 79*(9), 451–459.

Betz, C. L. (1991). The problem with case management. *Journal of Pediatric Nursing: Nursing Care of Children & Families, 6*(5), 295.

Jackson, B., Finkler, D., & Robinson, C. (1992). A case management system for infants with chronic illnesses and developmental disabilities. *Children's Health Care, 21*(4), 224–232.

Kanter, J. (1991). Integrating case management and psychiatric hospitalization. *Health and Social Work, 16*(1), 34–42.

Lourie, N. (1978). Case management. In J. Talbott (Ed.), *The chronic mental patient* (pp. 159–164). Washington, DC: American Psychiatric Association.

McClam, T., & Woodside, M. (1994). The practitioner's voice: Case management for effective service delivery. *Human Service Education, 14*(1), 39–45.

Mitchell, R. W. (1991). *Documentation in counseling records.* Alexandria, VA: American Association for Counseling and Development.

Sullivan, W., Wolk, J., & Hartmann, D. (1992). Case management in alcohol and drug treatment: Improving client outcome. *Families in Society: The Journal of Contemporary Human Services, 73*(9), 195–204.

Webster's New Collegiate Dictionary (1st ed.). (1973). Boston: G. & C. Merriam.

Wheeley, B. (1981). The human development liaison specialist: Theory, person, and process. *Journal of Community Psychology, 9*, 321–330.

Chapter Two

Historical Perspectives on Case Management

A lot of families fall through the cracks. . . . We come in and we do casework for them . . . we provide financial assistance . . . we assist with telephone bills . . . we go into the home and advocate services.

—Carolyn Brown, Third Avenue Family Service Center, Bronx, New York, personal communication, May 5, 1994

It is not competitive between the agencies. . . . We have case managers who are gatekeepers of the system. . . . We have the central and northwest part of the city. . . . Most of our clients come from a hospital discharge planner.

—Suzy Bourque, Family Counseling Center, Tucson, Arizona, personal communication, October 7, 1994

In one sense, we are not really state employees; but in another sense, we are. I think that it is a new trend in government where agencies are privatized so they can mainstream service delivery to youth. The office developed as a result of the governor's order and we are part of an agency that does a variety of things but we are now under the Department of Health.

—THOM PRASSA, East Tennessee Community Health Agency, Knoxville, Tennessee, personal communication, July 10, 1994

The purpose of this chapter is to establish a historical context for case management. It will describe four perspectives on case management that have evolved in the past 30 years: case management as a process; client involvement; the role of the helper; and utilization review and cost–benefit analysis. There follows a brief history of case management in the United States, including its evolution to broader service coordination responsibilities within the last decade, as well as an introduction to managed care.

By the end of each section of the chapter, you should be able to accomplish performance objectives listed below for that section.

PERSPECTIVES ON CASE MANAGEMENT

- Identify four perspectives on case management.
- Trace the evolution of case management.
- Describe the impact of managed care organizations on case management and service delivery.

THE HISTORY OF CASE MANAGEMENT

- Assess the contributions of the pioneers in the areas of advocacy, data gathering, recordkeeping, and cooperation.
- Using the Red Cross as an example, describe casework during World Wars I and II.
- Name the acts of federal legislation that further developed case management.

THE IMPACT OF MANAGED CARE

- List the goals of managed care.
- Summarize the impact of managed care on human service delivery.
- Differentiate between the types of managed care organizations: HMOs, PPOs, and POS.

EXPANDING THE RESPONSIBILITIES OF CASE MANAGEMENT

- Trace the shift in emphasis in case management.

Case management has long been used to serve clients in human services. Today, professionals are discovering new and more effective ways to deliver services, and there is no longer a standard definition. Modern-day case management does resemble the practice of the past, but many dramatic changes have occurred. Among them are the changing needs of individuals served; financial constraints on the human service delivery system; the increasing number of people in need of services; and the growing emphasis on client empowerment, evaluation of quality, and service coordination.

One consistent theme that runs through the study of human service delivery is diversity. The three helping professionals quoted at the beginning of this chapter describe the services of their agencies. The case worker from the Third Avenue Family Service Center describes the work of her agency as providing financial assistance and advocacy work for families for whom there is no other support. The Family Counseling Center in Tucson, on the other hand, gives patients just discharged from the hospital only the aftercare services they need. This often includes support to meet psychological, social, medical, financial, and daily living needs. In Thom Prassa's case, services to youth have changed recently. The state contracts with the agency where he works to deliver services. These services, like those of the other organizations, are delivered within the context of federal and state regulations and restrictions.

Much of the foundation of case management developed when it was used to serve people with mental illness who were deinstitutionalized in the 1970s. Illustrating our discussion here is the story of Sam, who was diagnosed as mentally ill and promptly institutionalized. Sam received many services since that first diagnosis, and his history with the human service delivery system reflects the evolution of service delivery from the traditional form of case management to the new paradigm that is applied today.

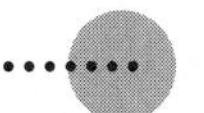

Perspectives on Case Management

This section will explore four different perspectives on case management, which together serve to illustrate the development of the process since the 1970s.

Case Management as a Process

In the 1970s, the mental health community was involved in the process of **deinstitutionalization:** the movement of large numbers of people from self-contained institutions to community-based settings such as halfway houses, family homes, group homes, and single residential dwellings. The following definition of case management was offered by Norman Lourie (1978), a member of the American Psychiatric Association's Ad Hoc Committee on the Chronic Mental Patient.

> My view is that it is a vital, perhaps the most primary, device in management for any individual with a disability where the requirements demand differential access to and use of various resources. Far from a new concept, it has long been the central device in every organized arrangement that heals, rehabilitates, cares for, or seeks change for persons with social, physical or mental deficits. . . . Case management is a key element in any approach to service integration. . . . A counselor manages assessment, diagnosis, and prescription

> . . . synthesizes information, emerges with a . . . treatment plan; and then purchases one or more interventions. (p. 159)

Many clients need assistance in gaining access to human services. Often they have multiple needs, limited knowledge of the system, and few skills to arrange services.

After several ear infections, Sam became severely hearing impaired when he was 3. He was the youngest of four children and lived in a small town with his mother, two sisters, and a brother. His mother took care of him, and he became dependent on her. They learned to communicate with each other using a sign language they devised themselves. None of his siblings learned to sign. Sam was often unruly and found that tantrums would get him what he wanted. The older he got, the more hard to handle he became. When Sam was 15, his mother died. None of his siblings would assume responsibility for him, so they decided to have him admitted to the state mental hospital in the capital. This occurred in the early 1950s; the exact date is unknown because a fire at the institution in the 1960s destroyed the records of those who were admitted previously. Sam was in the mental institution for many years before the deinstitutionalization movement started. At that point, Sam's long odyssey began. (See Table 2.1.)

Sam's first case manager was an employee of the institution, and his job was to identify those patients who could function in a community setting. Limited assessments of Sam's mental and emotional state indicated that he was not mentally ill, but simply hearing impaired. Unfortunately, his time in the institution had compounded his problems: He did not know American Sign Language, and he had begun to behave like other patients who did have mental illness. The case manager decided that Sam should be moved from the institution to another setting. The case manager located Sam's oldest sister, but he stayed with her for only one weekend. She returned Sam to the institution on Monday morning, saying that she couldn't handle him and his presence was too disruptive to her family. None of his other siblings was willing to help, so Sam remained in the institution while his case manager searched for a group home that had an opening. Eventually, Sam did move into a group home, but he lived there for just six days before returning to the institution. According to the home's director, no one could communicate with Sam, his behavior was inappropriate, and he needed constant supervision.

This account reflects the experiences of many clients who were institutionalized in the 1950s and 1960s and were later deemed appropriate for discharge during deinstitutionalization. The case management process illustrated here is an elementary one: limited assessment followed by placement. The responsibility of Sam's case manager was to find Sam a "least restrictive" environment that could foster his growth and development. Unfortunately, the search for such an

TABLE 2.1 SAM'S JOURNEY

Sam's age	*Provider*	*Activity*
3	mother	care at home
15	institution	residential care
26	first case manager sister group home institution again	limited assessment lived with sister for one weekend lived in group home for six days residential care
27	second case manager third case manager	assessment learning American Sign Language contact with deaf community planning involving Sam in process
28		independent living skills
29	day care program	socialization skills
30		fluent in ASL developed friendships
30–40	Lois Abernathy (care coordinator) interdisciplinary team	regular visits and assessment 3 halfway houses; 4 group homes; school for the day living with siblings living in apartment vocational training
40	rehabilitation counselor and care coordinator	vocational training living with friend in apartment
41	rehabilitation counselor	full job responsibility termination of services

environment was quite difficult, given Sam's institutional behavior and the limited assessment of his abilities. Sam is one of those clients who need access to multiple services before they can make the transition from an institution to a community setting.

Ozarin (1978) addresses the case management process by discussing the key elements for success: responsibility, continuity, and accountability. *Responsibility* entails one person or team assigned to assess the client's problem and then to plan accordingly. Linkages "must be established to form a network of service agencies which can provide specific resources when called upon without assuming total responsibility for the client, unless responsibility for carrying out the total plan is also transferred and accepted" (p. 167). In other words, there must be a clear line of responsibility for the case and the client.

Continuity is also a significant element of good case management. Planning is the key that ensures continuity. It is important not only during the intensive treatment phase but also in aftercare. In addition, to foster **accountability**, methods "must be in place to assure the patient is not lost. . . . The case management process must help the client increase the ability to function independently and to assume self-responsibility. The client should be involved in all aspects of decision making" (Ozarin, 1978, p. 168). Guided by these goals, organizations and professionals at every level work hard to develop systems that participants understand, working together to serve and involve the clients.

Similar to Ozarin's description of case management is that of Weil and Karls (1985). They describe case management as a "set of logical steps and a process of interaction within a service network which assure that a client receives needed services in a supportive, effective, efficient, and cost-effective manner" (p. 2). Case management is an important and necessary component of the human service delivery system, for it provides a focus and oversees the delivery of services in an orderly fashion. Moreover, they agree with Lourie (1978) that case management is needed because of the multiple needs of the clients and the complexity of the delivery system. As you read about Sam's case, you will see case management evolve into a more logical and complex process that focuses on client participation, integration of services, and cost effectiveness.

Client Involvement

In the 1980s, client involvement came to be emphasized more strongly. Dunst and Trivette (1989) proposed a model of case management based on the concept of enabling clients "to solve problems, meet needs, or achieve aspirations by promoting acquisition of competencies that support and strengthen functioning in a way that permits a greater sense of individual or group control over its developmental course" (p. 3). Sam's experience in the human service delivery system reflects the beginning of changes in service provision.

Sam remained institutionalized for the next four months because there was a shift in the case management process. The institution decided to contract with a local mental health agency for case management services. Case management was a new role and responsibility for this agency; it assigned two individuals 50 cases each, with few guidelines for performing this new function. Sam's new case manager, his second, spent some time assessing Sam's needs, getting to know him, and talking with the mental health professionals within the institution. In concert, they determined that Sam needed a very structured environment in the community if his deinstitutionalization was to succeed. He also needed to learn sign language and to begin to communicate with others using this medium. Because he was hearing impaired but did not have mental illness, he needed to be in contact with the deaf community, where he could find support and role models for independent living. At the age of 27, he had little ability to care for himself.

Unfortunately, his case manager left her position before she had the opportunity to implement the plan. A third case manager assumed responsibility for Sam's case, with much determination. Her own brother had been deaf since birth, and she recognized Sam's potential. She could communicate with him using American Sign Language (ASL). She was also committed to planning, documenting her work, involving Sam in the case management process, and following through on referrals and the involvement of other professionals.

The goals of the plan included having Sam learn ASL, teaching socialization skills and independent living skills, and introducing him to members of the deaf community. His case manager was able to find a day care program where Sam could learn independent living skills. Three times a week, he went to the local School for the Deaf for ASL lessons. Once a week, Sam and his case manager joined other hearing impaired adults for a special community program and social hour. Sam still lived at the institution. By the end of the second year, the case manager was able to include Sam in the process of setting priorities and planning for his treatment.

After two years, Sam had made considerable progress. He was able to use ASL to communicate his needs, and he had developed several friendships with people he met at the School for the Deaf and at the community programs. On his 30th birthday, he celebrated with his friends from the school. Tantrums continued to occur, but less frequently.

After a rocky beginning, Sam benefited from the service delivery process as it evolved. His third case manager assumed responsibility for his case, provided the continuity needed for him to make progress, and was accountable for his care. She used the process of logical steps described by Weil and Karls (1985) to establish goals and set priorities. She also established a partnership with Sam by involving him in problem identification, plan development, and service provision. By learning daily living skills, Sam reinforced his ability to care for himself, and his increasing mastery of ASL gave him a new medium for self-expression and communication.

The Role of the Case Manager

Chapter 1 listed an array of job titles that emerged to reflect the new goals of service delivery. Traditionally, terms such as *caseworker* and *case manager* had described the efforts of helpers. Today, job titles include *service coordinator, liaison worker, counselor,* and *case coordinator.* These new job titles represent not only the diversity of service delivery today but also the broader range of responsibilities and the different ways case managers perceive their roles. The change in job titles reflects the evolution of case management and, in a larger context, of service delivery. The emphasis shifted from what was previously understood to be case management, when it was considered to embody the skills of managing someone, to terminology reflecting a more equitable relationship—terms such as *coordinator* and *liaison.* There had occurred a change in philosophy regarding the role of the case manager: emphasizing working with other professionals, coordinating care and other services, and empowering individuals to use the system to help themselves. The focus became the strengths of the clients—whether individuals, families, or groups—to develop the skills needed to work within the human services network.

The next ten years were a struggle for Sam and for those who worked with him. His case manager of two years left her job for a promotion in a nearby city, and his case was transferred to Lois Abernathy, a care coordinator at a different agency. Because of increasing pressure to deinstitutionalize, it was decided to move Sam to a halfway house before helping him establish residence in the local community. Over the course of the decade, Sam lived in three halfway houses, four group homes, the School for the Deaf, with his siblings, and in an apartment with a roommate. Ms. Abernathy was the link between Sam and each of these placements. Her responsibilities included meeting with Sam regularly to review his needs, problems, and successes, and arranging any additional services for him. Often, she and Sam would meet with other professionals who were involved with his case. Ms. Abernathy was committed to giving Sam choices about his future. When he expressed the desire to work at a job, she helped him determine exactly what he would like to do. After exploring the options available to him, Sam decided that he would like to work with a local business vending program that Ms. Abernathy knew about. After much education and training, his responsibilities with this program came to include stocking machines, collecting money, and minor repair work.

The assignment of a *care coordinator* to Sam's case signaled a shift in the role of the case manager—from management to coordination. The client's participation in the process also became significant; there emerged a partnership to identify, locate, link, and monitor needed services.

Utilization Review and Cost–Benefit Analysis

One result of the spiraling cost of medical and mental health services and the push for health care reform is the growth of the managed care industry. The purpose of managed care is to authorize the type of service and the length of time care is provided and to monitor the quality of care. In the managed care environment, case managers function very differently from those described earlier.

What makes case management in managed care distinctive is the emphasis on efficient use of resources. Case managers are involved in utilization review and have the responsibility to authorize or deny services. Also, they must know how to interact with insurance providers and how to process claims through the insurance system.

This new case manager is also responsible for cost–benefit analysis. Such an analysis does not include the traditional reporting that is found in a case history as notes or recommendations and follow-up. It is focused on the financial matters of the case, specifically the cost and efficiency of service.

As we leave Sam at the age of 42, he is two months away from assuming responsibility for all the vending machines in a nearby neighborhood. It has taken two successive six-month training sessions to teach Sam the necessary job skills and the repair techniques required. Sam is living with a new friend in a small apartment near his vending area. This friend is also a client at the rehab center and is helping Sam train for his new job. Sam's rehabilitation counselor and his care coordinator from the mental health center have met and developed a coordinated plan, with input from Sam. Because of his recent success in rehabilitation, mental health services are no longer authorized for Sam.

Sam's experience with a care coordinator has led him in a new direction. The emphasis on tapping Sam's potential and coordinating care has given him a major voice in decision making. This requires coordination between the two systems of rehabilitation and mental health. Sam does make progress, and the managed care case manager decides to discontinue mental health services. As we leave Sam, his rehabilitation counselor supports him with regard to job training and housing. If he should again need professional mental health support, it is hoped that the rehabilitation counselor can arrange these services for him.

The History of Case Management

As important as various current perspectives on case management are, the historical roots are equally informative. The pages that follow trace the history of case management from its origins in institutional settings through the influences of early pioneers, the impact of the American Red Cross, and the influence of federal legislation. The chapter concludes with a discussion of case management as it is practiced today.

"The process of service coordination and accountability has a century-long history in the United States" (Weil & Karls, 1985, p. 1). As first used in institutional settings, case management included the responsibilities of intake, assessment of needs, and assignment of living space. These institutions provided residential services to people incarcerated for crimes, orphans, people with mental illness, people with disabilities, and elderly people. What professionals performed the case management function depended on the particular institution; among them were doctors, nurses, psychiatrists, psychologists, counselors, and teachers (Weil & Karls, 1985).

Early Organizations

One example of an institution with an early commitment to case management was the Massachusetts School of Idiotic and Feebleminded Youth, established in 1848. This school promoted the belief that people categorized as "idiotic" or "feebleminded" could improve if they were given appropriate clinical, social, and vocational services and support (Weil, 1985). In 1839, a child who had mental retardation as well as vision impairment was brought to the Institution for the Blind. It was clear that the child had needs beyond the expertise of the institution. The director, Samuel Howe, was determined to help this child and others with similar needs. He convinced the state that he could improve these children in three areas: bodily habits, mental capacities, and spirituality (Winsor, 1881).

The new institution founded under Howe's leadership was the Massachusetts School of Idiotic and Feebleminded Youth. It provided pioneer services in case management, although the school did not use the actual term. The early services at the school included observation and diagnosis of physical and mental behavior. The helping professionals tracked clients' progress, and they soon began differentiating and individualizing treatment: ". . . we cannot properly care for a young and helpless idiot in the houses devoted to the brighter moron children" (Trustees, 1920, p. 17).

When demand for the services increased, the school established outpatient clinics in several cities (Trustees, 1919). These clinics supported families who cared for children at home. Aftercare was also an important service, provided by trained "visitors" who helped plan the transition from the institution to the home or other setting. The visitor would gather information to determine whether the child should be sent home on trial or be released for vacation with family or friends. They would also follow up after transition to determine whether the release and placement were appropriate. It was assumed that aftercare (supported by parents, family, and the visitors) could "save much money to the state by helping to continue the custody and training of the troublesome defects, and to permit the liberty of those who can safely use such liberty" (Trustees, 1920, p. 19). This emphasis on aftercare was the forerunner of modern-day continuity of care, as well as today's commitment to provide services as the client makes a transition from the treatment setting to a less restrictive one.

Early in this century, the school made two improvements in the management of information, which is an important component of case management. In 1916, the institution began an evaluation of its services. This included a study of clients who had been discharged and those in aftercare. The information gathered included where patients lived, with whom they were living, whether they supported themselves, and (if so) how. The information was gathered by survey, and then patients, families, and friends were interviewed in their homes. In 1919, new legislation established a Registry for the Feebleminded in an effort to catalog the state's population of people with retardation (Trustees, 1920). These advances in recordkeeping and information management contributed to case management as we know it today.

In the 1860s, the state of Massachusetts created a Board of Charities. The board had responsibility to coordinate human services and to monitor public expenditures for the poor and the sick (Weil & Karls, 1985). In addition, Charity Organization Societies were established in New England to provide service coordination and interagency cooperation. Their primary responsibility was to register clients and needy families, and an effort was made to reduce duplication of services. The Charity Organization Societies' friendly visitors "had to be trained in investigation, diagnosis, preparations of case records, and treatment, all of which required guidance, counsel, supervision, and the knowledge of the scientific philanthropy" (Trattner, 1994, p. 238).

The next section discusses the work of some early pioneers in human services, who developed case management further, especially in coordination of services and interagency cooperation.

Early Pioneers

Early case management took either of two forms: a multiservice-center approach or a coordinated effort of service delivery. Jane Addams, Lillian Wald, and Mary Richmond were three early pioneers who contributed to the development of the emerging case management process.

HULL HOUSE

Hull House was founded in Chicago in 1889 by Jane Addams and Ellen Starr, classmates at the Rockford Female Seminary. In their twenties, while traveling in England, they visited Toynbee Hall, a university club that had established a recreation club for the poor. Addams and Starr were committed to increased communication between the classes, and the activities of Toynbee Hall inspired them. They returned to America and moved into a poor section of Chicago, hoping to improve that environment (Addams, 1910).

They bought Hull House, an older home on Chicago's West Side. Committed to sharing their home and their love of learning, they opened the house to the neighborhood. They acquired collections of furniture, art, and literature; soon they became involved with music and crafts. The purposes of Hull House were threefold: to provide a center for civic and social life; to improve conditions in the neighborhood; and to provide support for reform movements (Addams, 1910).

At first, the activities of Hull House involved only casual interactions with the neighborhood, but soon the bonds became more formal. Clubs enrolled members, and regular classes were held. This increased the need for effective administration and recordkeeping. Hull House was intended to be a gathering place that fostered a sense of community and friendship, so many participants would have been suspicious of formal files being kept, attendance logged, records of meetings maintained, or notes written about interactions or behavior. However, neighborhood people began using the staff members as references, and more detailed records had to be kept. Then information about households was gathered and recorded, just to maintain successful contact. This demographic information was used to generate mailing lists for announcements of classes and other activities. A card system was housed in the administrative office, and its contents were shown only to people who could establish a need to see it (Woods & Kennedy, 1911).

In addition to recordkeeping, advocacy was a case management function that was integrated into the work of Hull House. Jane Addams and her colleagues were involved in many efforts to improve the living and working conditions of the neighborhood and its inhabitants. In one such project, they worked with landlords and lawyers to improve existing housing and find better housing. First, they collected data about the types and characteristics of housing in the neighborhood. In order to understand the purchasing power of the renters, they also studied wage levels (Residents, 1985). Since the residents themselves served as the data gatherers, they became experts on the conditions of their environment (Brieland, 1990). Advocacy took many forms at Hull House; at one time, Jane Addams helped correct unsanitary conditions by becoming a health inspector for garbage collection (Polacheck, 1989). (See Box 2.1.)

HENRY STREET SETTLEMENT HOUSE

This organization was established in 1895 in New York by Lillian Wald and Mary Brewster, who were nurses. Early in their careers, they decided to provide services to New York City's Lower East Side, home to a large number of immigrants. Wald and Brewster lived in the neighborhood and provided health care services.

They established a system for nursing the sick in their own homes, promoting the dignity and independence of the patient. According to Wald (1915), "the nurse should be as ready to respond to calls from the people themselves as to calls from physicians, that she should accept calls from all physicians, and with no more red tape or formality than if she were to remain with one patient continuously" (p. 27). There was an explicit focus on accessibility.

The Henry Street Settlement House kept careful records of its services and made significant contributions to knowledge about the prevalence of disease, as well as treatment. For example, in 1914, the staff at Henry House cared for 3535 cases of pneumonia, and there was a mortality rate of 8.05% (Wald, 1915). This recordkeeping reinforced the idea that maintaining and managing information was an important part of the process of delivering services. The settlement house was committed to integrating its services with those of the school and the home. The staff kept data on schoolchildren and identified children who had traditionally been excluded from schools for medical reasons. They then served as advocates for educational services for these children.

The work at Henry Street led to two significant innovations: the designation of the visiting nurse and the development of the Red Cross. Both these services were important in promoting public health. One important function of the visiting nurse was to establish an organized system of care and instruction for people with tuberculosis and their families. Early in its history, the Red Cross facilitated the use of the public schools as recreation centers, taught housekeeping skills to women, and provided penny lunches for children (Wald, 1915).

BOX 2.1 A Hull-House Girl

Hilda Satt Polacheck was born in 1882 in Wolclawek, Poland. She was the eighth of twelve children born to a well-to-do Jewish family. Because of oppression by the Russian government, her family immigrated to America in 1892. During their early years, Hilda and her sister attended the Jewish Training School on Chicago's West Side. After the death of their father, however, they joined the ranks of the working poor. Hilda began working in a knitting factory when she turned 13. She first came to Hull-House for a Christmas party in 1896, and it soon became the center of her social life. She attended classes and club meetings, read literature, exercised, and performed in plays there. Later she worked there as a receptionist and a guide. Hilda grew up with Hull-House and spent much of her time there from 1895 until 1912. In her autobiography, she provides the following description of Jane Addams and her advocacy work.

> Bad housing of the thousands of immigrants who lived near Hull-House was the concern of Jane Addams. Where there were alleys in the back of the houses, these alleys were filled with large wooden boxes where garbage and horse manure were dumped. . . . When Jane Addams called to the attention of the health department the unsanitary conditions, she was told that the city had contracted to have the garbage collected, there was nothing it could do. . . . She was appointed garbage inspector for the ward. I have a vision of Jane Addams . . . following garbage trucks in her long skirt and immaculate white blouse. . . .
>
> Hull-House was in the Nineteenth Ward of Chicago. The people of the Hull-House were astonished to find that while the ward had 1/36 of the population of the city, it registered 1/6 of the deaths from typhoid fever. Miss Addams and Dr. Alice Hamilton launched an investigation that has become history in the health conditions of Chicago. . . . Whatever the causes of the epidemic, that investigation, emanating from Hull-House, brought about the knowledge of the sanitary conditions of the Nineteenth Ward and brought about the changes we enjoy today.

SOURCE: From *I Came a Stranger: The Story of a Hull-House Girl*, by H. Polacheck, pp. 71–72.

Mary Richmond Much of the work of Jane Addams and Lillian Wald was with the immigrant populations in the urban areas. Mary Richmond, a social reformer at the turn of the century, had a similar commitment to bettering the lives of this population. She, too, made significant contributions to the development of case management. Richmond promoted the idea that each person was a unique individual whose personality, family, and environment should be respected. Working with immigrants, she emphasized the need for social workers to resist the tendency to stereotype or overgeneralize (Lieberman, 1990). Richmond wrote, "the social adjustment cannot succeed without sympathetic understanding of the old world backgrounds from which the client came" (1922, p. 117). She also believed that professionals should work *with* clients rather than doing things *to* them.

One method Richmond developed to focus on the individual was **social diagnosis,** a systematic way for helping professionals to gather information and study client problems. She established a series of methods for gathering information about individuals, assessing their needs, and determining treatment. This process is often referred to as *social casework,* and it became a part of the case management process. Richmond contributed a case record form designed to focus on individuals and their unique problems (Pittman-Munke, 1985; Trattner, 1994).

She also recognized that data gathering is a complex process and urged the use of different methods for different individuals. According to Richmond (1917), some clients should be interviewed in an office; for others, the home was the preferred location. She believed in multiple sources of information and warned that data gathering was a complex and often incomplete process.

Cooperation was a continuing goal for Richmond in developing the theory of social casework. This emphasis on coordinated care extended to her work with individuals, and she later expanded that work to include families as well (Montalvo, 1982). She also fostered coordination among agencies. For example, her work in Philadelphia with the Charity Organization Society facilitated relationships between that society, the board of education, and the public education association (Pittman-Munke, 1985).

During the early part of the 20th century, the Red Cross emerged to meet the multiple needs of individuals. The subsection that follows describes its involvement in providing assistance, primarily to servicemen and their families in World War I and World War II.

The Impact of World Wars I and II and the American Red Cross

During the First World War, there was an increased interest in casework as developed by Mary Richmond. The American Red Cross, whose roots are found in the work of Clara Barton and the Civil War, used casework to address individuals' problems and their psychological needs. The use of a casework approach to assist individuals began during the Mexican civil war (1911–1917), when the American Red Cross provided a variety of services to support the daily living of civilians and troops along the Mexican border (Dulles, 1950). Subsequently, the services were extended to dependents of military personnel in Army installations throughout the country. These services, performed by the Home Corps of the Red Cross, helped address the needs of the families of military personnel.

In World War I, the Home Service Corps (later known as the Social Welfare Aide Service) provided help to families experiencing problems such as illness and marital difficulties. During the war, the Home Service Corps spent approximately $6 million and handled 2 million cases. During the Second World War, this work expanded: Expenditures climbed to $55 million, and the number of cases handled increased to 10 million (Hurd, 1959). The key was coordination with other welfare agencies, and the Red Cross entered into a formal agreement of cooperation with the Family Welfare Association of America.

The Home Corps faced a wide variety of problems. Dulles (1950) presents the following examples of messages received and sent by the Home Corps:

> Incoming: MILITARY AUTHORIZE INFORM FAMILY SERVICEMAN WELL ON ACTIVE DUTY NOT REPEAT NOT THE MAN THEY SAW IN NEWSREEL
> Outgoing: MESSAGE DELIVERED FAMILY MUCH RELIEVED MOTHER IMPROVING
> Incoming: SERVICEMAN REQUESTS MATERNITY REPORT WIFE EXPECTING CONFINEMENT EARLY JULY
> Outgoing: SON BORN JULY SEVEN BOTH WELL
> Incoming: SERVICEMAN INFORMED BY FRIEND MOTHER DIED TWO MONTHS AGO STILL RECEIVING LETTER FROM HER REGULARLY INVESTIGATE
> Outgoing: MOTHER DIED CANCER BREAST APRIL TWENTY SEVENTH SISTER FEARED SHOCK TO SERVICEMAN HAS BEEN WRITING IN MOTHERS NAME WILL WRITE IMMEDIATELY
> Incoming: SERVICEMAN BEGS WIFE DISREGARD HIS LAST LETTER MAILED JULY SEVEN RECEIVED ONE HUNDRED FIFTEEN LETTERS FROM HER YESTERDAY
> Outgoing: MESSAGE RECEIVED WIFE WILL WRITE
> Incoming: SERVICEMAN REQUESTS HEALTH CONFINEMENT REPORT WIFE
> Outgoing: WIFE DIED CHILDBIRTH JULY TWENTY SEVEN SON WELL WITH SERVICEMAN'S MOTHER GETTING GOOD CARE MOTHER WILL KEEP CHILD UNTIL SERVICEMAN'S RETURN (pp. 391–392)

The Home Corps workers made two contributions to the development of service delivery. First, extended help was offered to individuals and their families. The intervention was problem-focused but not time-bound. The Home Corps volunteer helped identify the problem and stayed with each family until it was resolved. Second, the volunteer did not just solve problems, but also became a broker of services. He or she would often coordinate communications and requests for services between the family and the agencies that could provide the help and support. The work often involved helping families communicate with the military (Dulles, 1950; Hurd, 1959).

After World War II, it became increasingly evident that many people needed assistance to improve their quality of life. In the early 1960s, the federal government became increasingly involved in helping people in need.

The Impact of Federal Legislation

This section describes several pieces of federal legislation, passed between the mid-1960s and the late 1980s, which fostered the use of the case management process to provide social services to people in need. What began as an implicit need for integrated services, with the Older Americans Act of 1965, emerged as mandated case management in the Family Support Act of 1988.

THE OLDER AMERICANS ACT OF 1965

The Older Americans Act of 1965 (Public Law 89-73) focused on providing services for older individuals in order to improve their quality of life. Among its contributions to the development of case management was an emphasis on the multiplicity of human needs. A summary of the act follows:

> To provide assistance in the development of new or improved programs to help older persons through grants to the States for community planning and services and for training, through research, development, or training project grants and to establish within the Department of Health, Education, and Welfare an operating agency to be designated as the "Administration on Aging." (Public Law 89-73, 1965)

This act advanced case management by recognizing the need to coordinate care. Section 101 of the act describes its goals and the services to be provided. (See Box 2.2.) The services are designed to meet a variety of needs—financial, medical, emotional, housing, vocational, cultural, and recreational. Faced with multiple needs, case managers must pay attention to the integration of services, so as to avoid fragmentation and duplication. The statute allowed the agencies implementing the act to define how services would be coordinated. Ten years later, the Education for All Handicapped Children Act of 1975 defined coordination of services more clearly.

EDUCATION FOR ALL HANDICAPPED CHILDREN ACT OF 1975

This act (Public Law 94-142) included an explicit case management process, so as to treat the client as a customer. The client was to be involved in identifying the problem; given complete information about the results of the assessment of needs; and empowered to help determine the type of services delivered. The client also participated in the evaluation of the helping process and in any decision to terminate or redirect the activities (Jackson, Finkler, & Robinson, 1992).

The passage and subsequent implementation of this act serves as an excellent example of how federal legislation applies the case management process (Weil & Karls, 1985). This public law, now known as the Individuals with Disabilities Education Act (IDEA), guarantees appropriate public education to all children who have educational, emotional, developmental, or physical disabilities. Congress found that the educational needs of more than 8 million handicapped children in the United States were not being met. According to the

BOX 2.2 The Older Americans Act of 1965

Title I—Declaration of Objectives: Definition

Sec. 101. The Congress hereby finds and declares that, in keeping with the traditional American concept of the inherent dignity of the individual in our democratic society, the older people of our Nation are entitled to, and it is the joint and several duty and responsibility of the governments of the United States and of the several States and their political subdivisions to assist our older people to secure equal opportunity to the full and free enjoyment of the following objectives:

(1) An adequate income in retirement in accordance with the American standard of living.
(2) The best possible physical and mental health which science can make available and without regard to economic status.
(3) Suitable housing, independently selected, designed and located with reference to special needs and available at costs which older citizens can afford.
(4) Full restorative services for those who require institutional care.
(5) Opportunity for employment with no discriminatory personnel practices because of age.
(6) Retirement in health, honor, dignity—after years of contribution to the economy.
(7) Pursuit of meaningful activity within the widest range of civic, cultural, and recreational opportunities.
(8) Efficient community services which provide social assistance in a coordinated manner and which are readily available when needed.
(9) Immediate benefit from proven research knowledge which can sustain and improve health and happiness.
(10) Freedom, independence, and the free exercise of individual initiative in planning and managing their own lives.

SOURCE: Older Americans Act of 1965, Public Law 89-73.

statute, state and local educational agencies would provide services to these children, and the federal government would assist with the funding. All handicapped children are to have access to free public education, including special education and other services to address their unique educational needs.

One tool to assist with planning, implementation, and evaluation was the *individualized educational program* (IEP):

> (19) The "individualized education program" means a written statement for each handicapped child developed in any meeting by a representative of the local educational agency or an intermediate

> education unit who shall be qualified to provide, or supervise the provision of, specially designed instruction to meet the unique needs of handicapped children, the teacher, the parents or guardian of such child, and whenever appropriate, such child, which statement shall include (A) a statement of the present levels of educational performance of such child, (B) a statement of annual goals, including short-term instructional objectives, (C) a statement of the specific educational services to be provided to such child, and the extent to which such child will be able to participate in regular educational programs, (D) the projected date for initiation and anticipated duration of such services, and (E) appropriate objective criteria and evaluation procedures and schedules for determining, on at least an annual basis, whether instructional objectives are being achieved. (Public Law 94-142, 1975)

Although educational agencies have implemented the IEP in numerous ways, it has always been critical that the case manager assume the leadership role of the team of participants. The following implementation procedure was developed in a school district in a city in the southeast.

- Children identified as possible candidates for services were referred to a school district.
- The district assigned a school psychologist to coordinate the testing and gathering of information about the child.
- The school psychologist made an assessment of the child's eligibility for receiving individualized services.
- Once eligibility was established, a team of professionals (including school counselor, teacher, social worker, and nurse) and the parents of the child met to discuss the assessment results.
- Based on those results, the team established goals for the coming year, strategies were developed to meet the goals, and appropriate services were determined.
- At the conclusion of the academic year, a final meeting was held to discuss the outcomes of the past year's work.
- Recommendations for summer activities were made.

In 1988, the Family Support Act was passed. As described here, the act mandated that case management be applied to the process of serving those who were deemed eligible. This marked new status for the case manager.

FAMILY SUPPORT ACT OF 1988

The act was passed with the expressed goal of increasing the economic self-sufficiency of families who receive AFDC. It increased the level of child support enforcement and added a new welfare-to-work program. A new Job Opportunity and Basic Skills program (JOBS) replaced the previously funded Work Incentive Program (Hagen, 1994). One notable characteristic of the act is the explicit authorization of the case management function in implementation. Education, training, employment-related services, and case management were to be available to JOBS clients and their children. Specifically, JOBS was "to assure that needy families with children obtain the education, training and employment that will help them avoid long term welfare dependence" (Family Support Act of 1988). The responsibility for implementing JOBS resides with the state welfare agency. Among the provisions of the act are the following:

- Mothers with children age 3 and older are required to participate.
- States have the option of including mothers who have children 1 year and older.
- Mothers younger than 20 must participate in the educational activities.
- States must guarantee child care.
- Participation of 7% of the eligible population was mandated for 1990, and 20% for 1995.
- Case management services are eligible for federal funds at 60% of the cost or at the state's Medicaid matching rate.

This legislation endorses the importance of the case management function in coordinating services for clients—in this case, young mothers and their children. In the legislation, case management is defined as a process that "must be responsible for assisting the family to obtain any services which may be needed to assist effective participation in the program." In addition, the act encouraged the development of new models and definitions of case management, since states could design their own case management systems (Hagen, 1994).

Just as social legislation was a major factor in the development of case management in the 1960s, 1970s, and 1980s, the advent of managed care during the 1980s has expanded the range of case management in the 1990s. The next section describes managed care and explores its impact on service delivery.

The Impact of Managed Care

The emergence of managed care as a model of health care delivery has increased the demand for case management services and provided new models and definitions of service delivery. To understand its impact, one must first grasp what managed care is. Therefore, we have included in Box 2.3 a list of definitions.

History of Managed Care

Until the 1930s, most medical care in this country was provided on a **fee-for-service** basis. In other words, a patient would be assessed a fee for each health or mental health service provided by a professional. For example, when Mrs. Fowler goes for her annual checkup, she receives a bill for the doctor's

BOX 2.3 Glossary of Managed Care Terminology

Case management Also referred to as Large Case Management. A method of managing the provision of health care to members with catastrophic or high cost medical conditions. The goal is to coordinate the care so as to both improve continuity and quality of care as well as lower costs. This generally is a dedicated function in the utilization management department.

HMO Health maintenance organization. The definition of an HMO has changed substantially. Originally, an HMO was defined as a prepaid organization that provided health care to voluntarily enrolled members in return for a preset amount of money on a per-member-per-month basis. With the increase in self-insured business, or with financial arrangements that do not rely on prepayment, that definition is no longer accurate. Now the definition needs to encompass two possibilities: a health plan that places at least some of the providers at risk for medical expenses, and a health plan that utilizes primary care physicians as gatekeepers.

Managed health care A regrettably nebulous term. At the very least, is a system of health care delivery that tries to manage the cost of health care, the quality of that health care, and access to that care. Common denominators include a panel of contractors/providers that is less than the entire universe of available providers, some type of limitations on benefits to subscribers who use non-contracted providers (unless authorized to do so), and some type of authorization system. Managed health care is actually a spectrum of systems, ranging from so-called managed indemnity, through PPOs, POS, open panel HMOs, and closed panel HMOs. . . .

POS Point of service. A plan where members do not have to choose how to receive services until they need them. The most common use of the term applies to a plan that enrolls each member in both an HMO (or HMO-like) system and an indemnity plan. . . . These plans provide a difference in benefits (e.g., 100% coverage rather than 70%) depending on whether the member chooses to use the plan . . . or go outside the plan for services.

PPO Preferred provider organization. A plan that contracts with independent providers at a discount for services. The panel is limited in size and usually has some type of utilization review system associated with it. . . .

SOURCE: From *The Managed Health Care Handbook,* Second Edition, by P. Kongstvedt, pp. 499–509.

consultation time, the tetanus injection, and the EKG. In the early 1930s, physicians implemented prepaid group plans or managed plans for medical services. This was an alternative way of organizing medical care. The basic concept of a prepaid plan was to guarantee a defined set of services for a negotiated fee. On such a plan, Mrs. Fowler would pay a yearly fee that covered a set of services such as those provided at her annual checkup.

The growth in prepaid group plans was relatively slow until the 1970s. Then the Health Maintenance Organization Act of 1973 (Public Law 93-222)

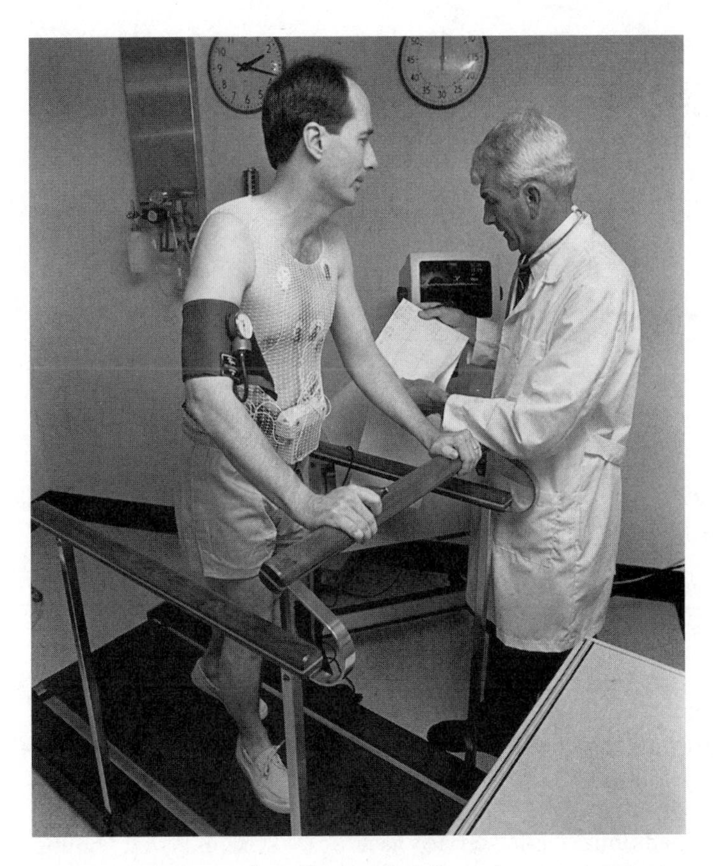

In recent years, managed care has greatly influenced case management and the human service delivery system.

allowed managed medical plans to increase in number and expand the numbers of patients being served (MacLeod, 1993). The prevalence of managed care is now commonly regarded as being connected to the rising cost and decreasing quality of health care and mental health care. The escalation of costs reflects many trends: improved technology; shifting of costs from nonpaying patients to paying patients; an older population; high expectations for a long and healthy life; increased administrative costs; and varying standards of efficiency and quality care (Kongstvedt, 1993).

Defining Managed Care

There are several ways to define **managed care.** First, the term may simply refer to an organizational structure that uses prepayment rather than fee-for-service payment. Second, it can designate the array of different payment plans, such as prepayment and negotiated discounts. It may also imply the inclusion of **quality assurance** practices, such as agreements for prior authorization and audits of performance (Hicks, Stallmeyer, & Coleman, 1993). Third, *managed* *care* may be used to refer to the policy of restricting clients' access to providers such as physicians and other health professionals. Instead, the providers or professionals are paid a flat fee to provide service to a certain group of patients or clients. Most simply stated, managed care is an agreement that health providers will guarantee services to clients within specified limits. The restrictions are intended to improve efficiency of services (Hicks, Stallmeyer, & Coleman, 1993). According to the same authors, the goals of managed care are as follows:

- To encourage decision makers (providers, consumers, and payers) to evaluate efficiency and priority of various services, procedures, and treatment
- To use the concept of limited resources in making decisions concerning services, procedures, and treatment
- To focus on the value received from the resources as well as the lower cost

Models of Managed Care

Three types of managed care models have evolved to meet the goals stated in the previous paragraph: health maintenance organizations (HMOs), preferred provider organizations (PPOs), and point of service (POS). Each has a particular strategy for maintaining cost and ensuring quality.

Health maintenance organizations This managed care model, which is the most structured and controlled, emphasizes positive health promotion. **HMO** is a generic term covering a wide range of organizational structures; it is distinguished from traditional fee-for-service health care systems by combining delivery and financing into one system. Five features of the HMO system are described by Luft, Feder, Holaman, and Lennox (1980):

- The HMO is responsible for developing and implementing contracts that define and state clearly the range of health care services. This can be as limited as preventative medicine or as inclusive as comprehensive long-term care.
- The HMO enrolls a defined population. One example might be all employees of an agency or a corporation such as a hospital, a university, or a state agency.
- With the HMO there are companies and individuals who voluntarily provide services for the HMO.
- The HMO receives a fixed periodic payment based upon the number of individuals who will enroll in the plan. This payment is assessed independent of the actual utilization of services.
- The HMO assumes the financial risk for providing the contracted services.

There are advantages and disadvantages to the HMO model. One immediate benefit is that both the services available and the cost of providing them are constantly monitored by the HMO. Physicians and other health professionals must establish a rationale for services, procedures, and treatment recommended, and their rate is monitored. Client spending is also monitored, since enrolled members can receive services only from the professionals participating in the plan. In some instances, the clients must obtain preauthorization from someone outside the plan. Through its leverage at the "site of services," the HMO can control utilization and improve the efficiency of service delivery (Hicks, Stallmeyer, & Coleman, 1993).

From the client's perspective, the "site of services" restrictions also represent the greatest disadvantage of an HMO. Clients do not like limits on their use of providers; they wish for more freedom to choose. In response to members' demands for freedom to choose their own providers, two other managed care systems have emerged: preferred provider organizations (PPOs) and point of service (POS) (Boland, 1993).

Preferred provider organizations The term **preferred provider organization (PPO)** does not describe any single type of managed care arrangement. Rather, this plan falls between the traditional HMO and the standard indemnity health insurance plan (Miller, 1989; Wagner, 1993). The following characteristics (Gannon, 1985) apply to PPOs:

- The PPOs, like the HMOs, select providers because of their ability and willingness to provide cost-effective and high-quality service.
- The PPO manages the choice of providers with the use of incentives rather than controls.
- The managed care group offers either to waive the deductible (this is the cost that the consumer pays before the insurance pays any fees), to reduce the deductible, and/or to reduce the percentage that the consumer is required to pay.
- The PPO includes a variety of services including hospitals, physicians, and other health professionals.
- The providers participate in the utilization review program and develop cost control information.
- This review board focuses on the volume of patients and the complexity of services provided. Providers can be withdrawn from the PPO if their review is unsatisfactory.

PPOs point with pride to their prompt payment of claims. The providers accept a negotiated discount, which represents the PPO fee, and they do not bill patients an additional amount. Based on the negotiated fee, both the clients and the PPO can anticipate their costs, and providers can anticipate their income. From the providers' perspective, they are assuming a business risk in terms of the fees that they agree to accept. On the other hand, they expect to increase the number of patients under their care. Many providers also maintain independent medical practices.

Point of service The third option of managed care offered today is the **point of service (POS).** This is often adopted by traditional HMO members who want more flexibility than the HMO or the PPO provide. The point of service plan is characterized by the following features:

- Customers are allowed to use out-of-plan providers, and they receive reduced coverage.
- To participate in the POS plan, clients pay higher premiums, higher deductibles, and a higher percentage of the medical fees (Curtiss, 1990).
- In this system, clients are encouraged to use the providers in the managed care system, but they do not use up all their benefits if they choose medical care outside the system.

Managed care has emerged as a response to the fact that employers, governments, payers, clients, and providers are all seeking ways of containing health care costs (Hicks, Stallmeyer, & Coleman, 1993). All three plans presented emphasize management of medical cases, review and control of utilization, and incentives for or restrictions on providers and clients to reduce costs and maintain quality. It is still too early to judge whether managed care systems are achieving these goals, but some clear advantages and disadvantages have emerged. Proponents of the system point to the following advantages of managed care.

- Providers, clients, and payers involved in health care are beginning to prioritize among services, procedures, and treatments provided.
- Providers are more thoughtful about any plan of action prescribed for the client, since they must provide justification for each component of the plan.
- Efficiency of service delivery improves, because services are to be provided in the shortest time needed to meet the goals established.
- From the client's perspective, there is a single, coordinated point of entry into the system.
- Resources are saved.
- Resources are spent according to priorities.

Those who question the virtue of the managed care system focus their concerns in two areas: the quality of services delivered and the efficiency of service. They see the following disadvantages of delivering services through managed care.

- Professionals do not believe that they are offering the best services available, since they are constrained by resource limitations.
- Managed care staff are making judgments about the suitability of proposed treatment without adequate training or professional knowledge.
- Services are not delivered in a timely manner because of the extra layer of bureaucracy.

- Managed care organizations require paperwork that limits the amount of time the professional can give to the client.
- Clients worry about the quality of the care they are receiving.
- Clients do not have access to all of the services that they believe they should have.
- Clients cannot choose their service providers.

Managed care is no longer just one alternative in the health care delivery system; it is part of the structure of service delivery. Managed care is also being adopted by the human service delivery system. This shift to managed care first influenced human services that were directly tied to health care, such as substance abuse and care for the elderly. In these areas, the social service or mental health service delivered has been restricted by the same provider agreements as described in the HMO, PPO, and POS plans. The use of managed care in human services will continue to influence the delivery of services.

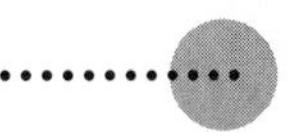

Expanding the Responsibilities of Case Management

Given the various perspectives on case management, its historical development, and the impact of managed care, it will be necessary to revise the case management process for the future. Shifts are evident in client involvement, the roles of the helper, and the emphasis on cost containment. Historically, the case management process has emphasized coordination of services, interagency cooperation, and advocacy, but the process of service delivery is expanding. Other trends have emerged from federal legislation, including coordination of care, integration of services, and the client as a customer. More recently, professionals providing services have been encouraged to empower clients, contain costs, and assure quality services. These shifts in emphasis are reflected in the roles and responsibilities of the people delivering service today.

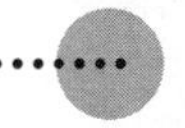

Chapter Summary

This chapter has explored four perspectives on case management and reviewed its historical development. The ramifications of federal legislation, the impact of managed care, and the evolution of expanded roles for the helper and the client add up to a process of service delivery that goes beyond traditional case management.

Chapter Review

Key Terms

Case management
Deinstitutionalization
Managed care
Advocacy
Social diagnosis
PPO
HMO
POS

Reviewing the Chapter

1. Trace the development of the case management function since the 1970s.
2. Describe the shift in the use of the term *case management* that occurred in the 1980s.
3. Why have *service coordination* and *care coordination* become common terms for case management?
4. How has the managed care movement influenced case management?
5. Describe the influence on case management of early organizations such as the Massachusetts School of Idiotic and Feebleminded Youth and the Board of Charities.
6. What emerging case management functions are illustrated by Hull House and the Henry Street Settlement House?
7. How did immigrant populations benefit from the form of case management practiced in the settlement houses?
8. Describe the case management functions of the American Red Cross.
9. What has been the impact of federal legislation on case management?
10. How did federal legislation contribute to the evolution of case management?
11. Trace the development of managed care in the 20th century.
12. Explain the various ways the term *managed care* is defined.
13. Identify and describe the three models of managed care.

Questions for Discussion

1. What part of the history of case management most influences the process today? Cite the reasons for your answer.
2. Why do you think case management developed in the United States over the past century?
3. If you could choose a managed care organization with which to work, what criteria would you use to make your decision?
4. What do you predict will be the future of case management in the United States?

References

Addams, J. (1910). *Twenty years at Hull House.* New York: New American Library.

Boland, P. (1993). Evolving managed care organizations and product innovation. In P. Boland (Ed.), *Making managed care work: A practical guide to strategies and solution* (pp. 151–192). Gaithersburg, MD: Aspen Press.

Brieland, D. (1990). The Hull House tradition and the contemporary social workers: Was Jane Addams really a social worker? *Social Work, 35*(2), 134–138.

Curtiss, F. (1990). Managed care: The second generation. *American Journal of Hospital Pharmacy, 47*(9), 2047–2052.

Dulles, F. (1950). *The American Red Cross.* New York: Harper.

Dunst, C., & Trivette, C. (1989). An enablement and empowerment perspective of case management. *Topics in Early Childhood Special Education, 8*(4), 87–102.

Gannon, J. (1985). PPOs: An alternative health care delivery system. *Dimensions in Health Care, 85*(1), 1–2.

Hagen, J. (1994). JOBS and case management: Developments in 10 states. *Social Work, 39*(2), 197–205.

Hicks, L., Stallmeyer, J., & Coleman, J. (1993). *Role of the nurse in managed care.* Washington, DC: American Nurses' Association.

Hurd, C. (1959). *The compact history of the American Red Cross.* New York: Hawthorne.

Jackson, B., Finkler, D., & Robinson, C. (1992). A case management system for infants with chronic illnesses and developmental disabilities. *Children's Health Care, 21*(4), 224–232.

Kongstvedt, P. (1993). *The managed health care handbook.* Gaithersburg, MD: Aspen Press.

Lieberman, F. (1990). The immigrants and Mary Richmond. *Child and Adolescent Social Work, 7*(2), 81–84.

Lourie, N. (1978). Case management. In J. Talbott (Ed.), *The chronic mental patient* (pp. 159–164). Washington, DC: American Psychiatric Association.

Luft, H., Feder, J., Holaman, J., & Lennox, K. (1980). Health maintenance organizations. In J. Feber, J. Holohan, & T. Marmor (Eds.), *National health insurance: Conflicting goals and policy choices* (pp. 129–180). Washington, DC: Urban Institute.

MacLeod, G. (1993). An overview of managed healthcare. In P. Kongstvedt (Ed.), *The managed health care handbook* (pp. 1–3). Gaithersburg, MD: Aspen Press.

Miller, J. (1989). New product imperative and the exclusive provider organization. In D. Cobbs (Ed.), *Preferred provider organizations: Strategies for sponsors and network providers* (pp. 147–162). Chicago: American Hospital.

Montalvo, F. (1982). The third dimension in social casework: Mary E. Richmond's contribution to family treatment. *Clinical Social Work Journal, 10*(2), 103–112.

Ozarin, L. (1978). The pros and cons of case management. In J. Talbott (Ed.), *The chronic mental patient* (pp. 165–170). Washington, DC: American Psychiatric Association.

Pittman-Munke, P. (1985). Mary E. Richmond: The Philadelphia years. *Social Case Work, 66*(3), 160–166.

Polacheck, H. (1989). *I came a stranger: The story of a Hull-House girl.* Chicago: University of Illinois Press.

Residents of Hull House. (1985). *Hull House maps and papers.* New York: Thomas Y. Crowell.

Richmond, M. (1917). *Social diagnosis.* New York: Russell Sage.
Richmond, M. (1922). *What is social casework?* New York: Russell Sage.
Trattner, W. I. (1994). *From poor law to welfare state: A history of social welfare in America.* New York: Free Press.
Trustees of the Massachusetts School for the Feebleminded. (1919). *Seventy-second annual report.* Boston: Wright and Potter.
Trustees of the Massachusetts School for the Feebleminded. (1920). *Seventy-third annual report.* Boston: Wright and Potter.
Wald, L. (1915). *The house on Henry Street.* New York: Dover Press.
Wagner, E. (1993). Types of managed care organizations. In P. Kongstvedt (Ed.), *The managed care handbook* (pp. 12–21). Gaithersburg, MD: Aspen Press.
Weil, M., & Karls, J. (1985). Key components in providing efficient and effective services. In M. Weil & J. Karls (Eds.), *Case management in human service practice* (pp. 29–71). San Francisco: Jossey-Bass.
Winsor, J. (Ed.). (1881). *The memorial history of Boston 1630–1880* (Vol. IV). Boston: Ticknor.
Woods, R., & Kennedy, A. (1911). *Handbook of settlements.* New York: Russell Sage.

Chapter Three

Models of Case Management

I did a lot of my work the last four years in rehab. I had a lady who had a stroke. She did well on rehab but she needed support. She couldn't go home independently. The son was not in the picture, so we developed a network like Meals On Wheels. . . . You know, we tried to come up with therapy in the home and access to transportation . . . that sort of thing. Then we sent her home and I would follow up with phone calls.

—JUDITH SLATER, Kennesaw State University, Kennesaw, Georgia, personal communication, October 14, 1995

I work with an agency that serves adults with developmental disabilities. Our age group ranges from 16 to at least 78 as our oldest. We have a residential program. We also have a day program. We have independent living. We do everything. I am learning that. I am a service coordinator, which we used to call case managers.

—KIM EHLERS, Mitchell Area Adjustment Training Center, Parkston, South Dakota, personal communication, October 14, 1995

Release plans are forms the inmates fill out about where they want to live and work once they get out. The parole officer investigates both of these. . . . We check out all of that to be sure there is room at the proposed residence and that there is not a conflict there. As far as the job, we want to make sure it is a legitimate business. . . . If we turn it down, we notify the institutional parole officer wherever the inmate is that the plan is rejected for whatever reason. We ask them to resubmit another plan . . . it is up to them to resubmit this.

—ANGELA LARUE, Office of Parole, Knoxville, Tennessee, personal communication, October 18, 1995

This chapter introduces the roles that helpers assume within the context of case management. We also present models of case management, reflecting the creative ways in which services are delivered.

The preceding quotations relate to both the roles that case managers perform in service delivery and the models used to deliver services. Judith Slater, working in hospital rehabilitation in Georgia, supports her clients as a coordinator, expediter, planner, and problem solver. Her agency provides services both in the hospital and after hospital discharge. The Mitchell Area Adjustment Training Center in South Dakota provides services in a residential program, a day program, and independent living. The program makes a long-term commitment to clients, regardless of their abilities or status. This requires flexibility of roles to accommodate the needs of the client. Working from a very different model, Angela LaRue of the parole office has a role that is defined by the state and shaped by her large caseload. Since the parolees assume most of the responsibility for themselves, Angela is primarily a recordkeeper, monitor, and problem solver.

For each section of the chapter, you should be able to accomplish the following objectives:

ROLES IN CASE MANAGEMENT

- Identify the roles in the case management process.

MODELS OF CASE MANAGEMENT

- List reasons why it is important to understand the different models of case management.
- Name the three models of case management.
- Illustrate each model with an example.
- Compare and contrast case management and problem solving.

Roles in Case Management

Each model of case management involves a set of roles to meet the goals established. Among the roles are linking clients with services, actually providing these services, and serving as advocate for the client (Lamb, 1980; Ross, Riffer, & Switchski, 1983). The roles discussed next are associated with service delivery.

Advocate

Advocacy is speaking on behalf of clients when they are unable to do so, or when they speak but no one listens. The case management process presents many opportunities for advocacy. Working at various levels, the case manager represents the interests of the client, to gain access to services or improve their quality. At the organizational level, the case manager influences the policies that control eligibility and access to services. At the legislative level, case managers can work to influence government policies and programs that serve the needs of their clients.

Jim was an advocate for Bryan, a 19-year-old recently admitted to an inpatient program for treating substance abuse. Bryan had participated twice before in a short-term inpatient program, and each time he had returned after six months. As Jim, his care coordinator, evaluated Bryan's past history, it became obvious that the previous treatment had not been effective. Jim petitioned the managed care team and the alcohol and drug treatment team to develop an individualized program for Bryan.

Coordinator

Many clients have multiple problems and need more than one service to meet their needs. In the role of coordinator, the case manager works with all the professionals involved to ensure that services are integrated. The case manager must know the current status of the client and the service delivered, and must assess what progress is being made. Monitoring the client's progress and interfacing with professionals is an important role for the case manager.

Jamie Wolfenbarger assumes the coordinator role for the local hospital's long-term care clients. In this role, she plans the aftercare for patients who will require long-term care. She coordinates previously unrelated services performed by professionals from different agencies. For Rose Woodson, a patient soon to be released from the Cardiac Observation Unit, Ms. Wolfenbarger has arranged home health care visits once a day, a housekeeper to clean twice a week, meals to be delivered at noon each day, and special ambulatory equipment. Ms. Wolfenbarger will contact these professionals each week for the next month for feedback on Ms. Woodson's progress.

Broker

As a **broker,** the case manager links the client with the needed service. Once the client's needs are clear, the broker helps the client choose the most appropriate service and negotiates the terms of service delivery. In this brokering role, the case manager is concerned with the quality of the service available and any difficulties the client may have in accessing it.

When Jo Sinclair assumed the brokering responsibilities for her elderly mother, the job was not an easy one. Jo had returned home from college at the end of her sophomore year to find her mother in poor mental health. Jo's father had left home the previous month, and her mom had sunk into deep depression. As the broker, Jo arranged appointments for her mother to see a psychologist, a physician, and a lawyer. When Jo needs professional help, she calls the Office on Aging; members of their client services staff are available for consultation.

Colleague and Collaborator

Helping professionals must establish and maintain good working relationships with other service providers. To perform the brokering role, linking clients to many services, it helps to know the human service community and to have an established network with other professionals. The case manager may also serve with other professionals on a case management team with the common goal of helping meet client needs. Each member of the team works with trust, mutual respect, and honest communication.

LaTonya Welch works for a disaster recovery agency. She has colleagues and collaborators all over the United States, and she has begun to gather professional contacts worldwide. Her primary responsibility is to provide services in her jurisdiction to individuals, families, neighborhoods, and regions that have experienced a crisis—especially one brought about by a natural disaster. Her secondary responsibility is to provide help in crisis situations elsewhere in the nation. Last month she even worked on an international relief effort that collected clothing and air-freighted it to Chile after a volcano erupted there. Ms. Welch has to have a network of colleagues who can help put together services at a moment's notice.

Community Organizer

Facilitating cooperation among agencies allows professionals who provide direct services to operate as colleagues and collaborators. Case managers involved with community organizing also help their agencies work together to assess the needs of the community and plan how the local human service delivery system will meet those needs. It is sometimes difficult for organizations to cooperate in an atmosphere characterized by competition and limited resources. However, many forward-looking organizations have come to serve their clients more effectively once they have established good working relationships with other agencies in the community.

In his role as chief of staff at the local hospital, Dr. Chin Lee is a community organizer. He is particularly interested in meeting the needs of mentally ill adults and teens. At present he chairs a task force to assess the size and characteristics of the population of young people and adults with mental illness in the community.

Consultant

Often an outside professional can help solve case management problems. An organization may need assistance with such matters as cost analysis, quality control, and organizational structure. A consultant may have the expertise to identify the problem, study it, and make recommendations. Consultants can also assist with case management of individual clients when special information or expertise is needed. This is especially true in small agencies that employ only generalist case managers.

Ann Marsella is a well-known expert on the treatment of young children with developmental disabilities. She is often called in for consultation on particularly challenging cases. Her expertise is in the legal and ethical aspects of serving these children and their families; and she is respected for her ability to make clear the ethical issues involved in a situation and the logical consequences of the proposed alternatives.

Counselor/Therapist

The case manager who is a counselor or therapist maintains a primary relationship with the client and his or her family. Having a thorough understanding of the client's mental health and medical history, this professional can tell what aspects of his or her current situation support or discourage progress.

David Tanaka maintains therapeutic relationships with 15 clients for whom he also serves as a caseworker. He sees each client once a week for an hour and also talks weekly with other professionals involved in each case. He expects to retain these 15 clients for the next two or three years, without increasing his caseload.

Evaluator

Evaluation is performed to determine the client's functioning and to assess service provision. In medical, psychological, financial, social, and vocational areas, the evaluator collects information from the client and from other professionals. The information will then be compiled and recommendations for treatment made.

Meena Shah is a case manager who works with a team of professionals, including a home health nurse, a psychiatrist, and a pharmacist, to conduct in-home assessments of elderly clients in the community. She is trained to complete a comprehensive assessment of functioning and to develop a treatment plan.

Evaluation also occurs during the implementation of the treatment plan. The case manager must evaluate the effectiveness of the plan in meeting the client's goals. This can result in minor or major changes to the plan when appropriate.

Renata Vennearo works in a school setting; she is an expert in working with children who have traumatic brain injury. One of her primary responsibilities is to manage the child's transition from the hospital to school—a difficult matter because the condition is not neurologically stable. Ms. Vennearo constantly monitors the plan and reassesses needs as the client's needs change. Educational programming for these children must be more flexible than for other special education students.

Expediter

Clients encounter many difficulties in the human service delivery system. As **expediter,** the case manager helps the client get through problems of duplicated services, ineligibility, seemingly closed doors, poor service quality, and irrelevant services. The expediter ensures that services are delivered efficiently and effectively. As an expediter, the case manager is performing the advocacy function described earlier.

Judy Yow just called her client, Ms. Vanderhoff, following up after a physical therapy session. She goes to therapy once a week to maintain flexibility and to promote muscular development in her arms and legs. For the second Monday in a row, Ms. Vanderhoff had gone there, only to find the center closed. Acting as expediter, Ms. Yow calls the therapy center and learns that the physical therapist had been in an automobile accident, and the center had been unable to find a replacement. Ms. Yow spends the next hour locating another physical therapist for Ms. Vanderhoff on a short-term basis.

Planner

One of the primary responsibilities of the case manager is planning the treatment that the client is to receive. This includes setting goals, determining outcomes, and implementing the plan with input from the client, other professionals, and other agencies. The case manager's planning role begins in the early stages of the helping process and continues until services are terminated. Planning may include a transition period until the client is able to manage his or her own case.

Tony Nix is in her first year as a case coordinator for the state parole board. She works with juvenile offenders after they have been paroled. Her first interaction with her clients occurs before they are released, and she develops a plan for their integration into their home environment. One of her greatest challenges is to plan for the first weeks after release, since these young people often believe that being released means they can do anything they wish. She includes them in the planning process to solidify their commitment to the stated goals and outcomes.

Problem Solver

The goal of problem solving is to make the clients self-sufficient, by helping them find alternatives to their current situations and learn to solve their own problems. The case manager is continually involved in problem solving; many of the problems arise unexpectedly, and time must be allotted each day for them.

Sonja McCreless has always admired her direct supervisor, Jim Fitzpatrick, because he is an expert problem solver. She remembers his work with a very difficult client, Sue D'Ambrosio, who was scattered and unfocused. Jim Fitzpatrick was able to modify the case management process to be more structured, clearly spelling out the responsibilities of the helper and the client. He made clear what the rules were, what behavior was acceptable, and what would constitute grounds for exclusion from the program. Sue responded positively to structured problem solving and eventually learned how to use the process without the guidance of her case manager.

Recordkeeper

Throughout service delivery, it is necessary to document assessment, planning, service provision, and evaluation. Detailed information is important for providing long-term care, communicating with other professionals and agencies, and monitoring and billing for services. Good documentation constitutes the linking element in the case management process. Many electronic systems of information management can help record, track, plan, monitor, and evaluate client progress, but the key is the quality of the data entered.

Eli Brawley works with families and children having severe medical problems. He makes detailed computer records of his activities. Each client has a file that contains a record of every interaction and action taken for or with that client. This serves as the official record of service delivery as well as the basis for accountability and quality assurance evaluations.

Service Monitor and System Modifier

In human service systems, case managers often give advice or alter the system so that agencies can work better together. They also evaluate how well the collaboration among agencies is helping the target populations. These responsibilities are usually assumed by those case managers who have administrative responsibilities. They possess the authority to change agency policy or redirect priorities, in addition to having the respect of the community and their colleagues.

Heather Lagenly is the director of the Office on Aging. She also serves on the board of the Professional Society for Geriatric Care, a group of community professionals committed to providing quality services for the elderly. One goal of the society is to make sure that the system offers a variety of services to the target population, does not duplicate services unnecessarily, and provides services in a coordinated way. In its sixth year, the group is deciding how to deal with the growing number of elderly people who wish to be cared for in their own homes.

TABLE 3.1 MODELS OF SERVICE DELIVERY

	Models		
	Role-based	*Organization-based*	*Responsibility-based*
Examples	Generalist	Comprehensive service center	Family
	Broker	Interdisciplinary team	Supportive care
	Primary therapist	Psychosocial rehabilitative center	Volunteer
	Cost containment		Client as case manager

Models of Case Management

This section introduces three modern models of case management. For several reasons, it is important to understand the different models. First, they demonstrate that service delivery can occur in a variety of ways; it is a flexible process. Second, the different goals that characterize each model give perspective on the case manager's responsibilities, roles, and length of involvement with the client. Third, each model has particular strengths and weaknesses. Knowing them makes it possible to determine in which particular settings each model is most relevant.

The three models presented are role-based, organization-based, and responsibility-based (see Table 3.1). The **role-based case management** model centers on the roles the case manager is expected to perform. One case manager may act primarily as a broker of services, whereas another is a therapist who has occasional brokering responsibilities. Another case manager may concentrate on cost containment and cost efficiency. Judith Slater's chapter-opening quote exemplifies the broker model; her major responsibility is linking the client to needed services.

Organization-based case management is a model that focuses on ways of configuring services to be comprehensive and meet the needs of clients with multiple problems. This model can be applied to work in a variety of situations, such as a comprehensive service center, a psychosocial rehabilitative center, or an interdisciplinary team approach. A quotation at the beginning of the chapter described an agency that serves individuals with developmental disabilities; the organization-based model applies well in this case. The agency provides a range of services, all are available in one location. Some clients live at the center full-time.

In the **responsibility-based case management** model, the case management function may be performed by family members, a supportive care network, volunteers, or the client. In the third chapter-opening quotation, Angela LaRue described how the the parolee assumes responsibility for a release plan. At her agency, the client has many responsibilities and is empowered to find a job and housing.

In the material that follows, each model is described according to the following characteristics: the goals of the process; the responsibilities and roles of the case manager; the length of the process; the strengths of the model; and its weaknesses. Illustrations of each model follow the description of its characteristics.

Role-Based Case Management

The role assumed by the designated case manager characterizes this model of case management. Roles may vary according to the function and the services provided (Austin, 1981; Jackson, Finkler, & Robinson, 1992; Mullahy, 1995; Weil, 1985; Weil & Karls, 1985). Following the characteristics of this model, we present four illustrations: the case manager as a generalist, a broker, a therapist, and a cost container.

Goal The case manager attempts to meet all the needs of the client through a single point of access. This may require the case manager to serve as the link to a variety of needed services, to be the provider of therapeutic care, and/or to monitor the efficiency and quality of services.

Responsibilities The case manager assumes a broad set of responsibilities, including intake interviewing, data gathering, planning, linking to services, coordinating and/or delivering services, referral, and evaluation.

Primary roles The roles in this model can include assessor, broker, collaborator, counselor/therapist, evaluator, expediter, planner, problem solver, recordkeeper, and service monitor.

Length of involvement The duration of case management varies according to the complexity of the client's needs, the limitations managed care places on service delivery, and the need for short-term or long-term care.

Strengths In many cases, there is a single point of access for the client. The case manager assumes the various roles as needed. Together they identify problems and develop a plan of services. The case manager remains closely involved, as both provider and monitor of services. This involvement promotes a strong relationship with the client, who has regular access. The case manager also provides assistance with the financial aspects of the medical and mental health systems.

Weaknesses Some agencies limit the services they provide; case managers may have large caseloads, no backup, and limited time for community involvement. Limited services may also mean more referrals, incomplete assessment, and a narrowing of the focus to only one perspective on service delivery. Where the case manager's role is oriented toward cost containment, there may be a focus on managed care concerns or a standard of care that does not correspond to the client's needs.

ILLUSTRATIONS

The first illustration of role-based case management involves the case manager as a generalist. This role, widely used in human services, focuses primarily on providing the services that can be delivered by a helper with knowledge and skills applicable to a range of clients in various settings. Brokering or linking the client to other services occurs infrequently.

> The Philadelphia Post Hospital Community Care Project in Philadelphia, Pennsylvania, provides short-term care to elderly individuals in their homes immediately after their discharge from acute-care hospitals. The goals of the service coordinators working in this project are to "help clients attain or maintain independence, prevent rehospitalization, improve functioning and quality of life, and prevent institutionalization." (Edelstein & Lang, 1991, p. 268)

Community Care Project case coordinators begin their work while the patients are still in the hospital. They conduct initial assessments, talk with the patient's support network, and prepare the home environment by restoring utilities, paying rent, and applying for entitlements for other services. Once the client returns home, the coordinators arrange for services such as light housekeeping. They also shop for groceries, cook meals, and buy medicines. In addition, the service coordinators provide counseling services. Clients express satisfaction with the services, and professionals involved with the project agree that the service coordination function is critical to the success of this aftercare. (Edelstein & Lang, 1991)

When the case manager's work focuses on the role of broker, the emphasis is on linking the client to other services. Interaction with the client is primarily to assess the person and the environment, link him or her to services, and monitor the services delivered by others. The case manager in this role provides fewer direct services than does the generalist.

For example, the broker role is especially relevant in working with elderly people in a rural setting where availability and accessibility of services are challenging. Clients are often "hardy individuals, extremely independent, with less formal education than urban residents" (Parker et al., 1992, p. 52). Such individuals tend to wait before seeking care, and many are in crisis by the time they enter the social service delivery system. One case manager's work in rural Wisconsin illustrates the broker role.

> Ann, a 76-year-old woman who lives alone in a rural area, has had Parkinson's disease for over 10 years. Nevertheless, she is determined to live alone. Her human service worker coordinates services with more than ten agencies and organizations. Over the years, this worker has

> established rapport with Ann and has convinced her of the need for special services. The helper has also coordinated the services described next that are common to long-term care for the elderly.
>
> Ann was in the hospital twice last year, once for a severe urinary tract infection and the other time for peripheral vascular problems that accompany Parkinson's disease. She receives Meals-on-Wheels daily, prepared by a center 20 miles away. The delivery person also has been trained to note any variation in Ann's condition that might require emergency medical or social assistance. Ann's daughter visits each weekend to bring groceries and to cook and freeze the week's meals. A housekeeper visits once a week to clean the home. During the winter, when this care is interrupted, a neighbor checks on Ann.
>
> In addition, Ann has assistance for lawn care, plumbing, and a communication system to use in emergencies. Ann's daughter takes her into town for medical care. Recently, the human service worker arranged for Ann to have a cushion lift chair, which has increased her mobility. She also receives fuel assistance. (Parker et al., 1992)

In some cases, the case manager is a counselor or a therapist, personally providing this service. Often in the mental health field, the client seeks the help of the counselor for a presenting problem. The counselor presents credentials and areas of expertise and works with the client to determine whether there is a match with the client's problem. Referral elsewhere occurs when there is not a good match, or if the client later experiences a crisis during the helping process.

Althea's family encouraged her to see a therapist in private practice. They believed she was suffering from depression. The therapist saw her once a week for five years and three months. At the beginning of treatment, the therapist conducted a mental health assessment using the Minnesota Multiphasic Personality Inventory (MMPI). On occasion, the therapist referred Althea to professionals for help with nutrition, exercise, and part-time employment. Althea and the therapist mutually agreed to conclude the treatment at the end of five years.

In many situations, clients need services because they become physically or mentally ill, sometimes catastrophically so. There is often a case manager responsible for cost containment. The case manager must make recommendations after considering outcomes, levels and quality of care, and the expense. The case manager gets the client, family, and professionals involved in the decisions. Negotiating services to find the best quality for the lowest price is one responsibility of the case manager. He or she also documents the negotiated arrangements, prepares a cost analysis statement, and tracks all interactions between the client and the providers. Box 3.1 lists the information included in a standard cost analysis; Box 3.2 gives an example.

BOX 3.1 Summary of Cost Containment Analysis Variables

1. Identifiers (file, case, or social security number; carrier, employer group; date case opened; date case closed; total weeks or months in case management; diagnosis)
2. Overview of case management intervention
3. Summary of intervention
4. Case management fees
5. Savings, such as avoided charges, potential charges, discounts and/or negotiated reductions, reductions in services, products, and equipment
6. Actual charges
7. Gross savings (potential charges minus actual charges)
8. Net savings (gross savings minus case management fees)
9. Status of case (open or closed)

SOURCE: From *The Case Manager's Handbook,* by C. Mullahy, pp. 283–284.

Organization-Based Case Management

In this model, the nature of case management is determined by the organizational structure of the agency or organization. In other words, it is the way in which the services are arranged that determines how services are delivered. The three examples of organization-based case management presented here illustrate the collaborative atmosphere that exists among many helping professionals. Each person has a specific assignment and responsibility. The services are organized so that the relationships between professionals are integrated to serve the clients' needs better.

Goal In this model, case managers meet multiple needs through a single point of access, with one location for service delivery. This comprehensive service delivery sometimes resembles that provided by the traditional extended family.

Responsibilities This model provides comprehensive case management: Each client receives an individual assessment and plan that may include social support, housing, recreation, work, and time to integrate into the community. The case manager's responsibilities range from coordinating services (supervision of intake, assessment, planning, brokering, monitoring, and termination) to leading a team of professionals who provide services to the client. Within the second scenario, the case manager role varies: Sometimes there is a professional whose primary responsibility is management of the case, and at other times the one who initiates services also assumes the management role.

BOX 3.2 Case Report

HCMC#: 5430
Group: Company X 123-100
Date of referral: 1/2/90
Diagnosis: Bronchopulmonary dysplasia
Status: Open

This child remains home vs. hospitalization. R.N. care is provided 16 hours/day × 7 days plus oxygen and medical supplies. Child has expended $100,000 of nursing charges this year on or about September 10.

Referral has been made and approval granted for Care at Home program retroactive to September 1, 1990. Care at Home will share cost of home care with carrier and family and referral to and approval for this program was through direct and ongoing intervention by Health Care Management Consultants.

Cost savings are represented by prevention of hospitalization, reduced cost for DME and supplies, and stabilization of condition. Parents' requests for increased services have not been deemed warranted.

Avoidance of potential charges:	
Potential hospitalization 90 days × $1,000	$90,000.00
Cost of P.T., O.T., S.T. $50 × 5 treatments/wk × 12 wk	3,000.00
	$93,000.00
Actual charges:	
R.N. care 16 hours/day × 7 days @ $520/day × 90 days	$46,800.00
DME and supplies	15,600.00
	$62,400.00
Case management fees	$ 1,072.00
Total actual charges:	$63,472.00
Net savings (potential charges minus actual charges):	$29,528.00

SOURCE: From *The Case Manager's Handbook,* by C. Mullahy, pp. 247–248.

Primary roles Advocate, broker, colleague/collaborator, coordinator, evaluator, expediter, planner, problem solver, recordkeeper, service monitor, service modifier.

Length of involvement In this model, the duration varies. If the case is complicated and several specialists are needed, services are provided longer. In other cases, short-term service is adequate.

Strengths Services are provided on an inpatient, outpatient, or residential basis, but all in one location. Client assessment is multifaceted, with a holistic approach. The plan is individualized and easily monitored. Staff members

Directory of a comprehensive service center

function as a team, with a common goal, regular meetings, and a common reporting scheme.

Weaknesses Resource availability may be a problem if the client needs services not available in the center. Service integration depends on clear organizational structure and lines of authority; the staff must agree on the problem, the plan, and the implementation. Resource availability can be a problem. The family of the client may be less engaged in the helping process than in other models. Also, the client can potentially become accustomed to the environment and never grow beyond it.

ILLUSTRATIONS

Many case managers see the multi-service center as the ideal. They believe it reduces the risk of people getting lost in the system, clients' feelings of frustration, and duplication of information, forms, and services. Clients can have a single intake interview and just one assessment. They have one case manager who has access to all the professionals involved in the delivery of services.

An example of the use of comprehensive service centers is a new approach to public welfare in the state of Massachusetts. In some service centers, this approach has transformed the eligibility worker into a case manager. The ultimate goal of this model is to move families toward self-sufficiency. Case managers' first responsibility is to collect information to determine eligibility; next, they must assess the client's use of other social support services, such as employment and training, child care, health services, housing services, and medical benefits. All these services are available in close proximity to the case manager. At the time of intake, the case manager and the client develop an individualized Family Independence Plan that reflects the needs of the client and the agencies involved in treatment.

This program is unique in that the services are located in close proximity. Another characteristic of the program is a definition of the case management team that includes all employees, managers, supervisors, case managers, specialists, and clerks. Everyone from the same office is trained together in the operations of case management, problem solving, and teamwork, and all are committed to helping clients find a route out of poverty (Pillsbury, 1989).

Another example of the organization-based model is the interdisciplinary team, which is particularly effective for clients with complex problems that require the involvement of several professionals. This approach brings specialists together, with mutual responsibility to help the client. Difficulties may arise when they do not agree, when the case manager and/or the specialists have too heavy a caseload, or when resources are insufficient for the services needed.

The Training in Community Living Model developed by Stein and Test (Solomon, 1992) is an example of the interdisciplinary team approach. It serves adults who have severe mental illness. This program provides all the services needed, with no brokering required. A team consisting of a coordinator, a nurse, and a psychiatrist delivers services in the client's own environment. These professionals try to teach the clients the daily living skills they need to live in the community, such as symptom recognition, medications, and money management. Each team has a manageable caseload and provides continuous care. Support is available 24 hours a day, seven days a week. No one is rejected from the program for lack of compliance, and all services are based on the client's needs.

Many professionals regard the psychosocial rehabilitative center as an excellent treatment approach. The care is comprehensive, and the client receives social support. Among the criticisms are that these centers may resemble institutions, and that they can be very expensive.

One such program is managed by the Eastern Panhandle Mental Health Clinic in a semirural area in Martinsburg, West Virginia (Pinney, 1991). The clinic has a total of about 700 clients, of whom 300 have mental illness. The program includes dispensing medication as a way to bring clients to the center. For example, medications are dispensed at 10:00 A.M., when snacks are served and special programs are provided. Then, the group work is expanded to include management of medications, management of symptoms, the development of self-care and social skills. Linkages are developed with other community resources, such as the public library, a vocational training center, and the department of human services. The clients complete a self-assessment each month and are actively involved in planning their treatment. The client's responsibility is emphasized.

Responsibility-Based Case Management

This model emphasizes the importance of having an individual or a team accountable for the care of the client. In too many instances, clients are shuttled among many human service agencies, with little or no coordination. Other clients have many needs that go unmet because no person or team takes responsibility to assess needs and find the necessary services.

Goal This model emphasizes long-term involvement of the case manager, the coordination of services, the help of volunteers, and the empowerment of clients.

Responsibilities The individual or group responsible for case management provides coordination, finds assessment services, and networks with others in the human service delivery system to access needed specialists and services. Problem identification, plan development, and implementation are other responsibilities. The case manager also provides support and assistance in making and maintaining other linkages.

Primary roles Broker, colleague/collaborator, consultant, evaluator, expediter, planner, problem solver, recordkeeper, service monitor, and service modifier.

Length of involvement The involvement may be short-term, during a crisis or developmental problem, or long-term, as with a physical or mental illness, a disability, or geriatric problems.

Strengths This model allows case management responsibilities to be assumed by various individuals or groups, including the family, neighbors, volunteers, and the client. In many cases, the designated case manager may already have an established relationship with the client. His or her involvement helps the client who does not have easy access to services. Under this model, service delivery is cost effective, the community is involved, and independence is encouraged.

Weaknesses In some cases, the person designated as case manager may not have the client's best interests at heart, may lack the necessary knowledge and skills, or may be ineffective in monitoring service provision. Training and supervision may be costly. Accepting family members or volunteers as case managers may be difficult for the client. Case managers who are not part of the human service delivery system may have trouble coordinating and gaining access to services.

ILLUSTRATIONS

It is a current trend in human services to ask families to act as case managers and then provide them with the support to do so. With costs escalating and institutional care being replaced by community care, it is cost effective for families to perform this central role. For such a system to be effective, the family must receive continuing education about the human service delivery system and must have professional help available when crises arise.

Project Continuity, a project supported by the University of Nebraska Medical Center in Omaha, Nebraska, involves families as care managers. This project has developed a care coordination effort that supports family-centered intervention for children with chronic illnesses and developmental disabilities. One goal of the project is to use service coordination to "support a family's capacities and competencies to identify, obtain, coordinate, monitor and evaluate resources and services to meet its needs" (Jackson, Finkler, & Robinson, 1992, p. 224).

In this project, infants and toddlers are identified and receive comprehensive services to meet their needs and those of their families. This includes the provision of services during hospitalization, assistance in transition from hospital to home, and follow-up when children return home. The service coordination functions include coordinating assessments, coordinating the Individual Family Service (IFS) Plan, assisting in securing services, and assisting in the necessary advocacy to meet the rights under Public Law 99-457. The ultimate goal of the IFS plan is to empower the family as care coordinators. (See Table 3.2.) Most of the activities specified in the IFS plan give the family the information they will need to assume responsibility for care (Jackson, Finkler, & Robinson, 1992).

The responsibility-based case management model also applies to supportive care, which is used most often when clients cannot travel to receive services. Most affected are people located in rural areas and those who live alone—often persons who are disabled or elderly. Sometimes the services provided are minimal, but they are helpful to people who otherwise would not be served at all.

One supportive care program was developed by Raymond Raschko of the Community Mental Health Center of Spokane, Washington (Emlet & Hall, 1991). The program uses "gatekeepers"—individuals who happen to have contact with elderly prospective clients in their day-to-day work.

TABLE 3.2 IFS PLAN FOR JANE SMITH

Bill & Linda Smith	Parents
Joanie Dinsmore	Nurse Specialist
Penny Lees	Child Life
Jean Tomasek	Social Work
Barb Jackson	Education

Identified outcomes	*Plan*	*Person responsible*
1. The Smiths will receive developmental support for Jane.	a. Developmental assessment will be completed prior to discharge.	B Jackson
	b. Referral will be made to Bellevue Public Schools.	Smiths
	c. Parents will be given suggestions for appropriate developmental intervention.	B Jackson
	d. Inpatient intervention times will be arranged with the parents.	P Lees
2. The Smiths will be directed to resources providing family support, to use at their discretion.	a. Parents will be given information on Nebraska Respite, Pilot Parents and Family Friends.	J Dinsmore
	b. Ms. Jones from Bellevue Public Schools will be contacted to identify other military families of children with special needs who may serve as a resource.	Smiths
3. The Smiths will have opportunity to review financial resources for medical care and equipment.	a. Arrange for consultation with Jean Tomasek, MSW.	J Dinsmore J Tomasek
4. The Smiths will receive supportive information regarding intervention with their daughter Jane.	a. Child Life will consult with the Smiths regarding information and visitation issues for Jane.	P Lees
	b. The Smiths will be provided with a list of available resource books for siblings.	P Lees J Dinsmore

(continued)

TABLE 3.2 IFS PLAN FOR JANE SMITH *(continued)*

Identified outcomes	*Plan*	*Person responsible*
5. Jane learns to communicate with others. Uses several ways to signal for more, e.g., smiles, vocalizes, increases movement, quiets.	It is important to set up a variety of situations where you can play simple games like rocking, talking to or pat-a-cake with Jane. Observe to see if Jane begins to signal for more. These signals might be in the form of a movement, a smile, or quieting, which is her way of saying that she wants more. If any indication is given, you immediately respond and play the game some more. Also see if she begins to anticipate any familiar games.	Parents Pediatric nursing staff
6. Jane anticipates familiar routines and games.		B Jackson P Lees J Bell
7. Jane sits balanced with little support from the adults.	Sit Jane on the mat between your legs facing away from you. Dangle toys in front of her to keep her attention and gradually reduce your support to see if she can sit for brief periods. If she falls back, your body will block the fall.	Smiths
8. Jane will pivot on her stomach to retrieve toys. Jane will bear weight on one hand/arm while she reaches and plays with a toy.	Present toys to Jane as she is on her stomach, slightly out of reach on her right or left side. See if she will move to attempt to retrieve it. If necessary, physically guide Jane's hips and shoulders in a pivoting movement to reach a toy. Also when Jane is on her stomach, offer a toy and see if she will shift her weight in order to reach for it.	Smiths

SOURCE: From *Family Centered Service Coordination for the 90s,* by J. Dinsmore and B. Jackson, Appendix C.

Elderly individuals who live alone are one type of client served by the Gatekeeper Program. Ms. G. is 80 years old and has been widowed for 27 years. Recently, she sprained her ankle on her weekly walk to the grocery store and has not left her apartment for four weeks. Her apartment manager, realizing that he had not seen her for a while, visited with her one afternoon. During the visit, it became clear to him that she was having difficulty caring for herself. She had lost weight and she was unkempt; her apartment was dirty and messy. After the visit, he called the Gatekeeper Program to get assistance for her.

Because of the rising cost of service delivery, many agencies and communities have developed strong volunteer programs. The agencies provide excellent training, ongoing education, and good supervision, thus allowing volunteers to assume case management responsibilities. In this way, the volunteers are able to contribute to the welfare of their local community.

An example of a volunteer program is at the Montefiore University Hospital, part of the University of Pittsburgh Medical Center. The program is designed to help citizens over 70 years of age remain in their own homes. Five neighborhoods participate in this project, which is directed by a team (nurse, social worker, environmental assessment specialist, volunteer coordinator, and program administrator). Since the case managers for the Department on Aging have caseloads of over 200 clients, this program supplements the Department on Aging Care.

The day-to-day contact with elderly clients is maintained by volunteers, who are supervised and assisted by community workers—paraprofessionals who themselves live in the neighborhoods. The services the volunteers provide might include grocery shopping, cooking, cleaning, companionship, and making others aware of any special needs or emergencies. One volunteer is Ms. F., an Italian immigrant, who was herself a client in the program when her husband was terminally ill. Now, she uses her language skills to help clients who speak only Italian (Bogdonoff, Hughes, Weissert, & Paulsen, 1981).

The responsibility-based case management model is most applicable when the client is case manager. This approach develops the maximum potential of the client. Self-determination—the right to establish one's own goals—is of primary importance here. The belief is that a client who has learned to act as manager can provide long-term care for himself or herself and for others. Indispensable in this approach, as in the others, is the availability of helping professionals to train, provide support, and (in times of crisis) provide services.

Clients served as case managers in one program developed by the Colorado Division of Mental Health. The goal of this project was to help four community health centers in the Denver area to employ 20 clients on a job-sharing basis, to fill 10 full-time case manager positions. Those selected received preliminary training and were then matched to their job sites, where they received on-the-job training. A network, including a weekly support group, was set up to give them continuing help. One of the case management aides wrote, "It is both rewarding and frustrating working with people with problems of mental illness. Some of the most gratifying aspects of this job are when the [professional] case managers come to us to ask us how we see things from our own perspectives, and seeing some of our clients really get a handle on their lives" (Sherman & Porter, 1991, p. 498).

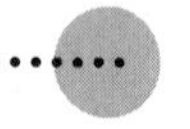

Case Management and the Problem-Solving Process

Problem solving, an integral part of the helping process, can be defined in many ways. For Brill (1995), problem solving is an orderly way of thinking and planning. Egan (1990) views it as a management strategy that can be used to develop opportunities. It is defined by Epstein (1985) as the ability to identify problems and solve them systematically.

We, the authors of this book, have developed a model for problem solving (McClam & Woodside, 1994). It includes three components: problem identification, decision making, and problem resolution. The goals of the model are twofold: solving the problem and teaching the problem-solving process. The roles a helper uses in the problem-solving process are data gatherer, broker, teacher/educator, evaluator, caregiver, planner, and advocate. The helper is a data gatherer and a broker during the problem-identification stage. During the decision-making stage, the roles of teacher/educator, evaluator, and caregiver are particularly important. Problem resolution calls for the roles of planner, advocate, evaluator, and teacher.

Much of the work in case management is directly related to problem solving. What is the relationship between these two processes? What are the distinctions between one process and the other?

Problem solving and case management have many similar characteristics. First, both processes can be outlined and described; they are not linear; they can be learned; both are applied in the helping process. Second, problem solving and case management use many similar roles to reach goals and outcomes—for example, data gatherer, broker, and evaluator. Third, both are grounded in concepts of helping, such as empowering clients and critically evaluating the delivery of services.

The difference between problem solving and case management is one of scope. Case management is the larger umbrella of providing services for clients, especially those with multiple needs. It includes assessment, the development of treatment plans, and the implementation of service plans through a process of providing and coordinating services. Problem solving is subsumed in case management as one of the responsibilities that a case manager might assume and one of the tools that can be used. It is especially appropriate that the case manager use the problem-solving process during the planning phase and the implementation of treatment.

A critical component of both case management and problem solving is creativity: finding innovative ways of approaching difficulties and opening up opportunities. When case management incorporates imagination and resourceful problem solving, case managers and clients are more likely to be satisfied with the process (McClam & Woodside, 1994).

Chapter Summary

Case managers perform many roles, which form the basis of their professional responsibilities. Three different views of case management are the role-based, organization-based, and responsibility-based models. Studying them illuminates the various methods of service provision, professional roles, and the flexibility of case management. Problem solving is another fruitful way of thinking about case management.

Chapter Review

Key Terms

Advocate
Broker
Expediter
Cost containment
Organization-based model
Role-based model
Responsibility-based model

Reviewing the Chapter

1. Describe situations in an agency setting where you might function in the following roles: planner, advocate, broker, expediter, recordkeeper.
2. Why is it important to understand the different models of case management?
3. How are the three models of case management different?
4. Describe a situation in which the role-based model of case management is applicable.
5. How does the helping professional function in the organization-based model?
6. Describe the characteristics of the responsibility-based model.

7. Compare and contrast the roles of case managers and clients in each of the three case management models.
8. What is the relationship between case management and problem solving?

◆ *Questions for Discussion*

1. Why do you think so many roles are involved in case management? Cite an example to illustrate your answer.
2. Now that you have completed your reading about the models of case management, what can you conclude about the definition of case management?
3. Suppose you wanted to use case management to deliver services to elderly people in your community. How would you determine which model would apply?
4. If you were working as a case manager in children's protective services, how would you determine which case management roles would be needed?

References

Austin, M. J. (1981). *Supervisory management for the human services.* Englewood Cliffs, NJ: Prentice Hall.

Bogdonoff, M. D., Hughes, S. L., Weissert, W. G., & Paulsen, E. (1981). *The Living-at-Home Program.* New York: Springer.

Brill, N. I. (1995). *Working with people: The helping process.* White Plains, NY: Longman.

Edelstein, H., & Lang, A. (1991). Posthospital care for older people: A collaborative solution. *Gerontologist, 31*(2), 267–270.

Egan, G. (1990). *The skilled helper: A systematic approach to effective helping.* Pacific Grove, CA: Brooks/Cole.

Emlet, C. A., & Hall, A. M. (1991). Integrating the community into geriatric case management: Public health interventions. *Gerontologist, 31*(4), 556–560.

Epstein, L. (1985). *Talking and listening: A guide to the helping interview.* St. Louis: Times Mirror/Mosby.

Jackson, B., Finkler, D., & Robinson, C. (1992). A case management system for infants with chronic illnesses and developmental disabilities. *Children's Health Care, 21*(4), 224–232.

Lamb, H. R. (1980). Therapist-case managers: More than brokers of services. *Hospital and Community Psychiatry, 31*(11), 762–764.

McClam, T., & Woodside, M. (1994). *Problem solving in the helping professions.* Pacific Grove, CA: Brooks/Cole.

Mullahy, C. (1995). *The case manager's handbook.* Gaithersburg, MD: Aspen.

Parker, M., Quinn, J., Viehl, M., McKinley, A. H., Polich, C. L., Hartwell, S., Van Hook, R., & Detzner, D. F. (1992). Issues in rural case management. *Family and Community Health, 14*(4), 40–60.

Pillsbury, J. B. (1989). Reform at the state level. *Public Welfare, 47*(2), 8–24.

Pinney, E. (1991). Combining approaches to care. *Hospital and Community Psychiatry, 42*(9), 956–957.

Ross, M., Riffer, N., & Switchski, T. (1983). The experience in New York State. In C. Sanborn (Ed.), *Case management in mental health services* (pp. 101–112). New York: Haworth.

Sherman, P. S., & Porter, R. (1991). Mental health consumers as case management aides. *Hospital and Community Psychiatry, 42*(5), 494–498.

Solomon, P. (1992). The efficacy of case management services for severely mentally disabled clients. *Community Mental Health Journal, 28*(3), 163–180.

Weil, M., & Karls, J. (1985). Key components in providing efficient and effective services. In M. Weil & J. Karls (Eds.), *Case management in human service practice* (pp. 29–71). San Francisco: Jossey-Bass.

Chapter Four

The Assessment Phase of Case Management

First I get the referral. Then I do a screening on the family to find out if they are suitable for our program. I set up an appointment to make a home visit. Sometimes they come into the office first and the home visit is made afterwards. I try first to verify their HIV status through their medical providers or clinic doctor. Then we do an assessment of the family to identify their needs. I follow up with the referral source to find out if all the needs are being met. If not, I make sure that things are put in place.

—Carolyn Brown, Third Avenue Family Service Center, Bronx, New York, personal communication, May 5, 1994

People fill out a preapplication. They do what we call "intake." It gives their status as to jobs, AFDC, or food stamps. The preapplication can tell you quite a bit: their job history, how many people are in their family, their ages, whether or not they have been in training. Some of the basic things like a physical disability will sometimes trigger vocational rehabilitation for us right off the bat. It gives us their education—if they quit in the tenth grade. The initial interview is real important to me because it gives me a lot of basic information.

—Janelle Stueck, Program Manager, Private Industry Council, Knoxville, Tennessee, personal communication, August 30, 1993

What we do first is introduce our agency and ourselves through a letter, and we send an appointment time to them. At that time, if we are lucky enough to get through the door, we do a complete assessment not unlike the Human Resource Administration assessment. It is a little more detailed, a little more personal than what the HRA DHS system does. We go into all areas with children and with the individual, and not only with their personal problems, but also problems with the systems . . . the welfare system and everything else.

—Marta Rivera, Caseworker, Casita Maria Settlement House, Bronx, New York, personal communication, May 4, 1994

Assessment is defined as "the appraisal of a situation and the people involved in it" (Brill, 1995, p. 114). As the initial stage in case management, assessment generally focuses on identifying the problem and the resources needed to resolve it. Specifically, the case manager identifies the initial presenting problem and makes an eligibility determination. As the three quotations show, data are gathered and assessed at this phase so as to see the applicant's problem in relation to the agency's priorities. Identifying possible actions and services and determining who will handle the case are also part of the assessment phase. These activities occur differently at each agency. Carolyn Brown at the Third Avenue Family Service Center does a preliminary screening after she has received a referral. At the Private Industry Council, the initial contact is a preapplication or intake interview. At Casita Maria Settlement House, the initial contact is by letter.

This chapter explores the assessment stage of case management—the initial contact with an applicant for assistance, the interview as a critical component in data gathering, and the case record documentation that is required during this phase. The assessment phase concludes with the evaluation of the application for services. For each section of the chapter, you should be able to accomplish the following objectives:

APPLICATION FOR SERVICES

- List the ways in which potential clients learn about available services.
- Compare the roles of the case manager and the applicant in the interview process.
- Define *interview.*
- Distinguish between structured and unstructured interviews.

- State the general guidelines for confidentiality.
- Define the case manager's role in evaluating the application.
- List the two questions that guide the assessment of the information gathered.

CASE ASSIGNMENT
- Compare the three scenarios of case assignment.

DOCUMENTATION AND REPORT WRITING
- Distinguish between process recording and summary recording.
- List the content areas of an intake summary.
- State the reasons for case or staff notes.

Application for Services

Potential clients learn about available services in a number of ways. Frequently, they apply for services only after trying other options. People having problems usually try informal help first; it is human nature to ask for help from family, friends, parents, and children. Some people even feel comfortable sharing their problems with strangers waiting in line with them or sitting beside them. Or a familiar physician or pastor might also be consulted on an informal basis. On the other hand, some people avoid seeking informal help because of embarrassment, loss of face, or fear of disappointment.

Previous experiences with helping agencies and organizations also influence the individual's decision to seek help. Many clients have had positive experiences with human service agencies, resulting in improved living conditions, increased self-confidence, the acquisition of new skills, and the resolution of interpersonal difficulties. Others have had experiences that were not so positive, having encountered helpers who had different expectations of the helping process, delivered unwanted advice, lacked the skills needed to assist them, were inaccessible, or never understood their problems. Increasingly, clients may also encounter local, state, and national policies that may make it difficult to get services. An individual's prior experiences play a role in the decision regarding whether to seek help.

Recently there has been a policy of relocating homeless families from other parts of New York City to the shelter system in the Bronx. Sandra Flecha, a caseworker at Casita Maria Settlement House, describes the potential recipients of the agency's services:

> The client has to be in agreement so we can work with him. We can't work with someone who refuses our help. We cannot obligate anybody to accept our services. It is a different client from the ones that come in and who already know what we are here for. These are clients that we are soliciting for services. It is a real big difference. Our name has been around for many years. Oh, I still hear, "Oh, my grandson used to go there." Casita Maria has been known for many years in the Latino

> community. The families who are coming to the Bronx don't know about us because they are not from the area. They come because they are going to a shelter system. The shelter system's obligation is to house them. So you might get a family who has lived in Brooklyn and now is placed in the Bronx. They know absolutely nothing about the Bronx. They do not want to be here to begin with. We are offering them services that they feel they don't need. They don't trust you. They don't know you. So it is very difficult for them and for us. (Personal communication, May 4, 1994)

These are families who are already in the human service delivery system. Their feelings about services are probably mixed, given their relocation to a new area. How that relocation has been handled will influence how they feel about involvement with Casita Maria Settlement House.

An individual who does decide that help is desirable can find information about available services from a number of sources. Informal networks are probably the best sources of information. Family, friends, neighbors, acquaintances, and fellow employees who have had similar problems (or who know someone who has) are people we trust to tell us the truth about seeking help. Other sources of information are professionals with whom the individual is already working, the media (posters, public service announcements, and advertisements), and the telephone book. Once people locate a service that seems right, they generally get in touch themselves (self-referral).

Other individuals are referred by a human service professional, if they are already involved with a human service organization but need services of another kind. Or they may be working with a professional, such as a physician or minister, who also makes referrals. These applicants may come willingly and be motivated to do something about their situation, or they may come involuntarily because they have been told to do so or are required to do so. The most common referral sources for mandated services are courts, schools, prisons, protective services, marriage counselors, and the juvenile justice system. These individuals may appear at the agency but ask for nothing, even denying that a problem exists.

The words of case managers at four different locations illustrate the different ways referral can happen. The first speaker works at a state institution.

> I pretty much work on a referral basis. Either referrals are from Leslie [another caseworker], the guidance counselors, teachers, cottage personnel, or the infirmary. Sometimes they are from department heads, although that is not frequent. And sometimes kids are referred during the preadmission interviews and sessions. Students must be referred by the local educational agency. We cannot solicit enrollment. So without a letter of referral from a school district, we can't evaluate them for admission. (Linda Smith, personal communication, October 27, 1993)

The second quote is from a caseworker at an institution that serves the elderly in a large metropolitan city.

> We get clients through referrals—word of mouth, professionals, people from Discharge Planning, agencies that serve the visually impaired, etc. We did do outreach in the community, but that hasn't seemed to net us the most clients. If I look through my book, I would find that most of our referrals in the past have come from the Jewish Guild for the Blind. This particular program, part of the day center, is a joint program of two agencies. Actually I am an employee of the Jewish Guild for the Blind, but I am on contract to the Jewish Home and Hospital, which is the administrator and holds the license for the day care program. The Guild contracts its staff to the Home to operate this part of the day care. So we get a lot of referrals from the parent agency. And again as I said, most of them come from community agencies. Among the city agencies, the access to services has a little more continuity to it. If certain people need certain services, they are required by law to pass them to that agency. If someone is in the Medicaid system and needs a certain type of care, they must be referred to long-term care. (Rosalyn Baum, personal communication, May 3, 1994)

The third speaker works at a local mental health center that coordinates services for clients with chronic mental illness.

> Our referrals began with the bed reduction [at the regional mental health institute] and have grown from that to in-house referrals from other departments within the Adult Center from our social workers, and some from the nurses who are the clients' primary clinicians. We get the referrals and I look over those, and if they are appropriate, then we distribute those out to the case managers. We also get referrals from the acute ward—people come in and out (revolving door). We can get referrals from other mental health agencies when the client wants to move to this area. They may end up being a transfer, and we can pick them up that way.
>
> When I get the referral, the patient's chart is attached. I review that chart to see if any of the criteria are met. Any concerns that I have I take back to the person who referred them. I will ask for more information, and in some instances, if we get referrals from Lakeshore [regional mental health institute], we will do a site assessment. This means I go out, talk with the consumer to find out what their plans are, and to see if they are interested in case management. There is a lot of consumer choice involved. (Paula Hudson, personal communication, April 27, 1995)

The final example is from a large hospital.

> Our committee, which is interdisciplinary—recreation, social work, nursing, and dietary—discusses the referral to see if we can meet some of that person's needs, and if that person can be helped in the program.

> Sometimes there is a question whether he or she can. We then decide if we can give it a try. The person may not have sufficient medical problems or service needs to be eligible for our program. It is our obligation to refer people to the appropriate level of care. (Roz Jaffe, personal communication, May 3, 1994)

These examples illustrate the different ways in which a referral occurs. In an institutional setting, referrals are made by other professionals and departments in the institution. In other settings, referrals can also be in-house as well as from other agencies or institutions. Sometimes the referral procedure may include an initial screening by phone (see Figure 4.1), a committee deliberation, or an individual interview. Usually, the individual who receives the referral checks that the necessary paperwork is included.

Not all applicants who seek help or are referred for services become clients of the agency. Only those who meet eligibility criteria are accepted for services. An intake interview is the first step in determining eligibility.

The Interview

An **interview** is usually the first contact between a helper and an applicant for services, although some initial contacts are by telephone or letter. The first helper an applicant talks with may be an intake worker who only conducts the initial meeting, or he or she may in fact be the case manager or service provider. If the first contact is an intake worker, the applicant (if accepted for services) will be assigned to a case manager, who will coordinate whatever services are provided. In this text, we will assume that the intake interviewer is the case manager.

The initial meeting with an applicant takes place as soon as possible after the referral. The interview is an opportunity for the case manager and applicant to get to know one another, define the person's need or problem, and give some structure to the helping relationship. These activities provide information that becomes a starting point for service delivery, so it is important for the case manager to be a skillful listener, interpreter, and questioner (skills that you will read about in Chapter 5). First, we will explore what an interview is, the flow of the interview process, and two ways to think about interviews that you may conduct as part of the intake process.

Interviewing is a critical tool for communicating with clients, collecting information, determining eligibility, and developing and implementing service plans—in all, a key part of the case management process. The two central objectives of interviewing are to help people explore their situation and to increase their understanding of it (Egan, 1990). The roles of the applicant and the case manager during this initial encounter reflect these objectives. The applicant learns about the agency, its purposes and services, and how they relate to his or her situation. The case manager obtains the applicant's statement of the problem and explains the agency and its services. Once there is an understanding of the problem and the services the agency offers, the case manager confirms the applicant's desire for services. The case manager is also responsible for

PROJECT LIVE INTAKE FORM

DATE ______________________________

NAME ______________________________ AGE __________

PHONE ____________________

ADDRESS AND DIRECTIONS ______________________________

____

____

____

REFERRAL SOURCE (AND REASON OR RELATIONSHIP) ______________

____

PROBLEM ______________________________

____

____

OTHER ______________________________

Figure 4.1 Phone screening form

recording information, identifying the next steps in the case management process, informing the applicant about eligibility requirements, and clarifying what the agency can legally provide a client. McGowan and Porter (1967) suggest that there are three desirable outcomes of the initial interview. (1) The applicant feels free to express himself or herself. (2) The applicant leaves confident of being able to work with the case manager toward a satisfactory solution. (3) Rapport is established between the two participants.

Exactly what is an interview? In case management, an interview is usually a face-to-face meeting between the case manager and the applicant; it may have a number of purposes, including getting or giving information, resolving a disagreement, or considering a joint undertaking. Epstein (1985, p. 2) describes it as directed conversation, "an event composed of a sequence of physical and mental experiences that occur when and where a helping professional practitioner and a client talk to one another." It has also been described as "professional conversation" that "involves communication between two people" (Garrett, 1972, p. 5).

An interview may also be an assessment procedure. It can be a testing tool in areas such as counseling, school psychology, social work, legal matters, and employment applications. One example of this is testing an applicant's mental status (discussed further in Chapter 6). We can also think of the interview as an assessment procedure in which one of the first tasks is to determine why the person is seeking help. An assessment helps define the problem, and the resulting definition then becomes the focus for intervention.

Another way to think about defining the interview is to consider its content and process (Enelow & Wexler, 1966). The *content* of the interview is what is said, and the *process* is how it is said. Analyzing an interview in these terms gives a systematic way of organizing the information that is revealed in the interaction between the case manager and the applicant. It also facilitates an understanding of the overall picture of the individual. This is particularly relevant in case management, since many clients have multiple problems. You will read more about process and content in the next chapter.

Ivey (1988) cautions that although the terms *counseling* and *interviewing* are sometimes used interchangeably, interviewing is considered the more basic process for information gathering, problem solving, and the giving of information or advice. An interview may be conducted by most anyone—business people, medical staff, guidance personnel, or employment counselors. Counseling is a more intensive and personal process, often associated with professional fields such as social work, guidance, psychology, and pastoral counseling. Interviewing is a responsibility assumed by most case managers, whereas counseling is not always the job of the case manager.

How long is an interview? Epstein (1985) states that an interview may occur once or repeatedly, over long or short periods of time. However, Okun (1992) limits the use of the word *interview* to the first meeting, calling subsequent meetings *sessions*. In fact, the actual length of time of the initial meeting depends on a number of factors, including the structure of the agency, the comprehensiveness of the services, the number of people applying for services (an individual or a family), and the amount of information needed to determine eligibility.

A case manager from a community action agency recalled that she had conducted an intake interview in 15 minutes. "When I was in school reading textbooks, I thought I would always have an hour and a half for every interview. And we would just ask questions to get all the information about this person.

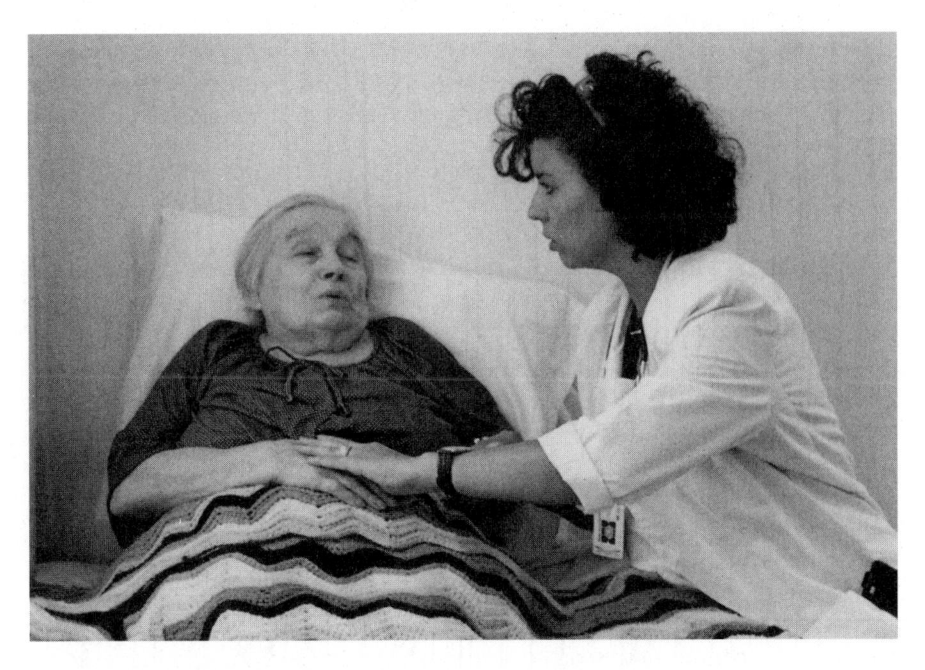

Caseworker making a home visit to a client

Not true!" (Janelle Stueck, Private Industry Council, Knoxville, Tennessee, personal communication, August 30, 1993).

Where does the interview take place? Interviews generally take place in an office at agencies, schools, hospitals, and other institutions. Sometimes, however, they are held in an applicant's home. In such cases, the case manager has the distinct advantage of observing the applicant in the home, which gives information about the applicant that may not be available in an office setting. An informal location such as a park, a restaurant, or even the street can also serve as the scene of an interview. Whatever the setting, it is an important influence on the course of the interview (Hutchins & Cole Vaught, 1997).

What do all helping interviews have in common? Kaplan and Saccuzzo (1993) have identified three factors that are common to all interviews in the human services. These should be present in any initial interview. First, there must be shared or mutual interaction: Communication between the two participants is established, and information is shared by both. The case manager may be sharing information about the agency and its services, while the applicant may be describing the problem. No matter what the subject of their conversation is, the two participants are clearly engaged as they develop a relationship.

A second factor is that the participants in the interview are interdependent and influence each other. Each comes to the interaction with attitudes, values, beliefs, and experiences. The case manager also brings the knowledge and skills of helping, while the person seeking help brings the problem that is causing distress. As the relationship develops, whatever one participant says or feels triggers a response in the other participant, who then shares that response. This type of exchange builds the relationship through the sharing of information, feelings, and reactions.

The third factor is the interviewing skill of the case manager. He or she remains in control of the interaction and clearly sets the tone for what is taking place. Perlman (1979) stresses the helper's *authority* as a characteristic of the helping relationship. This authority stems from knowledge and expertise of the case management process which distinguishes the case manager from the applicant and from any informal helpers who have previously been consulted. Since the helping relationship develops for a specific purpose and often has time constraints, it is important for the case manager to bring this authority to the interaction, in addition to providing information about the agency, its services, the eligibility criteria, community resources, and so forth.

THE INTERVIEW PROCESS

The interview's *structure* refers to the arrangement of its three parts: the beginning, the middle, and the end. The beginning is a time to establish a common understanding between the case manager and the applicant. The middle phase continues this process, through sharing and considering feelings, behaviors, and events. At the end, a summary provides closure by describing what has taken place during the interview and identifying what will follow. Let's examine each of these parts in more detail.

The beginning Several important activities occur at the beginning of the interview: greeting the client, establishing the focus by discussing the purpose, clarifying roles, and exploring the problem that has precipitated the application for services. The beginning is also an opportunity to respond to any questions that the applicant may have about the agency and its services and policies. The following questions are those most frequently raised by clients (Weinrach, 1987).

- How often will I come to see you?
- Can I reach you after the agency closes?
- What happens if I forget an appointment?
- Is what I tell you confidential?
- What if I have an emergency?
- How will I know when our work is finished?
- What will I be charged for services?
- Will my insurance company reimburse me?

Answering these questions can lead to a discussion of the applicant's role and his or her expectations for case management.

The middle The next phase of the interview is devoted to developing the focus of the relationship between the case manager and the applicant. Assessment, planning, and implementation also take place at this time. Assessment occurs as the problem is defined in accordance with the guidelines of the agency. This activity includes consideration of the applicant's eligibility for services in light of the information that is collected. These activities lead to initial planning and implementation of subsequent steps, which may include additional data gathering or a follow-up appointment.

The end At the close, the case manager and the applicant have an opportunity to summarize what has occurred during the initial meeting. The summary of the interview brings this first contact to closure. Closure may take various forms, including the following scenarios. (1) The applicant may choose not to continue with the application for services. (2) The problem and the services provided by the agency are compatible; the applicant desires services; and the case manager moves forward with the next steps. (3) The fit between the agency and the applicant is not clear, so it is necessary to gather additional information before the applicant is accepted as a client.

STRUCTURED AND UNSTRUCTURED INTERVIEWS

Interviews may be classified as structured or unstructured. A brief overview of these two types of interviews is given here; the next chapter will provide more in-depth information about the structured interview, in connection with the discussion of the skills for intake interviewing.

Structured interviews are directive and focused; they are usually guided by a form or a set of questions that elicit specific information. The purpose is to develop a brief overview of the problem and the context within which it is occurring (Long, Paradise, & Long, 1981). It can range from a simple list of questions to soliciting an entire case history.

Agencies often have application forms that the applicant can complete before the interview. If the forms are completed with the help of the case manager during the interview, the interaction is classified as a structured one. Of course, this is a good way to establish rapport, but case managers must be cautious. The interview can easily become a mere question-and-answer session if it is structured exclusively around a questionnaire. Asking yes/no questions and strictly factual questions limits the applicant's input and hinders rapport.

The **intake interview** and the **mental status examination** are two types of structured interviews, each of which has standard procedures. Generally, an agency's intake interview is guided by a set of questions, usually in the form of an application. The mental status examination (which typically takes place in psychiatric settings) consists of questions designed to evaluate the person's current mental status by considering factors such as appearance, behavior, and general intellectual processes. In both situations, the case manager has responsibility for the direction and course of the interview, even though the areas to be covered are predetermined.

In contrast, the unstructured interview consists of a sequence of questions that follow from what has been said. This type of interview can be described as broad and unrestricted. The applicant determines the direction of the interaction, while the case manager focuses on giving reflective responses that encourage the eliciting of information. Helpers who use the unstructured interview are primarily concerned with establishing rapport during the initial conversation.

CONFIDENTIALITY

Our discussion of the initial meeting would be incomplete if we did not address the issue of **confidentiality.** Human service agencies have procedures for handling the records of applicants and clients and for maintaining confidentiality; all case managers should be familiar with them. An all-important consideration is access to information.

Every communication during an interview should be confidential, in order to encourage the trust that is necessary for the sharing of information. Generally, human service agencies allow the sharing of information with supervisors, consultants, and other staff who are working with the applicant. The client's signed consent is needed if information is to be shared with staff employed by other organizations. The exception to these general guidelines is that information may be shared without consent in cases of emergency such as suicide, homicide, or other life-threatening situations.

Applicants are frequently concerned about who has access to their records. In fact, they may wonder whether they themselves do. Legally, an individual does indeed have access to his or her record; the Federal Privacy Act of 1974 established principles to safeguard clients' rights. In addition to the right to see their records, clients have the right to correct or amend the records.

There are two potential problem areas related to confidentiality. First, it is sometimes difficult in large agencies to limit access of information to authorized staff. Support staff, visitors, and delivery people come and go, making it essential that records are secure and conversations confidential (Hutchins & Cole Vaught, 1997). The second problem has to do with **privileged communication,** a legal concept under which clients' "privileged" communications with professionals may not be used in court without client consent (Corey, Corey, & Callanan, 1988). State laws determine which professionals' communications are privileged; in most states, human service workers are not usually included. Thus, the helper may be compelled to present in court any communication from the applicant or client. Sometimes it can be a challenge for the case manager to explain this limitation to an applicant while trying to gather essential information. It can also be perplexing to the applicant.

Evaluating the Application for Services

During this phase, the case manager's role is to gather and assess information. In fact, this process may actually start before the initial meeting with the applicant, when the first report or telephone call is received, and continue through and beyond the initial meeting. The initial focus on information gathering and assessment then narrows to problem identification and the determination of eligibility for services. This process is influenced to some extent by guidelines and parameters established by the agency or by federal or state legislation. At this point in the process, the case manager must pause to review the information gathered for assessment purposes.

Part of the assessment of available information is responding to the following questions:

- Is the client eligible for services?
- What problems are identified?
- Are services/resources available that relate to the problems identified?
- Will the agency's involvement help the client reach the objectives and goals that have been established?

Reviewing these questions helps the case manager determine the next steps. To answer the questions and evaluate the application for services, the case manager engages in two activities: a review of information gathering and an assessment of the information.

Review of information gathering Usually the individual who applies for services is the primary source of information. During the initial meeting, the case manager forms impressions of the applicant. The problem is defined, and judgments are made about its seriousness—its intensity, frequency, and duration (Hutchins & Cole Vaught, 1997). As the case manager reviews the case, these impressions are considered, along with the application for services, case notes that summarize the initial contact with the client, and any case notes that report subsequent contacts. The case manager learns more about the applicant's reasons for applying for services, his or her background, the problem that is causing difficulty, and what the applicant wants to have happen as a result of service delivery. The case manager also uses information and impressions from other contacts. Other information in the file that may contribute to an understanding of the applicant's situation comes from secondary sources, such as the referral source, the client's family, school officials, or an employer. Information from secondary sources that can be part of the case file might be medical reports, school records, a social history, and a record of services that have previously been provided to the client.

An important part of the review of information gathering is to ascertain that all necessary forms, including releases, have been completed and signatures obtained where needed. It is also a good idea at this point to make sure that all necessary supervisor and agency reviews have occurred and are documented.

Assessing information Once the case manager has reviewed all the information that has been gathered, the information is assessed. Many case managers have likened this part of case management to a puzzle. Each piece of information is part of the puzzle; as each piece is revealed and placed in the file, the picture of the applicant and the problem becomes more complete. Margaret Mikol at Sick Kids Need Involved People of New York (SKIP) described it as "a large body of information from all kinds of different people, then you have to sort through that and say, 'Now what are the real issues here?' in a concrete fashion" (personal communication, May 4, 1994). Margaret Lanning, Director of Programs at SKIP, added that workers "need to gather a large quantity of information and break that down, and be tremendously proactive" (personal communication, May 4, 1994). "One of the most difficult pieces of case management," suggests Janelle Stueck, a program manager with the Private Industry Council, "is doing that continual assessment and not resting on what you think things were two weeks ago, because they are different now, you know. And responding differently, differentially each time. It requires an incredible amount of judgment" (personal communication, August 30, 1993).

Knowing what information is in the file, the case manager's task shifts to assessing the information. Two questions guide this activity. Is there sufficient information to establish eligibility? Is additional information necessary? In addition to answering these questions, the case manager also evaluates the information in the file, looking for inconsistencies, incompleteness, and unanswered questions that have arisen as a result of the review.

Is there sufficient information to establish eligibility? In order to answer this question, the case manager must examine the available data to determine what is relevant to the determination of eligibility. The quantity of data gathered is less important than its relevancy. Human service organizations usually have specific criteria that must be met to find an applicant eligible. The data must correspond to these criteria if the applicant is to be accepted for services.

The criteria for acceptance as a client for vocational rehabilitation services is a good example. Vocational rehabilitation is a state–federal program whose mission is to provide services to people with disabilities so as to enable them to become productive, contributing members of society. Essentially, there are three criteria for acceptance for services. First, the individual must have a documented physical or mental disability. Second, the disability must be a substantial handicap to employment. Finally, there must be a reasonable expectation that vocational rehabilitation services will render the applicant fit for gainful employment.

During the assessment phase, a vocational rehabilitation counselor assesses the information gathered to determine whether the applicant has a documented disability. The next step is to document that the disability is a handicap to employment. Does the disability prevent the applicant from returning to work? Or, if the person has not been employed, does the disability prevent him or her from getting or keeping a job? If the answers are yes, the counselor's final task is to find support for a reasonable expectation that, as a result of receiving services, the applicant can be gainfully employed. This brings us to the second main question in the information assessment activity.

Is additional information necessary in order to determine eligibility for services? If the answer is no, the case manager and the client are ready to move to the next phase of case management. If the answer is yes—additional information is necessary to establish eligibility—then a decision is made about what is needed and how to obtain it. In the vocational rehabilitation example, the counselor examines the file for the documentation of a disability. Specifically, the counselor is looking for a medical report from a physician or specialist that will establish a physical disability, or a psychological or psychiatric evaluation that will establish a mental disability. If the needed information is not in the file, the helper must make arrangements to obtain the necessary reports.

Establishment of eligibility criteria is not the sole purpose of this phase. Data gathered at this time may prove helpful in the formulation of a service plan. Certainly the case manager does not want to discard any information at this point; neither does he or she want to leave unresolved any conflicts or inconsistencies. Relevant, accurate information is an important part of the development of a plan. Ensuring relevance and accuracy at this point in the process saves time and effort and allows the helper and the client to move forward without delay.

Case Assignment

Once eligibility has been established and the applicant accepted for services, there are three possible scenarios, depending on the particular agency (Austin, 1981). In all three, the applicant becomes a client who is assigned a case manager to coordinate services.

In many instances, the case manager is the same person who handled the intake interview and determination of eligibility. In some agencies, however, there are staff members whose primary responsibility is conducting the intake interview. After a review, the case is then assigned to a case manager—the helping professional who assumes primary responsibility for the case and is accountable for the services given to the client, whether provided personally, by other professionals at the agency, or by helpers at a different agency.

A second scenario involves the *specialized worker,* a term that may refer to either level specialization or task specialization. Level specialization has to do with the overall complexity and orientation of the client and the presenting problem. Is the case under consideration simple or complex? Is it a case of simply providing requested information, or is it a multiproblem situation? Task specialization focuses on the functions needed to facilitate problem resolution. Does the case require highly skilled counseling, or is coordination sufficient?

The third scenario occurs most often in institutions where a team of professionals is responsible for a number of clients. For example, in a facility for children who are mentally retarded, clients interact daily with a staff that includes a teacher, a nurse, an activity coordinator, a cottage parent, and a social worker. These professionals work together as a team to provide services to each client.

Documentation and Report Writing

Documentation and report writing play critical roles in the assessment phase of service coordination. The main responsibilities facing the case manager in this phase are identifying the problem or problems and determining the applicant's eligibility for services. Documentation of these two responsibilities takes the form of **intake summaries** and **staff notes.** Most agencies have guidelines for the documentation of information gathered and decisions made. In this section, we will discuss the forms of documentation, their purposes, and how to write them. Before we discuss intake summaries and staff notes, let's distinguish between process recording and summary recording.

Process Recording and Summary Recording

A **process recording** is a narrative telling of an interaction with another individual. In the assessment phase of case management, a process recording would show what each participant said by an accurate account of the verbal exchange, a factual description of any action or nonverbal behavior, and the interviewer's analysis and observations. The person making the recording should imagine that a tape recorder and a camera are taking in everything that is heard or seen. Of course, since records are required to be brief and goal-oriented, the helper would not attempt an exhaustive description. But this approach helps focus the reporter's attention on accuracy and impartiality.

We believe, as does Wilson (1980), that process recording is a useful tool for helping professionals in training, especially those who are learning to be case managers. Tape recorders, video cameras, and VCRs are readily available today, but many agencies and organizations don't have them, and case managers may not have time or authorization to use the equipment. Process recording is still an effective way to hone one's skills of direct observation.

Process recording is most often used with one-on-one interviews. It includes the following elements.

- Identifying information: names, date, location, client's case number or identifying number, and the purpose of the interview.

  Paulette Maloney saw the client, Rosa Knight, for the first time on Monday, November 5 at the agency. Ms. Knight is applying for services, and the purpose of the interview was to complete the application form and inform her of the agency's services.

- Observations: description of physical and emotional climate, any activity occurring during the interaction, and the client's nonverbal behavior.

  Ms. Knight appeared in my office on time, dressed very neatly in a navy dress. I asked Ms. Knight to come in, introduced myself, shook her hand, and asked her to sit down. Although there was little eye contact, she smiled shyly with her head lowered. In a soft voice, she asked me to call her Rosa. Then she waited for me to speak.

- Content: an account of what was said by each participant. Quotes are helpful here, to the extent that they can be remembered.

I explained to Rosa that the agency provides services to mothers who are single parents, have no job skills, and have children who are under the age of 6. She replied, "I am a single parent with two sons who are 18 months old and 3 years old. I have worked briefly as a domestic." I asked her how long she had worked and where. She replied that she worked about 5 months for a woman a neighbor knew. She quit "because the woman was always cancelling at the last minute." She related that one time when she went to work, the house was locked up and the family was out of town. During this exchange, Rosa clasped her hands in her lap and looked up.

- Recorder's feelings and reactions: Sometimes this is called a *self-interview,* meaning that the recorder writes down feelings about and reactions to what is taking place in the interview.

I was angry about the way Rosa had been treated in her work situation. She appears to be a well-mannered, motivated young woman who genuinely wants a job so that she can be self-supporting. Her shyness though may prevent her from asserting herself when she needs to. Her goal is to get out of the housing project. I wonder if my impressions are right.

- Impressions: Here the recorder gives personal impressions of the client, the problem, the interview, etc. It may also be appropriate to make a comment about the next step in the process.

Rosa Knight is a 23-year-old single mother of two sons, ages 18 months and 3 years. She has worked previously as a domestic. Her goal is to receive secretarial training so that she can be self-supporting and move from the projects. Based on the information she has provided during this interview, she is eligible for services. I will present her case at the next staff meeting to review her eligibility.

This type of recording is particularly useful in learning interviewing skills. Austin (1981) suggests that as students are conducting an interview, they should use legal-size paper divided into three columns: Supervisory Comments, Content—Dialogue, and Gut-Level Feelings. This format is beneficial both to supervisors and to those who are learning how to interview. First, the supervisor can make comments next to the interaction or feelings recorded. This format also helps the instructor or supervisor identify areas that have been mastered as well as those requiring additional practice. Second, writing everything that is said and everything that happens provides a complete record for the recorder. Initially, the recorder may leave out something that is considered relatively unimportant but may actually be a critical piece of information. The description should include as much information as the recorder can remember. Finally, recording how one felt as something was being said or happening increases self-awareness and helps the learner differentiate among facts, feelings, and impressions.

The other style of recording, summary recording, is preferred in most human service agencies. It is a condensation of what happened, an organized presentation of facts. It may take the form of an intake summary or staff notes (both discussed later in this section). Summary recording is also used for other types of reports and documentation. For example, a diagnostic summary presents case information, assesses what is known about the client, and makes recommendations. (See the Report for Juvenile Court in Chapter 7.) A second example is problem-oriented recording, which identifies problems and treatment goals. This type of recording is common in an interdisciplinary setting or one with a team structure. (An example is the Psychological Evaluation—also in Chapter 7.)

Summary recording differs from process recording in several ways. First, a summary recording gives a concise presentation of the interview content rather than an extensive account of what was said. The focus remains on the client, excluding the case manager's feelings about what transpired. Summary recordings usually contain a Summary section, which is the appropriate place for the writer's own analysis. Finally, summary recording is organized by topics rather than chronologically. The case manager must decide what to include and omit under the various headings: Identifying Information, Presenting Problem, Interview Content, Summary, and Diagnostic Impressions.

Summary recording is less time-consuming to write, as well as easier to read. It is preferred for these reasons—not to mention the fact that it uses less paper, thereby reducing storage problems. Where computerized information systems are used, a standard format makes information easy to store, retrieve, and share with others. A reminder is in order here: Agencies often have their own formats and guidelines for report writing and documentation. Generally speaking, however, the basic information presented here applies across agency settings.

Intake Summaries

An intake summary is written at some point during the assessment phase. It is usually prepared following an agency's first contact with an applicant, but it may also be written at the close of the assessment phase. For purposes of illustration, assume that it is written after the intake interview. After the case manager conducts the intake interview, he or she assesses what was learned and observed about the applicant. This assessment takes into account the information provided by the applicant, forms that were completed, and any available information about the presenting problem. The case manager also considers any inconsistencies or missing information. While integrating the information the case manager also considers the questions that are presented in the previous section. Is the applicant eligible for services? What problems are identified? Are services/resources available that relate to the problems identified? Will the agency's involvement improve the situation for the applicant?

This information is organized into an intake summary, which usually includes the following data.

- Worker's name, date of contact, date of summary
- Applicant's demographic data—name, address, phone number, agency applicant number
- Sources of information during the intake interview
- Presenting problem
- Summary of background and social history related to the problem
- Previous contact with the agency
- Diagnostic summary statement
- Treatment recommendations

Figure 4.2 is a sample intake summary from a treatment program for adult women who are chemically dependent. To qualify for the residential program, applicants must have a child who is 3 years of age or less and has been exposed to drugs. The day program is available to mothers who have a child older than 3 years. The applicant whose intake is summarized in Figure 4.2 is applying to the residential program, which lasts one year. During this time children live with their parent in one of ten agency apartments. To graduate from the program, the client must be employed or enrolled in school and be free of substance abuse. Some individuals enter the program voluntarily, and others are ordered to come as part of their probation. Upon admission, a staff member conducts a 30-minute interview, which is written up and placed in the client's file within two days. The client then receives an orientation to the program and is assigned a care coordinator in charge of that case. The care coordinator and the client then have a more extensive interview, lasting approximately 90 minutes.

Staff Notes

Staff notes, sometimes called *case notes*, are written at the time of each visit, contact, or interaction that any helping professional has with a client. Staff notes usually appear in a client's file in chronological order. They are important for a number of reasons.

- Confirming a specific service. The helper wrote, "John missed work 10 of 15 days this month. He reported that he could not get out of bed. I referred him to our agency physician for a medication review."
- Connecting a service to a key issue. The helper may write, "I observed the client's interactions with peers during lunch," or "I questioned the client about his role in the fight this morning."
- Recording the client's response. "Mrs. Jones avoided eye contact when asked about her relationship with her family," "Janis enthusiastically received the staff's recommendation for job training," or "Joe resisted the suggestion that perhaps he could make a difference."

INTAKE SUMMARY
OAK HILL CENTER

CLIENT'S NAME Katie Dunlap ADMISSION DATE ____________

INTERVIEWER ______________ DATE 9-11-95

Katie Dunlap was admitted to Oak Hill Center September 11, 1995. The client's date of birth is May 22, 1974. She is from Nashville. This is her first admission to this center. Prior to admission the client resided with her sister at 1010 Western Avenue in Memphis, TN. Her drug of choice is crack/cocaine. The client started using at age 14 and the date of last use was September 9, 1995. She has been in four rehabilitation centers; however, she has not graduated from any of them. The client has 2 daughters: Candy who is 4 years of age, and Kristy who is 2. They have been in foster care for one year. The youngest child was chemically exposed. The children will be returned to the client one month from the admission date. The client's father and mother were also chemically dependent. Her father died when she was 11. The client stated that her mother had mental health problems and that they do not have a good relationship. The client has not been treated for being emotionally abused by her mother and physically abused by the father of her children. Currently she has no means of income, nor does she have health insurance. A psychological evaluation is being completed at the present time. She stated that she views herself as happy, easygoing, and appreciative.

______________________ ______________________

INTAKE WORKER'S SIGNATURE DATE

Figure 4.2 Intake summary

- Describing client status. Case notes that describe client status use adjectives and observable behaviors: "Jim worked at the sorting task for 15 minutes without talking," and "Joe's parents were on time for the appointment and openly expressed their feelings about his latest arrest by saying they were angry."
- Providing direction for ongoing treatment. Documenting what has occurred or how a client has reacted to something can give direction to any

treatment. "During our session today, Mrs. Jones said she felt angry and guilty about her husband's illness in addition to the feelings of sadness she expressed at our last meeting. These feelings will be the focus of our next meeting."

The format of case notes depends on the particular agency, but they are always important. For instance, one substance abuse treatment facility uses a copy of its form for each client every day. A worker on each of the three shifts checks the behavior observed and makes chronological case notes on the other side. These notes allow the case manager and others working with a client to stay up to date on treatment and progress and provide the means of monitoring the case.

Another format for case notes is illustrated by the notes presented below, from a residential treatment facility for emotionally disturbed adolescents. Recorded January 22, 1995, these notes are in the file of a 15-year-old female client who is thought to have a borderline personality disorder. At this facility, a staff person from each shift is required to make a chart entry; the notations DTC, ETC, and NTC refer to day, evening, or night treatment counselor. Any other staff member who has contact with the client during the day (such as the therapist, teacher, nurse, or recreation specialist) also writes a staff note. The word *Level* refers to the number of privileges that a client has. For example, a client at Level 1 has a bedtime of 9 P.M.; a client at Level 3 would go to bed at 10.

1/22/95, 8:15 A.M. DTC: Ct. awoke on time this morning. She was showered, dressed, and ready for school on time. She had positive interactions with her peers this morning. Ct. received one cue for being loud. Ct. completed her chores in a timely manner and responded appropriately to all staff requests. Ct. remains on Level 2 and under constant supervision. _______ Jan Allen

1/22/95, 2:45 P.M. TEACHER: Ct. completed all of her classroom assignments in a timely manner. She received one C for cursing. She accepted the C appropriately and maintained a good attitude. Ct. had positive interaction with peers throughout the day. She had a slight confrontation with one peer, but the two of them worked it out in a positive manner. Ct. has been very talkative and cheerful today. Ct. responded appropriately to all staff requests. Ct. remains on Level 2 and under constant supervision. _______________ Carlos Chaney

1/22/95 4:30 P.M. THERAPIST: Ct. requested and received her Level 3 today. She appeared very excited and stated that she has worked very hard for this and feels that she deserves it. We discussed the fact that being on Level 3 requires displaying Level 3 behavior. Ct. was receptive to this and seems to want to make the effort to do so. We also mutually decided that her new goal will be to take responsibility for her own actions and not blame other people. She understands that this is essential to her treatment and was receptive to doing so. _______________ Kim Stuck

1/22/95 9:30 P.M. ETC: Ct. participated appropriately in all group activities this evening. Went outside for structured activity time and participated appropriately. Interacted positively with her peers. Ct. participated actively and appropriately in group. Discussed her new goal and how much she desires to achieve it. Gave positive and appropriate feedback in group. Ct. completed her chore in a timely manner without being prompted by staff. She discussed with staff how much she desires to go home and how she wants to work the program as quickly as possible so that she can go home. She seemed a little quieter than usual this evening. Ct. required no cues for the evening and responded appropriately to all staff requests. Ct. remains on Level 3 and under constant supervision. ______________ B. Greer

1/23/95 5:00 A.M. NTC: Ct. slept through the night with no problems and was present for each 15-minute bed check. ______________ Phil Thress

---

This residential treatment facility gives its staff members strict guidelines for charting and consistent abbreviations: "ct" to mean client, "cue" to mean a warning, and "C" to signify a consequence. Also, there must be no blank lines or spaces in a case note, so the recorder puts in the line and signature at the end of each entry. Note also that no names of other clients or staff members appear in any entry. Notes such as these are routinely reviewed by the case manager.

You will also note that the word *appropriate* appears often. In this facility, it is an important word, because these clients often behave and interact inappropriately. It is common for them to act out sexually, exhibit interpersonal difficulties, and rebel against any rules or authority figures.

Chapter Summary

In the assessment phase of case management, an individual applies for services, or they are requested for him or her by others. The helping professional gathers preliminary information and determines eligibility for services, at which point the applicant becomes a client. He or she participates in each part of the process. Documentation records what has occurred. Once the individual is accepted for services, a case manager at the agency assumes responsibility for the case and begins work with the client on the next phase of case management—developing a plan for services.

Chapter Review

Key Terms

Applicant
Client
Interview
Structured interview
Intake interview
Mental status examination
Confidentiality
Privileged communication

Process recording
Content recording
Intake summary
Staff notes

Reviewing the Chapter

1. How does a person's previous experience with human service agencies influence his or her decision to seek help?
2. In what different ways do people learn about available services?
3. Distinguish between the terms *applicant* and *client.*
4. What purposes does the interview serve?
5. Describe the roles of the helper and the applicant during the initial meeting.
6. Discuss different ways to define *interview.*
7. What is the difference between interviewing and counseling?
8. List the three factors common to all interviews.
9. Describe what takes place in each of the three parts of an interview.
10. Give three examples of issues that applicants may raise during the initial interview.
11. Why are intake interviews and mental status examinations structured interviews?
12. What does a helper need to know about confidentiality in the assessment phase?
13. Discuss the two problem areas related to confidentiality.
14. What questions guide the assessment of the information gathered during this first phase of service coordination?
15. Describe case assignment.
16. Compare process recording and summary recording, and provide an example of each.
17. What purposes do staff notes (case notes) serve?

Questions for Discussion

1. Why do you think the interview is an important part of the case management process?
2. What evidence can you give that a good assessment is vital to the case management process?
3. Speculate on what could happen if a client's confidentiality were violated.
4. Develop a plan to interview a client who has applied for public housing.

References

Austin, M. J. (1981). *Supervisory management for the human services.* Englewood Cliffs, NJ: Prentice-Hall.
Brill, N. I. (1995). *Working with people: The helping process.* White Plains, NY: Longman.

Corey, G., Corey, M. S., & Callanan, P. (1988). *Issues and ethics in the helping professions.* Pacific Grove, CA: Brooks/Cole.

Egan, G. (1990). *The skilled helper: A systematic approach to effective helping.* Pacific Grove, CA: Brooks/Cole.

Enelow, A., & Wexler, M. (1966). *Psychiatry in the practice of medicine.* New York: Oxford Press.

Epstein, L. (1985). *Talking and listening: A guide to the helping interview.* St. Louis: Times Mirror/Mosby.

Garrett, A. (1972). *Interviewing: Its principles and methods.* New York: Family Service Association of America.

Hutchins, D. E., & Cole Vaught, C. (1997). *Helping relationships and strategies* (3rd ed.). Pacific Grove, CA: Brooks/Cole.

Ivey, A. E. (1988). *Intentional interviewing and counseling: Facilitating client development.* Pacific Grove, CA: Brooks/Cole.

Kaplan, R. M., & Saccuzzo, E. P. (1993). *Psychological testing: Principles, applications and issues.* Pacific Grove, CA: Brooks/Cole.

Long, L., Paradise, L. V., & Long, T. J. (1981). *Questioning: Skills for the helping process.* Pacific Grove, CA: Brooks/Cole.

McGowan, J. F., & Porter, T. L. (1967). *An introduction to the vocational rehabilitation process.* Washington, DC: U.S. Department of Health, Education, and Welfare.

Okun, B. F. (1992). *Effective helping: Interviewing and counseling techniques.* Pacific Grove, CA: Brooks/Cole.

Perlman, H. H. (1979). *Relationship: The heart of helping people.* Chicago: The University of Chicago Press.

Weinrach, S. G. (1987). The preparation and use of guidelines for the education of clients about the therapeutic process. *Psychotherapy in Private Practice, 5*(4), 71–83.

Wilson, S. J. (1980). *Recording: Guidelines for social workers.* New York: Free Press.

Chapter Five

Effective Intake Interviewing Skills

Questions are the best way to get the information you need. To be able to sit back and say, "Well, she is an introvert. I am going to have to get her to open up." Sometimes humor works, and sometimes it is just having insight.

—JANELLE STUECK, Private Industry Council, Knoxville, Tennessee, personal communication, August 30, 1993

With your first contacts with the family, you need to be able to ask the right questions. We have families who are very articulate and very forthcoming about what they are struggling with and what they need. And then we have many families who are not. . . . They don't know exactly what they need, or they can't articulate it for whatever reason. Our job then is to examine the whole situation, to gather every bit of information possible.

—Margaret Lanning, Sick Kids Need Involved People, New York City, personal communication, May 4, 1994

Listening is the most important tool. Not so much what you can tell them, but what they can tell themselves. Often a client will start talking to you and they will ask their questions and answer them in one breath. You have nothing to say because they will resolve their own problems. All you have to do is to be there and listen to them.

—Sandra Flecha, caseworker, Casita Maria Settlement House, Bronx, New York, personal communication, May 4, 1994

Interviewing was described in the previous chapter as directed conversation or professional conversation. Many helpers consider it an art as well as a skilled technique that can be improved with practice. In case management, the intake interview is a starting point for providing help. Its main purpose is to obtain an understanding of the problem, the situation, and the applicant. A clear statement of the intentions of the interview helps the case manager and the client reach the intended outcomes (Kelly, Hall, & Miller, 1989).

The chapter-opening quotations illustrate some important skills that are needed during the interviewing process. For the case manager at the Private Industry Council, asking questions is a key method of gathering information. This case manager uses techniques such as humor and insight to achieve positive results. Margaret Lanning begins her contact with families by using questions to grasp the big picture. She makes a distinction between questions in general and the *right* questions. Sandra Flecha, a caseworker at Casita Maria Settlement House, emphasizes listening as a critical skill in the interviewing process. It is her belief that clients will tell their problems if given an opportunity. Each of these professionals describes interviewing in a different way, but a common thread is respect for the considerable skills involved in using the interview to gain an understanding of the client's situation.

A number of factors influence interviewing in the helping professions. Some factors apply directly to the interviewer, such as attitudes, characteristics, and communication skills. Others are determined by the agency under whose auspices the interview occurs: the setting, the purpose of the agency, the kinds of information to be gathered, and recordkeeping. This chapter will explore many of these factors.

The intake interview is usually the first face-to-face contact between the helper and the applicant. In some agencies, the person who does the intake interview will be the case manager; others have staff members whose primary responsibility is intake interviews. Interviews are also a part of the subsequent case management process, and some of the skills used in the intake interview apply there, too. This chapter uses the term *case manager* to refer to the helping professional who is conducting the interview.

This chapter is about effective interviewing in case management: the attitudes and characteristics of interviewers, the skills that make them effective interviewers, how these skills are used in structured interviews, and the pitfalls to avoid when interviewing. For each section of the chapter, you should be able to accomplish the following objectives:

ATTITUDES AND CHARACTERISTICS OF INTERVIEWERS

- List two reasons why the attitudes and characteristics of the case manager are important to the interview process.
- Name five characteristics that make a good interview.
- Describe a physical space that encourages positive interactions between the client and the case manager.
- List barriers that discourage a positive interview experience.

ESSENTIAL COMMUNICATION SKILLS

- List the essential communication skills that contribute to effective interviewing.
- List three interviewing skills.
- Support the importance of listening as an important interviewing skill.
- Offer a rationale for questioning as an art.
- Write a dialogue illustrating responses that a case manager might use in an intake interview.

INTERVIEWING PITFALLS

- Name four interviewing pitfalls.

Attitudes and Characteristics of Interviewers

The case manager's attitudes and characteristics as an interviewer are particularly important during the initial interview, because this meeting marks the beginning of the helping relationship. Research supports the view that the personal characteristics of interviewers can strongly influence the success or failure of helping (Brown & Srebalus, 1988). In fact, Brammer (1993) concluded after a review of numerous studies that these personal characteristics are as significant in helping as the methods that are used.

One approach to the attitudes and characteristics of interviewers is a framework based on the work of Brammer (1993) and Combs (1969). It suggests that two sets of attitudes are important: one related to self and the other related to how one treats another person. Those related to self include self-awareness and personal congruence, whereas respect, empathy, and cultural sensitivity are among the attitudes related to treatment of another person. Elsewhere in the literature, other perspectives on helping attitudes and characteristics have as common themes the ability to communicate, self-awareness, empathy, responsibility, and commitment (Woodside & McClam, 1994).

The case manager communicates helping attitudes to the applicant in several ways, including greeting, eye contact, facial expressions, and friendly responses. The applicant's perceptions of the case manager's feelings are also important in his or her impression of the quality of the interview (Morales & Sheafor, 1980). Communicating warmth, acceptance, and genuineness promote a climate that facilitates the exchange of information, which is the primary purpose of the initial interview. The following dialogue illustrates these qualities.

INTERVIEWER:	*(stands as applicant enters)* Hello, Mr. Johnson *(shakes hands with a smile)*. My name is Clyde Dunn—call me Clyde. I'll be talking with you this morning. Please have a seat. Did you have any trouble finding the office?
APPLICANT:	No, I didn't. My doctor is in the building next door, so I knew the general location.
INTERVIEWER:	Good. Sometimes this complex is confusing because the buildings all look alike. Have you actually been to the Hard Rock Cafe in Cancún? *(pointing to the applicant's shirt)*
APPLICANT:	No, I haven't. A friend brought me this T-shirt. I really like it.
INTERVIEWER:	They certainly are popular. I see them all over the place. Well, I'm glad you could come in this morning. What seems to be the problem?

The case manager communicates respect for the applicant by standing and shaking hands. It is also easy to imagine that Clyde Dunn is smiling and making eye contact with Mr. Johnson. Clyde takes control of the interview by introducing himself, suggesting how Mr. Johnson might address him, and asking him to have a seat. His concern about Mr. Johnson finding the office and his interest in the T-shirt communicate warmth and interest in him as a person. Clyde also reinforces Mr. Johnson's request for help in a supportive way. All of these behaviors reflect an attitude on Clyde's part that increases Mr. Johnson's comfort level and facilitates the exchange of information.

The positive climate created by such a beginning should be matched by a physical setting that ensures confidentiality, eliminates physical barriers, and promotes dialogue. It is disconcerting to the applicant to overhear conversations from other offices or to be interrupted by phone calls or office disruptions. He or she is sharing a problem, and such events may lead to worries about the confidentiality of the exchange. Physical barriers between the client and the case manager (most commonly, desks or tables) also contribute to a climate that can interfere with relationship building. As much as the physical layout of the agency allows, the case manager should meet applicants in a setting where communication is confidential and disruptions are minimal. A furniture arrangement that places the case manager and the applicant at right angles to one another without tables or desks between them, facilitates eye contact, positive body language, and equality of position are preferable.

A sensitive case manager is also cognizant of other kinds of barriers, such as **sexism, racism, ethnocentrism,** and **ageism.** Problems inevitably arise if the case manager allows any biases or stereotypes to contaminate the helping interaction. To help you think about your own biases and stereotypes, indicate whether you believe each of the following statements is true or false.

T	F	Boys are smarter than girls when it comes to subjects like math and science.
T	F	Men do not want to work for female bosses.
T	F	Mothers should stay home until their young children are in school.
T	F	Women cannot handle the pressures of the business world.
T	F	Asians are smarter than other ethnic groups.
T	F	People on welfare do not want to work.
T	F	People who do not attend church have no moral principles.
T	F	A mandatory retirement age of 65 is necessary because people at that age have diminished mental capacity.
T	F	The older people get, the lower their sexual interest and ability.

How did you respond to these statements? Each statement reflects an unjustified opinion that is based solely on a stereotype of gender, race, or age.

Sensitivity to issues of ethnicity, race, gender, and age is important for the case manager when conducting interviews. Many clients and families will have backgrounds very different from that of the case manager. In the United States today, one-fourth of the population originates from non-European backgrounds, a large number of clients are women, and the population proportion of elderly people is increasing rapidly. For many people in these populations, life is difficult, and they have few places to turn for help. Many of them live in poverty, have inadequate education, have a disproportionate chance of getting involved in the criminal justice system (either as a victim or a perpetrator), possess few useful job skills, are unemployed, and suffer major health problems at a disproportionate rate (Smith, 1995a).

Case managers should ask themselves, "How do I become sensitive to my clients and relate to them in a way that respects and supports their race, culture, gender, and age?" The following suggestions are derived from Kirst-Ashman (1993).

Expect clients to be unique individuals It is easy to stereotype cultural, racial, gender, or age groups, but clients cannot be understood strictly in terms of their particular culture. For example, many African Americans do share values and experience similar life events, but not all African Americans are the same. During interviews, case managers must take special care to get to know each individual client rather than categorizing him or her as a member of one particular group. For example, Marta Rivera, a caseworker at Casita Maria Settlement House in New York City, explained the complexities their workers face when working with Latino clients. "With Hispanics, it is just no longer the Puerto Rican family coming, but people from Nicaragua, Colombia, and Honduras come for help. All we share is that we are labeled Hispanics or Latino. . . . The dialects, the beliefs, the family lifestyles are totally different" (personal communication, May 4, 1994).

Remember that differences in language can be confusing Do not assume that words mean the same to everyone who is interviewed. When the case manager asks interview questions, clients sometimes do not understand the terminology. Likewise, words or expressions that clients use may have a very different meaning for the interviewer. For example, questions about family and spouse are familiar subjects in an intake interview. When clients talk about "partners" or "family," these terms can have various meanings, depending on the cultural background and life experiences of the individual being interviewed. For example, in the Native American culture, the *family* is an extended one that includes many members of the clan. For gay men and lesbian women, the word *partner* has the special meaning of "significant other."

Explain to the client the purpose of the intake interview and the case manager's role Clients may show up for the interview without understanding its purpose or the role of the interviewer in the helping process. Confidentiality may also be an important issue for them—sharing information about themselves and others may be contrary to the rules of their culture. For example, for many people raised in Asian cultures, to describe a problem to someone who is not in the family implies making the matter public, which is considered to bring shame to the family.

Expect that others will be different from you It is easy to make the mistake of expecting the clients we serve to be like us. We begin the interview process wanting to find similarities as a way of building a bridge to them. When clients prove to be very different, or we cannot understand them, we often want them to change so that they will be easier to "manage." In the United States, we often like to think of our country as a melting pot in which all cultures mix in together and lose their original identities. When individuals do not want to lose their own culture, there is a tendency to blame them for being difficult. Case managers must take special care in the interview process to let clients know that there is respect for differences.

We have assembled some suggestions for developing sensitivity in interviewing individuals with certain cultural backgrounds (Baruth & Manning, 1991; Bernal et al., 1983; Gilligan, 1982; Kirst-Ashman, 1993; Smith, 1995a, 1995b). These are meant to be guidelines and points of awareness; they should be used with caution. As we mentioned earlier, individuals seldom exhibit all of the characteristics of their cultural group.

Interviewing clients of Native American origin In many Native American cultures, sharing information about oneself and one's family is difficult. It is important not to give others information that would embarrass the family or imply wrongdoing by a family member.

Listening behaviors such as maintaining eye contact and leaning forward are considered inappropriate and intrusive in some Native American cultures.

For many Native Americans, trust increases as you become more involved in their lives and show more interest in them. Making home visits and getting to know the family can significantly improve an interviewer's chances of getting relevant information.

Native Americans tend not to make decisions quickly. This slow process could influence how soon the client is willing to share information or make judgments.

Native American cultures sometimes incorporate a fatalistic element—a belief that events are predetermined. During the initial stages of the process, the client may not understand how his or her responses and actions can influence the course of service delivery.

Interviewing clients of Latino origins In many Latino cultures, informality is an important part of any activity, even the sharing of information. Taking time to establish rapport with the client before direct questioning begins is helpful.

Some people of Latino origin may be perceived as submissive to authority because they appear reticent or reluctant to answer questions. This behavior, in fact, may be shyness or the natural response to a language barrier.

In Latino cultures, the father may be seen as aloof as he performs his roles of earning a living for the family and establishing the rules. The mother and other members of the family tend to assume more nurturing roles. Questions that do not take these roles into consideration may be misinterpreted by the clients or may suggest to them that the interviewer is an outsider incapable of understanding the culture or of helping them.

Also in Latino cultures, fatalism often plays a role. Latino clients may not see any point in discussing the future, preferring to talk about the present.

Interviewing African Americans Given a long history of being denied access to services, African Americans may not want to share information during the intake interview. Or they may be angry about discriminatory treatment that they have previously experienced in the human service delivery system. During the intake interview, it is important to focus on concrete issues that can be connected to services. This will show respect for the client's right to expect fair treatment and quality services.

When being interviewed by a white professional, an African American may feel powerless because of previous experiences in white society. The interviewer will need to show that the client's input does matter.

Interviewing women Many women do not know how to talk about the difficulties that they are experiencing, and they may not know how to respond to the questions they are asked. Some have had few opportunities to discuss their problems, and they may believe that they do not have the right to complain. Listening carefully is very important.

Anger may play a part in the initial interview. Many women come to the helping process frustrated, either because their efforts have been unrecognized or because they believe that others expect them to be perfect. Often this anger must be expressed before any information can be gathered.

Women often feel powerless and do not expect the bureaucracy to serve them well. They may be reluctant to communicate and doubtful that the interview or the process as a whole can make a difference.

Women may be overly dependent and assume that the case manager will take complete control of the interview. They may want the interviewer to be the one to identify problems and possible goals. In such cases, care should be taken to give the woman opportunities and encouragement to respond more fully.

Interviewing elderly clients In this society, elderly people are often disregarded and devalued. During the interview, the case manager must show respect for the elderly client's answers and opinions about the issues discussed. Such a client needs to be assured that his or her responses are important and have been heard by the helper.

Pay special attention to the elderly client's description of support in his or her environment. Many live in an environment of decreasing support (changing neighborhood, death of friends) and/or decreasing mobility. Others live with limited family support. These clients may not realize how their environment has changed.

Elderly clients may be reluctant to share their difficulties, for fear of losing much of their independence. They may understate their needs or overstate the amount of support they have, hoping to avoid changes in their living conditions such as being removed from their home or relinquishing their driving privileges.

These are only a few of the differences that helpers may encounter during the intake interview with individuals of various ethnic, racial, gender, and age groups. In several ways, case managers can continue to learn more about how to interview culturally diverse clients. Among them are becoming knowledgeable about other cultures; reading professional articles that focus on ways to modify the interviewing process to meet the needs of certain client groups; and talking with other helpers whose own cultural origins give them insight into cultural barriers. Gaining an understanding of diversity is a process that continues throughout the professional life of every effective helper. Such an understanding enhances the interviewing environment for both parties.

Essential Communication Skills

Communication forms the core of the interviewing process. In interviewing, communication is the transmission of messages between applicant and helper. As the first face-to-face contact, the interview is a purposeful activity for both participants. In many cases, the motivation is a mutual desire to decide whether the applicant is in the right place for the needed services. This is a negotiation that is facilitated by **effective communication** skills.

An important skill that promotes the comfort level of the applicant and lays the foundation for a positive helping relationship is using language the person understands. This means avoiding the use of technical language. For example, terms such as *eligibility, resources,* and *Form 524* may not mean much to an applicant who has not become familiar with the human service system. To take another example, imagine that the interviewer is discussing the benefits of taking a vocational or interest test. Rather than going into detail about the validity or reliability of the test, the case manager should discuss how it might help in establishing a vocational objective. Using language or words the applicant does not understand tends to create distance and disengagement.

Congruence between verbal and nonverbal messages is another way to facilitate the interaction between an applicant and a case manager. A major part of the meaning of a message is communicated nonverbally so when conflict is apparent between the verbal and nonverbal messages, the applicant is likely to believe the nonverbal message. An example of this that many of us have experienced is the person who says: "Yes, I have time to talk with you now." Unfortunately, as this is said, the case manager is dialing the phone or looking through her desk drawer for a folder. The lack of eye contact or any other encouraging nonverbal message communicates to us that the person is indeed busy or preoccupied with other matters.

Another skill that facilitates the interview process is **active listening**—making a special effort to "hear" what is said, as well as what is not said. An interviewer who is sensitive to what the applicant is communicating, verbally as well as nonverbally, gains additional information about what is really going on with the individual. This ability is particularly helpful in situations where the presenting problem may differ from the underlying problem. Later in the chapter, we present a more detailed discussion of listening as it relates to the intake interview.

A popular way to elicit information is asking questions. Questioning is an art as well as a skill. Unfortunately, case managers don't often develop their questioning skills, relying instead on questioning techniques that have served them well in informal or friendly encounters. Typically, this means asking questions that focus on facts, such as "What happened?" "Who said that?" "Where are you?" "Why did you react that way?" Questions such as these usually lead to other questions, placing the burden of the interview on the case manager and allowing the applicant to settle into a more passive role. The applicant's participation is then limited to answering questions, so the interview may begin to feel like the game "Twenty Questions." Skillful questioning combined with effective responding helps elicit information and keep the interaction flowing. Appropriate questioning and responding techniques are also introduced later in this chapter.

Patterns of communication vary from culture to culture, according to religion, ethnic background, gender, and lifestyle differences. In the dominant culture in the United States, it is effective to use a reflective listening approach where feelings are important. Many of the techniques that are useful in this approach are not appropriate for all cultures. For example, eye contact is inappropriate among some Eskimos. The sense of space and privacy is different for Middle Easterners, who often stand closer to others than Americans do. Some people from Asian cultural backgrounds may prefer more indirect, subtle approaches of communication. Thus, a single interviewing approach may have different effects on people from various cultural backgrounds. The skillful and sensitive case manager must be aware of these differences.

Both spoken language and body language are expressions of culture. Even if the client's native language is the same as the interviewer's, cultural assumptions and ways of structuring information may differ. Both talking and listening provide many occasions for misunderstanding (Kolanad, 1994). Marta Rivera, a caseworker at Casita Maria Settlement House, illustrates the importance of word choice as she describes a counseling referral: "In the Hispanic community, you tell someone they need counseling, and right away they say, 'I'm not crazy.' So you don't tell someone they need counseling. You always say, 'Well, maybe it would help to talk to someone' " (personal communication, May 4, 1994).

Assigning great significance to any single gesture by the applicant is also risky, but a pattern or a change from one behavior to another is meaningful (Sielski, 1979). Once again, the key is the case manager's awareness during the interview process.

Now that you have read about general guidelines for essential communication, let's focus on the specific skills of listening, questioning, and responding.

Interviewing Skills

Interviewing skills aim to enhance communication, which involves both words and nonverbal language. Spoken language varies among individuals and cultures. Understanding spoken language is challenging, because it is always changing, it is usually not precise, meanings vary, and it is ambiguous. Body language, which is also important and challenging to understand, includes body movement, posture, facial expression, and tone of voice. Knowing the ways in which body language varies culturally can help the interviewer fathom the thoughts and feelings of the applicant.

In talking with an applicant, the case manager must strive for effective communication—making sure that the receiver of the message understands the message in the way the sender intended. In the intake interview, the case manager listens, interprets, and responds. To understand the applicant's problem as fully as possible, the case manager constantly interprets the meanings of behaviors and words. He or she should always have a "third ear" focused on this deeper interpretation.

At the same time, the applicant is interpreting the words and behaviors of the case manager. A case manager who is an effective interviewer can help the applicant make connections and interpretations. Also contributing to correct interpretations and connections are a good working relationship between the two of them, good timing, and sensitivity to whether the material being discussed is near the applicant's level of awareness.

Zulma Resto, a caseworker at Casita Maria Settlement House, describes what the initial meeting is like at her agency:

> Normally when clients come in, they are already tired of the system, and so when they meet you, they are like ticked off and probably hate you. One thing that I try to get across right away is that I have nothing to do with the welfare system, which is what most of them are against. I let them know that I am here for them and that whatever we say is confidential. I try to keep the first session free of paperwork. Keep it friendly and not get too much into their space the first day. Then the second time they see me, they know who I am and we can talk a little bit more. It's important to bond with them, get their confidence and trust, so that they will tell you what their real problems are. (Personal communication, May 4, 1994)

Linda Smith, a caseworker at the School for the Deaf, works hard at establishing a relationship during the initial meeting. She sees much of her job as trying to keep communication open between the school and the family.

> When there are problems, I already have a relationship with the family. You know, there can be a lot of confusion and a lot of misunderstanding because of the lack of communication. . . . You have to be available to listen to what the other person is saying and not saying. I think you have to be very accepting and nonjudgmental. You need good communication skills. You also have to meet the client where he/she is rather than imposing your values upon him/her. And patience, a lot of patience. (Personal communication, October 27, 1993)

Both of these helping professionals are experienced at intake interviewing. They value the helping relationship and recognize its importance in the service delivery that is to follow. To establish the relationship, they use such communication skills as listening, questioning, and responding. These are discussed and illustrated next, with excerpts from intake interviews.

LISTENING

Listening is the way in which most information is acquired from applicants for services. The case manager listens to the applicant's verbal and nonverbal messages. "Listening with the eyes" means observing the client's facial expressions, posture, gestures, and other nonverbal behaviors, which may signal his or her mood, mental state, and degree of comfort. Verbal messages communicate the facts of the situation or the problem and sometimes the attendant feelings. Often, however, feelings are not expressed verbally, but nonverbal messages provide clues. A good listener should be sensitive to the congruence (or lack of it) between the client's verbal and the nonverbal messages. The case manager must pay careful attention to all of these forms of communication.

Good listening is an art that requires time, patience, and energy. The case manager must put aside whatever is on his or her mind—whether it is what to recommend for the previous client, the tasks to be accomplished by the end of the day, or making a grocery list—to focus all attention on the applicant. The case manager must also be sensitive to the fact that his or her behavior gives the applicant feedback about what has been said (Epstein, 1985). During the interview, the case manager must also be cognizant of cultural factors that play into the interpretation of body language. For example, the proper amount of eye contact and the appropriate space between case manager and applicant may vary according to the cultural identity of the applicant. As you can see, listening is indeed complicated. What behaviors characterize good listening? How are attentiveness and interest best communicated to the applicant?

Attending behavior, responsive listening, and *active listening* are terms that indicate ways in which case managers let applicants know that they are being heard. Egan (1990, pp. 108–110) lists the following five behaviors as a set of guidelines for the interviewer. They can be easily remembered by the acronym S-O-L-E-R.

- **S:** Face the client *squarely.* This is a posture of involvement. To face away from the client or even at an angle lessens the degree of involvement.
- **O:** Adopt an *open posture.* This is usually perceived as nondefensive. Crossing arms or legs may not communicate openness or availability.
- **L:** *Lean toward* the other person. A natural sign of involvement, this posture is a slight forward inclination. Moving forward or backward can frighten a client or communicate lessened involvement.
- **E:** Maintain *eye contact.* This is normal behavior for two individuals who are involved in conversation. It is different from staring.
- **R:** Try to be *relaxed.* This means avoiding nervous habits such as fidgeting or tapping a pencil. Behaviors such as these can distract the client.

Attending behavior is another term for appropriate listening behaviors. Eye contact, attentive body language (leaning forward, facing the client, facilitative and encouraging gestures, etc.), and vocal qualities such as tone and rate of speech are ways for the interviewer to communicate interest and attention (Ivey, 1988). Attending behavior also means allowing the applicant to determine the topic.

Other guidelines for good listening are provided by Epstein (1985, pp. 18–19):

1. Be attentive to general themes rather than details.
2. Be guided in listening by the purpose of the interview in order to screen out irrelevancies.
3. Be alert to catch what is said.
4. Normally, don't interrupt, except to change the subject intentionally, to stop excessive repetition, or to stop clients from causing themselves undue distress.
5. Let the silences be, and listen to them. The client may be finished, or thinking, or waiting for the practitioner, or feeling resentful. Resume talking when you have made a judgment about what the silence means, or ask the client if you do not understand.

A skillful listener also hears other things that may help him or her understand what is going on. A shift in the conversation may be a clue that the applicant finds the topic too painful or too revealing, or it might indicate that there is an underlying connection between the two topics. Another consideration is what the applicant says first. "I'm not sure why I'm here" or "My probation officer told me to come see you" give clues about the applicant's feelings about the meeting. Also, the way in which the applicant states the problem may indicate how he or she perceives it. For example, an applicant who states, "My mother says I'm always in trouble" may be signaling a perception of the situation that differs from the mother's. Concluding remarks may also reveal what the applicant thinks has been important in the interview. The skilled interviewer also listens for recurring themes, what is not said, contradictions, and incongruencies.

Good listeners make good interviewers, but as you have just read, listening is a complex activity. It requires awareness of one's own nonverbal behaviors, sensitivity to cultural factors, and attention to various nuances of the interaction. It is further complicated by the fact that people seeking assistance don't always say what they mean or behave rationally (Garrett, 1972). However, the use of good listening skills always increases the likelihood of a successful intake interview.

QUESTIONING

Questioning, a natural way of communicating, has particular significance for intake interviews. It is an important technique for eliciting information, which is a primary purpose of intake interviewing. Many of us view questioning as something most people do well, but it is in fact a complex art. This section will elaborate on questioning skills, introduce the appropriate use of questions, identify problems that should be considered, and explore the advantages of open inquiry as one way to elicit information.

Questioning is generally accepted by some as low-level or unacceptable interviewer behavior (Carkhuff, 1969; Egan, 1990; Gordon, 1974). Others view it as a complex skill with many advantages (Ivey, 1988; Long, Paradise, & Long, 1981). Let's explore its complexity and its advantages. Long, Paradise, and Long (1981) give three reasons why questioning is a complex skill: questioning may assist *and* inhibit the helping process; it can establish a desired as well as an undesired pattern of exchange; and it can place the client in the one-sided position of being interrogated or examined by the helper.

For these reasons, we may consider questioning an art form. The wording of a question is often less important than the manner and tone of voice used to ask it (Garrett, 1972). Suzy Bourque at the Family Counseling Agency in Tucson says: "I think that people have to be detectives. . . . They have to enjoy walking into a new setting and seeing what is there. . . . It is not just going out with your 12-page assessment form and asking alienating questions" (personal communication, October 7, 1994). Linda Smith at the School for the Deaf concurs: The "skills are those involved with being a private eye, nosy in a tactful way" (personal communication, October 27, 1993).

Also, too many questions will confuse the applicant or produce defensiveness, whereas too few questions place the burden of the interview on the client, which may lead to the omission of some important areas for exploration. The pace of questions influences the interview, too. If the pace is too slow, the applicant may interpret this as lack of interest, but a pace that is too fast may cause important points to be missed. A delicate balance is required.

What are the advantages of questioning? One is that questioning saves time. If the case manager knows what information is needed, then questioning is a direct way to get it. Questioning also focuses attention in a particular direction, moves the dialogue from the specific to the general as well as from the general to the specific, and clarifies any inaccuracies, confusion, or inconsistencies. Let's examine some examples of the appropriate uses of questioning. After each example, you are asked to provide two relevant questions.

To begin the interview Could you tell me a little about yourself? What would you like to talk about? Could we talk about how I can help?
You work at the county office on aging. A woman comes in with her elderly mother. List two questions that you might use to begin the interview.

____

____

____

____

To elicit specific information How long did you stay with your grandmother? What happens when you refuse to do as your boss asks? Who do you think is pressuring you to do that? Can you give me an example of a time when you felt that way?
A client tells you about mistreatment by her boss at her new job. She claims that she is being sexually harassed. What two questions would you ask to help you understand what happened?

____

____

____

____

To focus the client's attention Why don't we focus on your relationship with your daughter? What happens when you do try to talk to your husband? Of the three problems you've mentioned today, which one should we discuss first?
A client is worried about how her surgery will go, who will care for her children while she is in the hospital, and whether she will be fired for

missing so much work. She wrings her hands and seems ready to burst into tears. What are two questions you could use to focus her attention?

____

____

____

____

To clarify Could you describe again what happened when she left? How did you feel about that conversation compared with others you have had with him? What is different about these two situations?
A young man shares his anguish over his mother's death a year ago. You notice that he is smiling, and you are confused about what he is really saying. Write two questions that would help you clarify what is going on.

____

____

____

____

These are examples of interview situations in which the case manager might legitimately use questions. In all of them, the general rule of questioning applies: question so as to obtain information or to direct the exchange into a more fruitful channel (Garrett, 1972).

Although questioning may seem to be the direct path to information, sometimes this strategy can have negative effects. Long, Paradise, and Long (1981) suggest that interviewers not rely on questions to carry the interaction or interview. This is particularly problematic for beginning helpers, because people generally have a tendency to ask a question whenever there is silence. (This will be discussed later in this section.) Questions may also be inappropriate when the case manager does not know what to say. Unfortunately, to do so may lead to more questions, which can put the case manager in the position of focusing on thinking up more questions rather than listening to what the client is saying.

An overreliance on questioning can create other problems for both interviewers and clients. For the client, too many questions can limit self-exploration, placing him or her in a dependent role in which the only responsibility is to respond to the questions. A client may also begin to feel defensive, hostile, or resentful at being interrogated. Using too many questions may place the case manager in the role of problem solver, giving him or her most of the responsibility for generating alternatives and making decisions. In the long term, overreliance on questioning leads to bad habits and poor helping skills. Using questions to the exclusion of other types of helping responses eventually results in the withering of these other skills (as discussed in the next section).

In conclusion, questioning is an important strategy for effective interviewing, but it is more than a strategy for obtaining information. Because of the subtleties of questioning, the matter of its appropriate uses in interviewing, and the potential problems, questioning is an art that requires practice. The skillful case manager who uses questioning to best advantage knows when to use open and closed inquiries to gather information during the intake interview. These types of questions are discussed next.

CLOSED AND OPEN INQUIRIES

The questions used in intake interviews can be categorized as either open or closed inquiries. Determining which one to use depends on the case manager's intent. If specific information is desired, closed questions are appropriate: "How old are you?" "What grade did you complete in school?" "Are you married?" If the case manager wants the client to talk about a particular topic or elaborate on a subject that has been introduced, open questions are preferred: "What is it like being the oldest of five children?" "Could you tell me about your experiences in school?" "How would you describe your marriage?"

Closed questions elicit facts. The answer might be yes, no, or a simple factual statement. An interview that focuses on completing a form generally consists of closed questions like those in the previous paragraph. However, the interviewer must be cautious, for a series of closed questions may cause the client to feel defensive, sensing an interrogation rather than an offer of help. One suggestion is to save the form until the end of the interview, review it, and complete the unanswered questions at that time. If the completion of an intake form is allowed to take precedence in the interview, the case manager misses the opportunity to influence the client's attitudes toward the agency, getting help, and later service provision. Perhaps equally important, information that could be acquired through listening and nonverbal messages may be missed if the interviewer is focused on writing answers on the intake form.

Open inquiries, on the other hand, are broader, allowing the expression of thoughts, feelings, and ideas. This type of inquiry requires a more extensive response than a simple yes or no. The exchange of this type of information contributes to building rapport and explaining a situation or a problem. Consider the following example.

FATHER: I'm having trouble with the oldest boy, William. He's in trouble again at school.

INTERVIEWER 1: How old is William?

INTERVIEWER 2: Could you tell me more about what's going on?

Interviewer 1's response is a closed question that asks for a simple factual answer. Interviewer 2's response is an open inquiry that asks the father to elaborate on what he thinks is happening with William. This allows William's father to determine what he wishes to tell the interviewer about the situation. Such an open inquiry emphasizes the importance of listening—to what the individual says first, how he or she perceives the problem, and what is considered important.

An intake interview

You can see how valuable open inquires can be in intake interviewing. They also provide an opportunity for the client to introduce topics, thereby putting them at ease by allowing discussion of their problems in their own way and time. Besides providing the information that the case manager needs, open inquiries encourage the exploration and clarification of the client's concerns.

Four methods are commonly used to introduce an open inquiry (Evans, Hearn, Uhlemann, & Ivey, 1979). Each is presented here with an example of a client statement, the interviewer's response, and the kind of information that the client might volunteer in response to the open inquiry.

- "What" questions are fact-oriented, eliciting factual data.

  MR. CAGLE: I'm here to get food stamps. Here's my application.

  INTERVIEWER: Let's review it to make sure you've completed it correctly. What's your income?

  MR. CAGLE: Well, I make minimum wage at my job, and my wife don't make much either. We have three children and we live in a low-income apartment.

- "How" inquiries are people-oriented, encouraging responses that give a personal or subjective view of a situation.

  SUZANNE: My boyfriend doesn't like my parents, and when we are all together, nobody agrees with any one about anything.

  INTERVIEWER: How do you feel about that?

SUZANNE: I hate it. Everyone is so uncomfortable. I want everyone to get along, but I dread the times we have to be together. Sometimes I feel like somebody will yell at someone else or even hit somebody else.

- "Could," "could you," or "can you" are the kinds of open inquiries that offer the client the greatest flexibility in responding. These inquiries ask for more detailed responses than the others.

MIKE: I hate school. My teacher doesn't like me. She's always on my case about stuff.

INTERVIEWER: Could you describe a time when she was on your case?

MIKE: Well, I guess. Like yesterday, she was mad at me because I was late to class . . . but I was only five minutes late. Then she called on me to answer a question. Well, I hadn't read the stuff because I lost the book, so how could I answer the question? I mean, give me a break.

The fourth type of open inquiry is the "why" question, which experienced interviewers often avoid because it may cause defensiveness in clients. Examples of "why" questions that may do this are "Why did you do that?" and "Why did you think that?" Phrased this way, these responses may be perceived as judgments that the client should not have done something, felt a certain way, or had certain thoughts. Less risky "why" questions are those phrased less intrusively: "Why don't we continue our discussion next week?" "Why don't we brainstorm ways that you could handle that?"

In what follows, we analyze some excerpts from an intake interview that occurred at juvenile court. Tom Rozanski is the case manager who was assigned to court on that particular day. In some such cases, the juvenile is remanded to state custody that very day—he or she can leave the courthouse only to go to a local or state facility. The juvenile in this case, Jonathan Douglas, has been charged with breaking and entering. He has a history of substance abuse and school truancy and is well known to the judge, who finds him guilty and remands him to state custody. The case then comes under the jurisdiction of an Assessment, Care, and Coordination Team (ACCT), which takes responsibility for assessing the case, developing a plan of services, and coordinating the needed services among the agencies that are involved with the plan. Tom finds on this day that court is very crowded. Once Jonathan Douglas has been remanded to state custody, Tom asks him to follow him into the hall, and the initial intake interview occurs there. Jonathan's parents also join them, as do two officers—who suspect that Jonathan will run if he gets the chance. They stand together in the hall for a brief interview so that Tom can gather enough information to arrange a placement that afternoon. Here's what happens:

TOM: Jonathan, my name is Tom. *(Shakes hands)* I work for the Assessment, Care, and Coordination Team. We are responsible for assessing your case and planning services for you.

JONATHAN: *(Limply shakes hands and looks everywhere but at Tom)*

TOM: Jonathan, are you listening? Please look at me. Are you on any drugs right now?

JONATHAN: *(Unintelligible response)*

Tom realizes that it is futile to try to talk with Jonathan now and hopes that in a few hours he will be down from whatever drugs he has taken.

TOM: Mr. and Mrs. Douglas, I am Tom Rozanski, a case manager for the Assessment, Care, and Coordination Team in this county. Let me review for you what has happened. The judge found Jonathan guilty of breaking and entering. Because of his prior record, he is in state custody, and it is my job to find a place for him to stay while we evaluate his case. I need some basic information right now. Can you help me?

MRS. DOUGLAS: Yes, we want to help him any way we can.

TOM: Does Jonathan live with either of you?

MRS. DOUGLAS: He stays with me once in a while, but mostly he stays with his dad.

TOM: Mr. Douglas, could you describe his behavior when he stays with you?

MR. DOUGLAS: Well, I guess he goes to school sometimes. Leastways, when I leave for work, I try to get him up. I don't know if he goes though. Sometimes he's here when I get home and sometimes he isn't. He's a big boy now, and I can't do much with him, so I just let him be.

TOM: Do either of you have health insurance?

BOTH PARENTS: No.

The interview lasts approximately 5 more minutes, and Tom obtains some key information about the family situation. He has very little time and needs specific information, so he hurriedly asks closed questions. "What is your address, Mr. Douglas?" "What grade is Jonathan in?" "Has he had a medical examination recently?" Finally, Tom has enough information that he can complete most of the intake form. That afternoon, he meets with Jonathan and makes another attempt to talk with him. He is relieved to find Jonathan more communicative at this meeting. Here's an excerpt; note Tom's use of open inquiries.

TOM: Jonathan, I would like to talk with you about what's going to happen. I'd also like you to tell me your side of what's going on.

JONATHAN: *(Looks at Tom but makes no comment)*

TOM: When we finish talking, Deputy Johnston will take you to Mountainview Hospital, where you will spend the next two weeks. During that time, we will talk again, you will take some tests, and you will meet with a group of patients who are your age. At the end of that time, we will develop a plan of services for you. Now, could you tell me about yourself?

JONATHAN: Well, I'm 15. I don't like school and I don't get along with either of my parents. My mother doesn't want me since she moved, and my dad don't care if I'm at home or not.

TOM: This is the first time you have been in trouble for breaking and entering. What happened?

JONATHAN: Well, I was with these guys and we needed money for some dope. It looked easy. I think I made a mistake.

TOM: Yeah. It seems so. Let's talk about what you can do now. What kind of changes would you like to see?

JONATHAN: Well, I don't want to go to jail and I don't want to go to Red River [a juvenile correctional facility]. I can't stay home though. They don't care about me and I don't care about them.

TOM: How would you describe your relationship with your parents?

JONATHAN: We don't have no relationship. They don't care about me. Sometimes I stay with my mom, but she's looking for another husband and she don't want me around. My dad, he just don't want to be bothered.

TOM: Hmm. Sounds as though you're not sure if there's a place for you with them. What changes would you like to see in your relationship with your parents?

JONATHAN: I wish they . . . I wish . . . I wish they liked me.

TOM: I see. Could you give me an example of what they would do if they liked you?

JONATHAN: I don't know.

TOM: Can you describe a time when you did something they liked?

JONATHAN: My mom likes it when I come in early. My dad, he don't care.

TOM: What have you done to please your mom?

JONATHAN: *(Pauses)* I cleaned up the kitchen once.

In this excerpt, Jonathan mentions his family and school in his first response. Tom picks up on the family situation and decides to explore it with Jonathan. He has talked with the parents, and although they are not living together, he senses that both are interested in Jonathan and willing to help him but don't seem to know what to do, and they feel that Jonathan rebuffs any overtures they make. Tom is trying to discover what kind of support may be available to Jonathan from his parents and how receptive he would be to it. Tom uses open inquiries in his conversation with Jonathan to elicit the boy's thoughts and feelings about this issue. The use of "what" questions gets at factual information, and the "how" questions are aimed at people-oriented information.

In summary, case managers who are good interviewers use both open and closed inquiries, although open inquires are preferred whenever possible. They are also careful to ask one question at a time and to avoid asking consecutive questions of a kind that might create the feel of a cross examination. What other types of responses do interviewers use? The next section suggests other ways of responding to clients in an interview situation.

RESPONDING

A case manager might use various kinds of responses during the course of an intake interview. Of course, the type of response depends on the intent at that particular point. Let's review some of the most common responses. In the following material, each response is followed by an example of its use. Joe Barnes, a recent parolee, has returned home and is having a difficult time with his wife. His parole officer, sensing that the relationship is in trouble, suggests that Joe see a counselor at the Family Service Center.

Minimal responses Sometimes called *verbal following,* minimal responses let the client know that you are listening. "Yes," "I see," "Hmm," and nodding are minimal responses. Using them is important when getting to know the applicant.

JOE BARNES: I'm here because my probation officer thought it would be a good idea for me to talk with someone about things at home. Things haven't been very good since I came home.

MIKE MATSON: I see.

Paraphrase This response is a restatement (in different words) of the main idea of what the client has just said. It is often shorter and can be a summary of the client's statement. Paraphrasing lets the client know that the case manager has absorbed what was said.

JOE: I just don't know what the trouble is. I was glad to get home, and I thought my wife would be glad to have me there. But we fight about everything—even stuff like when to feed the dog. I don't know what to do.

MIKE: You don't know what's happening between you and your wife since you got home. Sounds like it's pretty unpleasant for both of you . . . and you're wondering what to do about it.

Reflection Sometimes people get out of touch with their feelings, and reflection can help them become more aware. The feelings may not be named by the individual, but rather communicated through facial expression or body language. For example, a flushed face or a clenched fist may show anger. The case manager's reflective response begins with an introductory phrase ("You believe," "I gather that," "It seems that you feel"), then clearly and concisely summarizes the feelings the case manager perceives.

JOE: Yes. I don't know how we can continue to live like this. I know she is really angry about me getting in trouble with the law, but I've paid my dues, learned my lesson. I don't plan to ever get in that mess again.

MIKE: I gather that you really do feel bad about what you did, but you would like to put the past behind you and focus on the future and how to make your marriage work.

Reflection is a response that facilitates a discussion of the client's feelings, particularly when he or she may feel threatened by such a discussion. It is also helpful as a way to check and clarify the case manager's perception of what was said during the interview.

Clarification Clarifying helps the case manager find out what the client means. When the case manager is confused or unsure about what has taken place, it is more productive to stop and clarify at the time than to continue.

JOE: I got so angry last week because she wouldn't listen to me and she didn't seem to care that I was home. I was yelling, she was yelling, she threw a bowl at me, and I almost hit her.

MIKE: Sounds to me like you got so angry and frustrated that you were almost out of control.

Summarizing With this response, the interviewer provides a concise, accurate, and timely summing up of the client's statements. It also helps organize the thoughts that have been expressed in the course of the interview. Summarizing is used to begin an interview when there is past material to review. It is also useful during the interview when a number of topics have been raised. Summarizing directs the client's attention to the topics and provides direction for the next part of the interview.

From the summary, the client can choose what to discuss next. Summarizing is also useful when the client presents a number of unrelated ideas or when his or her comments are lengthy, rambling, or confused; such a response can add direction and coherence to the interview. Finally, summarizing is a way to close the interview: The case manager goes over what has been discussed. Prioritizing next steps or topics becomes easier at this point.

JOE: I told her I didn't care what she thought. I'm sure she knew what I meant even though *I* didn't know what I meant. She won't give me a chance. I am trying hard, so what does it mean to her that I have been gone? She has no idea what I have been through.

MIKE: Let me see if I can summarize what we've talked about today. Returning home has been very difficult for you and you're confused about your relationship with your wife. She still seems angry about your trouble with the law, and the two of you just can't seem to communicate.

JOE: I guess that's about it.

MIKE: Let's focus on the communication problems at our next meeting.

Other kinds of responses help us increase the client's understanding during the interview. Most often, these responses are used after the presenting problem is clearly defined and explained. Among them are interpretation, confrontation, and informing.

Interpretation The case manager adds something to what the client has said, based on the meanings gleaned from his or her verbal and nonverbal communications.

JOE: I may not be able to stay with her, and that is okay given how awful everything is at home. I'm ready to get out.

MIKE: It's almost a relief to think about not having fights all the time. But it's sad at the same time to think about leaving your home and ending your marriage.

Confrontation This type of response is used to point out inconsistencies, discrepancies, or distortions.

JOE: Yeah, I do feel relieved—but sad, too. We've been married ten years. I have done the best that I can do. I told you I paid my dues. I really would like to start over and find me another woman.

MIKE: This is really different information than you gave me at the beginning of our conversation when you told me that you were glad to get home to your wife.

Informing The case manager uses informing to provide information about such topics as agency services, resources, and problems. In the following example, the presenting problem has been identified, and the case manager shares information about a possible service.

JOE: I guess we do need some help trying to talk to each other about what is happening to us. I guess I'm not ready to end it yet. And there isn't any other woman. How do we get back what we had?

MIKE: I think you have two options here. One is for both you and your wife to meet with me to discuss what is happening. A second option is for the two of you to attend a group that is made up of couples just like you. There are four couples in the group now, talking about their marriages, how to communicate better, and how to resolve disagreements.

The following is an excerpt from an intake interview that incorporates all the responses that you have just read about: open and closed inquiries, minimal responses, paraphrases, reflection, clarification, summarization, interpretation, confrontation, and informing. Notice how and when the case manager uses each response and the client's reaction to it.

Mathisa walked into the AIDS Community Center one Wednesday evening about 8 o'clock. She had come to talk to a counselor because she had just discovered that her best friend had AIDS. She had told Mathisa and no one else, and Mathisa was scared. She did not know what to tell her friend, and she did not know what to do. Mathisa had passed by the center on her way to school each morning, but she had barely noticed it. And now she was here.

A young man came up to her and introduced himself. She said "Hi," but did not want to tell him her name. In fact, she really did not want anyone to know that she was there. He asked her if she had come to talk and she nodded "yes." He led her into a small room that had three comfortable chairs. He sat in one and pointed to one where she could sit.

The young man, Dean, started by telling Mathisa about the agency and about his job as a service coordinator. He also talked to her about the confidentiality policies of the agency.

DEAN: *I'm glad you're here.*

MATHISA: *I'm not sure I'm glad to be here. I've never been in this place before.*

DEAN: *It's scary to be in a place for the first time. We're always glad to welcome newcomers and visitors.* (Smiles) *Can you tell me why you're here?*

MATHISA: (Pauses) *I'm here for a friend.*

DEAN: *Your friend is very lucky that you could come for him or her. How did you decide to come here?*

MATHISA: *Well, this is a place I pass every morning on my way to school. Sometimes I wonder what it's like here. And today I knew that I needed to come. Can I be sure that nobody will find out what I tell you?*

DEAN: *Yes, what you tell me stays between the two of us. Confidentiality is very important to you.*

MATHISA: *I have some information, and I don't want anyone else to know. I don't know what I can do.*

DEAN: *Umm . . .* (Nods)

MATHISA: *You need to know what* before you can help, I guess.

DEAN: *Could you describe the event that brought you here?*

MATHISA: *I'm just so scared and I don't know what to do.*

DEAN: *It's scary having information and not having any idea what to do with it. How do you think I can help you?*

MATHISA: *I don't know for sure. But I do know that you understand AIDS and you help people with AIDS. I only know what they taught us in school.* (Mathisa is obviously in distress; she is almost in tears and is choosing her words carefully.)

DEAN: *Your quiet voice and your tears let me know that the reason you came is very upsetting to you.*

MATHISA: (Nods)

DEAN: (Silence)

MATHISA: *My best friend just told me that she has AIDS. She got tested when she was on a trip a month ago. She went to a state that does not ask your real name. She just found out yesterday. She's really blown by this. No one else knows—not even her parents.*

DEAN: *She told you and you don't know what to do.*
MATHISA: *I don't really know anything about it. I don't want her to die, and I don't want to die. Her boyfriend doesn't know, and I don't know what she'll tell her parents. She may even run away or kill myself, but what if the tests are wrong? And seeing the really sick people here makes me think that I don't want to live.*
DEAN: *Mathisa, I'm not sure what you said just then; you said that you didn't want your friend to die, and then you said that you didn't want to die.*
MATHISA: *She doesn't want to die, I mean, I don't want her to die.*
DEAN: *I hear you say that you want to kill yourself.*
MATHISA: *I must have been confused, I mean. . . .*

In this interview, Dean promoted good rapport with Mathisa by providing a good physical setting. It was simple, without distractions; they sat in close proximity, with no barriers between them, in comfortable chairs. Perhaps most important, it was an environment that was private. Dean introduced himself and assured her of confidentiality so that Mathisa felt comfortable beginning to talk.

Dean used a combination of open inquiries and responses. His first open inquiry was "Can you tell me why you're here?" This was designed to elicit a fact from Mathisa. She did not elaborate, but she did give enough information to continue the conversation. Dean also used "how" and "could" questions to encourage Mathisa to provide more information.

Dean's responses also included a paraphrase—"Confidentiality is very important to you"—as well as reflection—"Your quiet voice and your tears let me know that the reason you came is very upsetting to you." Both of these responses helped Mathisa understand that Dean was actively listening to her and had heard what she had said. He had also interpreted her nonverbal messages.

At the conclusion of this excerpt from the interview, Dean used confrontation—"I'm not sure what you said just then; you said that you didn't want your friend to die, and then you said that you didn't want to die"—and clarification—"I hear you say that you want to kill yourself"—to try to sort through the information that Mathisa has given. In the remainder of the interview, Dean will continue to find out more about the problem and its implications for Mathisa and her friend.

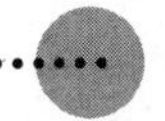

Interviewing Pitfalls

Clearly, interviewing requires a great deal of skill. An effective interviewer is one who listens attentively, questions carefully, and uses other helpful responses to elicit information and promote client understanding. However, caution is necessary. The desire to be helpful and the anxiety of conducting that first interview can lead to a number of pitfalls. Four of them will be discussed here.

Premature problem solving This arises from a desire to be helpful to the applicant by removing the pain, the discomfort, or the problem itself as soon as possible. Unfortunately, if the interviewer suggests a change, strategy, or solution before the problem has been fully identified and explored, this may address a symptom of the presenting problem rather than the actual problem. Premature problem solving may cause the client to lose confidence in the case manager's knowledge and skills or to become impatient. Also, premature problem solving undermines the client's self-determination and can lead to false assumptions, misinterpretation of what the client says, and steering him or her in the wrong direction. In the case of mental illness, misdiagnosis can result.

Giving advice In attempting to solve the problem or offer a solution, the case manager may mistakenly give advice. When given hurriedly and before the problem has been explored sufficiently, advice may be seen as indicating a lack of interest or thoroughness. The client may also feel misunderstood, or he or she may superficially agree, without having an intention of following through. Advice giving also tends to diminish the client's level of responsibility, self-determination, and partnership in problem solving.

Overreliance on closed questions This pitfall has been discussed elsewhere in this chapter. Remember that closed inquiries are usually directive and focused on facts; they rarely provide the opportunity for exploration. A series of closed inquires may also make the client defensive. Once this feeling is established, it is difficult to overcome.

Rushing to fill silence Because silence is often awkward in everyday social situations, beginning helpers as well as seasoned professionals are sometimes uncomfortable with pauses and rush to fill them, believing that silence indicates that nothing is happening. In fact, silence does have meaning. The client may be waiting for direction from the interviewer, thinking about what has transpired so far, or just experiencing an emotion. Constant dialogue can be a false signal that something is "happening." Skillful case managers learn to listen to silence.

Chapter Summary

This chapter has focused on the attitudes, characteristics, and skills that contribute to effective intake interviewing, which is an art. Communication is at the core of interviewing. Listening, questioning, and responding are communication skills that are used to obtain information, establish rapport, develop a relationship, and promote client understanding. These skills are critical for the case manager during the initial phase of the process. Among the pitfalls that may arise from the desire to be helpful or the anxiety of a first meeting are premature problem solving, giving advice, using too many closed questions, and rushing to fill silence.

Chapter Review

Key Terms

Sexism
Racism
Ageism
Effective communication
Attending behavior
Active listening
Open inquiries

Reviewing the Chapter

1. What attitudes and characteristics facilitate the development of a helping relationship?
2. Write a dialogue representing the beginning of an intake interview, to illustrate desirable attitudes and characteristics of the helper.
3. Describe an office setting that facilitates relationship building.
4. Discuss the problems that are created by stereotypes based on gender, race, and age.
5. Give general guidelines for essential communication skills in interviewing.
6. Discuss the importance of listening in the intake interview.
7. What is attending behavior (active listening)?
8. What are the five listening behaviors represented by the acronym S-O-L-E-R?
9. Why are listening and questioning both complex skills and arts?
10. State the advantages and disadvantages of questioning.
11. Describe the four situations in which questioning is appropriate.
12. Distinguish between closed and open inquiries.
13. What are the four commonly used methods of introducing an open inquiry?
14. Name four pitfalls of interviewing and tell how each may be avoided.

Questions for Discussion

1. Do you think that you will be able to conduct a good interview? What skills will you need to strengthen your competence as an interviewer?
2. Speculate about how interviewing will change if computers are used in the process.
3. Discuss the kinds of activities that might help you practice your listening skills.
4. Do you believe that certain communication skills are essential to effective interviewing? If your answer is no, why not? If yes, what are they, and why are they important?

References

Baruth, L., & Manning, M. (1991). *Multicultural counseling and psychotherapy.* New York: Merrill.

Bernal, G., Martinez, A. C., Santisteban, D., Bernal, M. E., & Olmedo, E. E. (1983). Hispanic mental health curriculum for psychology. In J. C. Chunn, P. J. Dunston, & R. Ross-Sheriff (Eds.), *Mental health and people of color* (pp. 65–96). Washington, DC: Howard University Press.

Brammer, L. M. (1993). *The helping relationship.* Englewood Cliffs, NJ: Prentice-Hall.

Brown, D., & Srebalus, D. J. (1988). *Introduction to the counseling profession.* Boston: Allyn & Bacon.

Carkhuff, R. R. (1969). *Helping and human relations: A primer for lay and professional helpers. Vol 1: Selection and training.* New York: Holt, Rinehart & Winston.

Combs, A. W. (1969). *Florida studies in the helping professions.* Gainesville: University of Florida Press.

Egan, G. (1990). *The skilled helper: A systematic approach to effective helping.* Pacific Grove, CA: Brooks/Cole.

Epstein, L. (1985). *Talking and listening: A guide to the helping interview.* St Louis: Times Mirror/Mosby.

Evans, D. R., Hearn, M. T., Uhlemann, M. R., & Ivey, A. E. (1979). *Essential interviewing: A programmed approach to effective communication.* Pacific Grove, CA: Brooks/Cole.

Garrett, A. (1972). *Interviewing: Its principles and methods.* New York: Family Service Association of America.

Gilligan, C. (1982). *In a different voice.* Cambridge, MA: Harvard University Press.

Gordon, T. (1974). *Teacher effectiveness training.* New York: David McKay.

Ivey, A. E. (1988). *Intentional interviewing and counseling: Facilitating client development.* Pacific Grove, CA: Brooks/Cole.

Kelly, K. R., Hall, A. S., & Miller, K. L. (1989). Relation of counselor intention and anxiety to brief counseling outcome. *Journal of Counseling Psychology, 36,* 158–162.

Kirst-Ashman, K. K. (1993). *Understanding generalist practice.* Chicago: Nelson Hall.

Kolanad, G. (1994). *Culture shock! India.* Singapore: Time Books International.

Long, L., Paradise, L. V., & Long, T. J. (1981). *Questioning: Skills for the helping process.* Pacific Grove, CA: Brooks/Cole.

Morales, A., & Sheafor, B. W. (1980). *Social work: A profession of many faces.* Boston: Allyn & Bacon.

Sielski, L. M. (1979). Understanding body language. *Personnel and Guidance, 57*(5), 238–242.

Smith, S. (1995a). Family theory and multicultural family studies. In B. B. Ingoldsby & S. Smith (Eds.), *Families in multicultural perspective* (pp. 5–35). New York: Guilford Press.

Smith, S. (1995b). Women and households in the Third World. In B. B. Ingoldsby & S. Smith (Eds.), *Families in multicultural perspective* (pp. 235–267). New York: Guilford Press.

Woodside, M., & McClam, T. (1994). *Introduction to human services.* Pacific Grove, CA: Brooks/Cole.

Chapter Six

Service Delivery Planning

We call our clients "consumers" because they are here for our services. They are buying our services. There are other agencies doing exactly what we are doing, so they can certainly go to another agency that they find much more amicable or friendlier. So they are definitely consumers.

—Yolanda Vega, CSW, Director of Agency Services, Casita Maria Settlement House, Bronx, personal communication, May 4, 1994

It is the responsibility of the service coordinator and the referring case manager to work together on making the plan. You also need a Plan B. Part of the planning process is planning in case your first plan doesn't work. And sometimes Plan B is not agreeable to the participant.

—Janelle Stueck, Program Manager, Private Industry Council, Knoxville, Tennessee, personal communication, August 30, 1993

In the interview, I tell clients what we expect in our program. For example, they have to have a 9.5 in math and 10.0 in reading on their TABE test. They have to be able to type at least 30 words a minute. They know their expectations at the very beginning. And they are only allowed to miss up to 8 days in this training program. . . .

—Cathy Lowe, Private Industry Council, Knoxville, Tennessee, personal communication, August 30, 1993

At this point in the process, the agency has determined that the applicant meets the eligibility criteria and can now receive services. At Casita Maria Settlement House, the applicant becomes a "consumer," whereas the Private Industry Council uses the term *client.* An agency in South Dakota that serves adults with developmental disabilities calls the service recipients *individuals,* explaining that "they are not clients any more or consumers. They are just people." Other agencies or organizations use the term *customer.* The change in status from applicant to recipient of services marks the move into the second phase of case management: planning service delivery.

The quotes that introduce this chapter identify some of the activities that occur during this phase. The workers at the Private Industry Council stress the need for not only a plan of services for a client, but also a fallback plan. They also indicate the importance of expectations or goals for clients and the uses of test data. Margaret Mikol at SKIP in New York City summarizes this phase of case management: "You have to be able to procure and digest a large body of information from all kinds of different people, then you have to sort through the information and say, 'Now, what are the real issues here?' Then finally you have to be able to say, 'Okay, if this is the real problem, what do I have to do to solve it?' " (personal communication, May 4, 1994).

Linda Smith, a caseworker at the School for the Deaf, also notes the importance of gathering information in order to see the big picture: "Probably one of the most important things in my job is being able to communicate with all the people who are involved with the student because . . . so many people have different pieces of information, and if we can pull all the information together, we can get a better picture of the way to work with our kids or their families. I think being able to see the whole in addition to the individual little narrow parts is a very important characteristic for somebody to be able to see that big picture" (personal communication, October 27, 1993).

This chapter explores the planning phase of case management, wherein the helper and the client together determine the steps necessary to reach the desired goal. The activities involved in this phase include the review and continuing assessment of the problem, the development of a plan, the use of an information system, and the gathering of additional information. Running through our discussion in this chapter are two critical components of the case management process—client participation and documentation.

For each section of the chapter, you should be able to accomplish the objectives listed here.

REVISITING THE ASSESSMENT PHASE
- List the two areas of concern that are addressed when reviewing the problem.

DEVELOPING A PLAN FOR SERVICES
- Identify the parts of a plan.
- Write a plan.

IDENTIFYING SERVICES
- Locate available services.
- Create an information and referral system.

GATHERING ADDITIONAL INFORMATION
- Compare interviewing and testing as data collection methods.
- Identify the five types of interviews.
- Show how sources of error can influence an interview.
- Illustrate the role of testing in case management.
- Define *test*.
- Categorize a test.
- Identify sources of information about tests and the information that each provides.
- Analyze the factors to be considered when selecting, administering, and interpreting a test.

Revisiting the Assessment Phase

This phase of case management begins with a review of the problems identified during the assessment phase. Before moving ahead with the process, the case manager will need to know if the problem has changed, if the same client resources are available, and if any shift in agency priorities has occurred. According to Janelle Stueck at the Private Industry Council, "one of the most difficult pieces of case management is doing that continual assessment and not resting on what you think things were two weeks ago. They may be different. It requires an incredible amount of judgment" (personal communication, August 30, 1993). In order to complete the review quickly before moving into a planning mode, the case manager examines two aspects of a case.

The first area of concern involves a review of the relevant facts regarding the problem. At this point, the case manager revisits the identification of the problem. The initial question the helper asks can help determine whether the problem still exists. Working with people requires an element of flexibility; clients' lives change, just as ours do. Thus, the problem may have changed in some way, the client may have a different perspective on it, the participants may be different, or assistance may no longer be needed or wanted. Once the case manager has confirmed that the problem still exists and has documented any changes that have occurred, the problem itself is revisited. Is the problem an unmet need such as housing or financial assistance, or is it stress that limits the client's coping abilities or causes interpersonal difficulties? Or is the problem a combination of several factors? This activity is best accomplished by talking with the client and reviewing his or her file.

A second area of concern in the review of the problem requires an examination of available information to answer three questions:

- What do I know about the source of the problem?
- What attempts have been made previously (before agency contact) to resolve the problem?
- What barriers may affect the client's attempts to resolve the problem?

Information in the case file will help the case manager answer these questions. Another important source of information is the client. Questioning the client can reveal what he or she has thought about doing, what has been tried, and what some possible solutions are. Such questions indicate that the process of case management continues to be a partnership between the client and the system.

Other techniques that are helpful in reviewing the problem are observations, documentation, and intuition. In the course of receiving the application, conducting the intake interview, and/or making a home visit, the case manager has had opportunities to observe the client. These observations may be richer if they occur in the home or in the office, or if the client is accompanied by family members or a significant other. Information available from such observations includes the client's thoughts, feelings, behaviors, and relationships.

Documentation in the case file also provides facts and insights about the client. Case notes, reports from other professionals, and intake forms help the case manager pin down past occurrences and pertinent facts about the present situation. Finally, the case manager's intuition about the client and all the known facts also provide direction to case planning. Case managers who have a long history in service delivery may call on knowledge and experience from the past to understand a current case. Sometimes, knowledge comes from a case manager's "own perception, instinct, kind of experience, street know-how" (McClam & Woodside, 1994, p. 41).

Once the case manager has revisited the problem, confirmed its existence, documented any changes, and reaffirmed the client's desire for assistance, the two of them move to the next step of the planning phase, which addresses the need to determine the steps necessary to reach the identified goal(s). This is the plan that will guide service provision.

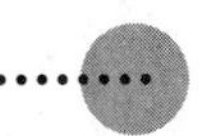

Developing a Plan for Services

The plan is a document setting forth the goals and objectives of service delivery and directs the activities necessary in order to reach them. The plan also serves as a justification for services by showing that they meet the identified needs and will lead to desired outcomes. More specifically, a plan describes the service to be provided, who will be responsible for its provision, and when service delivery will occur. If there are financial considerations, the plan may also identify who will be responsible for payment. Sometimes financial support is available from outside sources, including the client and the family. Usually, the completed plan is signed by the client and the case manager as the representative of the agency. It may then be approved by someone else in the agency before the authorization to provide services.

Clearly, the plan is a critical document, since it identifies needed services and guides their provision. How is it developed? What is included? What are goals and objectives? These questions will be answered as you read this section.

Plan development is a process that includes setting goals, deciding on objectives, and determining specific interventions. The process begins with the synthesis of all the available data. This information is scrutinized carefully for as complete a picture of the case as possible. It is analyzed so as to identify desirable outcomes. In Roy Roger Johnson's case in Chapter 1, the information available at the time of plan development was derived from Roy's application for services, the intake interview, reports from his orthopedic surgeon, case documentation, a general medical examination report, a psychological evaluation, and a vocational evaluation report. All of this was reviewed and considered as the counselor and Roy formulated the plan.

Roy had a back injury and needed assistance finding a job; he also met economic eligibility criteria. His Individualized Written Rehabilitation Plan (IWRP) is reproduced in Figure 6.1. It included a program objective and intermediate objectives. For each objective, a service was identified, as well as a method of checking progress toward the achievement of the objective. The form also provided space to describe any other client, family, or agency responsibilities or conditions. Because this agency values client participation, Roy's view of the program was also noted. Then both Roy and the counselor signed the plan.

Exactly what a plan looks like varies from agency to agency. However, if you are employed by an agency that provides case management or client services, you can be sure that a plan will guide your work. Let's examine the components of a plan of services.

Service plans are goal directed and time limited, so they should include both long-term and short-term goals. Long-term goals state the situation's ultimately desired state. Short-term goals aim to help the client through a crisis or some other present need. Whatever the time constraints, goals establish the direction for the plan and provide structure for evaluating it.

INDIVIDUALIZED WRITTEN REHABILITATION PROGRAM

1. NAME JOHNSON, ROY — PROGRAM TYPE ☒ INITIAL ☐ AMENDMENT

2. YOU ARE ELIGIBLE FOR: ☒ VOCATIONAL REHABILITATION SERVICES ☐ EXTENDED EVALUATION SERVICES ☐ PAST EMPLOYMENT SERVICES

BECAUSE:
☒ A. YOU HAVE A PHYSICAL OR MENTAL DISABILITY WHICH CONSTITUTES A SUBSTANTIAL HANDICAP TO EMPLOYMENT AND:
☒ B. YOU CAN REASONABLY BE EXPECTED TO BENEFIT IN TERMS OF EMPLOYABILITY FROM SERVICES.
☐ C. IT CANNOT BE DETERMINED WHETHER OR NOT YOU CAN BENEFIT IN TERMS OF EMPLOYABILITY FROM REHABILITATION SERVICES.
☐ D. POST EMPLOYMENT SERVICES ARE NEEDED FOR YOU TO MAINTAIN EMPLOYMENT.

3. PROGRAM OBJECTIVE: Business Communications — ANTICIPATED DATE OF ACHIEVEMENT: MONTH 1 YEAR 92

4. INTERMEDIATE OBJECTIVE, SERVICES METHODS OF CHECKING PROGRESS.

ESTIMATED DATES TO REACH OBJECTIVE & RECEIVE SERVICES

	RESPONSIBILITY	FROM	TO
OBJECTIVE To correct physical impairment so that client might			
SERVICES reach vocational objective	Client	1/91	1/92
Possible office visit with the doctor			
METHOD OF CHECKING PROGRESS Medical information			
OBJECTIVE To provide background information and educational skills			
SERVICES so that client might reach vocational objective	VR	1/91	1/92
A. Tuition/UT Knoxville		1/91	1/92
B. Miscellaneous Educational Expenditures		1/91	1/92
METHOD OF CHECKING PROGRESS R-II, Grade Reports			
OBJECTIVE To follow client's progress and develop plan amendment if needed so that client might			
SERVICES reach objective	Client	1/91	1/92
A. Possible RP-B		1/91	1/92
B. Client/Counselor Contacts		1/91	1/92
METHOD OF CHECKING PROGRESS R-11			

5. CLIENT OR FAMILY AND AGENCY RESPONSIBILITIES AND CONDITIONS: I. Client is responsible to maintain contact with counselor twice each semester by mail, phone or in person. II. Client is responsible to furnish VR Counselor with a copy of grades at the end of each term. III. Client is responsible to maintain an average load of classes and average grades throughout his program. IV. Client is responsible to file for any similar benefits which might help him pay for his program. V. Client is responsible to furnish VR counselor with a resume and a list of potential employers to interview with during the first part of his senior year. VI. Client is responsible to notify counselor of any significant change of address, health, phones number of financial status.

6. CLIENT'S VIEW OF PROGRAM The client and I have discussed the services necessary to help him reach his vocational objective and we are in mutual agreement with his plan.

I HAVE PARTICIPATED IN THE DEVELOPMENT OF THIS PROGRAM AND I UNDERSTAND IT.
I UNDERSTAND AND ACCEPT THE STATEMENT OF UNDERSTANDING WHICH HAS BEEN EXPLAINED TO ME.

Roy Johnson	5-6-91	Susan Fields	5/6/91
Client's Signature	Date	Supervisor Signature	Date

Figure 6.1 Roy's IWRP

"A **goal** is a statement describing a broad or abstract intent, state, or condition" (Mager, 1984, p. 33). For clients, a goal is a brief statement of intent concerning where they want to be at the end of the process—for example, "Learn daily living skills in order to live independently," "Acquire knowledge and skills for a career in business communications," or "Develop a support network for help coping with phobias."

Having written goals helps us focus on what we are trying to accomplish before we take action or provide any services. Action is often easy, but sometimes relating actions to outcomes is not. For accountability reasons, service provision is tied to outcomes. This makes writing goals a critical step in plan development. Remember that these broad statements of intent can be achieved only to the degree that their meaning is understood, so well-stated, reasonable goals are essential to problem resolution (Dixon & Glover, 1984).

How does one write goals that are well stated and reasonable? Three criteria help us achieve this. First, the goal should be expressed in language that is clear and concise; second, the goal statement should be unambiguous; and third, the goal must be realistic and achievable. These criteria are illustrated in the following goals, which were established for a 74-year-old woman who will attend the Daily Living Program at the Oakes Senior Citizens Center.

Draft 1 is a goal statement for Ms. Merriweather; Draft 2 improves the statement by making it more clear and concise.

> *Draft 1:* Ms. Merriweather will participate often in many of the Oakes programs that relate to sports, games, music, communication, exploring other cultures, and other educational programs as they are developed by the creative staff in the activities area.
> Draft 2: Ms. Merriweather will increase her social opportunities by participating in center activities.

A description of the plan is presented in Draft 1. In Draft 2, it is restated less ambiguously by defining who will help with medications and what the help entails.

> *Draft 1:* They will work with Ms. Merriweather and her numerous family members to help with medications.
> *Draft 2:* Nursing staff will develop a plan to administer Ms. Merriweather's medication.

The goal in Draft 1 establishes general physical goals for Ms. Merriweather. Draft 2 restates these goals in realistic and achievable terms.

> *Draft 1:* Ms. Merriweather will increase her range of motion, physical strength, and stamina.
> *Draft 2:* Ms. Merriweather will participate four times a day in an exercise program that includes walking, weightlifting, and stretching.

Thus, goals are an important part of the service plan. They increase the chance of solving the problem by providing direction and focusing attention on well-expressed, reasonable statements. Since formulating goals also requires collaboration between the client and the case manager, writing them also highlights the shared responsibility for the case. Once a broad statement of intent has been agreed upon, it is time to identify the activities that will lead to the desired outcomes. This process continues as a cooperative effort between the client and the case manager. Activities are identified as objectives.

An **objective** is an intended result of service provision rather than the service itself. It tells us about the nuts and bolts of the plan—what the person will be able to do, under what conditions the action will occur, and what the criteria for acceptable performance are—so that we can know whether the objective has been accomplished. Objectives are useful for several reasons. First, they tell us where we are going. Second, they give the client guidance in organizing his or her efforts. Third, they state the criteria for acceptable performance, thereby making evaluation possible. Objectives are all-important for the case manager, since they provide the standards by which progress is monitored. As progress is made, the case manager adjusts the plan as needed.

Writing clearly defined objectives benefits the client, the case manager, and the agency. Boserup and Gouge (1977, p. 111) provide the following guidelines for writing and evaluating service objectives:

1. The statement of objective should begin with the word "to" followed by an action verb. The achievement of an objective must come as a result of action of some sort. Therefore, the commitment to action is basic to the formulation of an objective.

2. The objective should specify a single key result to be accomplished. In order for an objective to be effectively measured, there must be a clear picture of when it has or has not been achieved.

3. The objective should specify a target date for its accomplishment. It's fairly obvious that to be measurable, an objective must include a specific completion date, either stated or implied. If the objective is of a continuing nature, the target date could be assumed to be the end of the eligibility period. A situation of this nature may occur when services are being provided to a client whose situation is such that prospects for improvement seem very slim.

4. An objective should specify the "what" and "when"; it should avoid venturing into the "why" and "how." Once again, an objective is a statement of "results to be achieved." The "why bridge" should have been crossed before the actual writing of the objective has started. The means of achieving an objective should not be included in the objective statement.

5. Objectives should be realistic and attainable but still represent a significant challenge. Since an objective can and should serve as a strong motivational tool for the individual worker and client, it must be one that is within reach. This simply means that resources must be available to achieve the objective.

6. Objectives should be recorded in writing. Each of us, whether consciously or unconsciously, has a convenient "memory." We tend to remember the things that turn out the way we want them to and either forget or modify those things that are less than we wish. If objectives were not put in writing, it would be relatively easy to look on accomplishments as if they were in fact planned objectives. On the other side of the coin, one of the sharpest areas of conflict among case manager, client, and supervisor is illustrated by such

phrases as "I thought you were working on something else!" or "That's not what we agreed to do" or "You didn't tell me that's what you expected." Having objectives in writing will not eliminate all of these problems, but it will provide something more tangible for comparison. Furthermore, written objectives serve as a constant reminder and an effective tracking device for the case manager, the client, and the supervisor in order to measure progress.

7. A statement of objective must be consistent with the resources available or anticipated.

8. Ideally, an objective should avoid or minimize dual accountability for achievement when joint effort is required.

9. Objectives must be consistent with basic agency policies and practices.

10. Objectives must be willingly agreed to by the client without undue pressure or coercion.

11. The setting of an objective must be communicated not only in writing but also in face-to-face discussions with the client and the resource persons or agencies contributing to its attainment.

The following case example illustrates the development of goals and objectives with a client who is elderly and needs assistance.

Mary Sue Davis is an 86-year-old white married female. Her husband has recently been placed in a nursing home facility so that he can be provided with full-time care. Mrs. Davis has a severe heart condition and has been ordered by her physician to rest every 2 hours and not to travel by herself because of dizzy spells. The nursing home is now receiving her husband's Social Security income. Mrs. Davis lives in a two-bedroom apartment and receives $10 a month in food stamps. They have one son who lives an hour away and also has a heart condition. Mrs. Davis is requesting assistance with transportation in order to visit her husband on a more regular basis.

An interview with Mrs. Davis at her apartment revealed that her income consists solely of her Social Security checks. She does have Medicare to help with the costs of treatment for her severe heart condition. She currently uses public transportation (bus) to get where she needs to go. During the interview, the service coordinator identified additional problems: the affordability of her current apartment, the availability of affordable housing, the need for an escort for travel, and possible grief issues regarding her husband's condition and placement in a nursing home.

Mrs. Davis agrees that she cannot afford her apartment and needs to seek more affordable housing. She is willing to apply for CAC (Community Action Committee) transportation that will pick her up at her door. She is very realistic regarding her husband's condition. Although she wishes he could come home, she has accepted that he will most likely remain at the nursing home. She realizes that she has to take care of her own health, but at the same time she has to get things done, and there is not always somebody around to help.

Council on Aging
Client Plan

CLIENT Mary Sue Davis

DATE 5/3/97

GOAL 1 Locate affordable housing

Objective 1: To contact KCDC this week for application for rent-controlled apt.—service coordinator.

Objective 2: To review list of apts., decide which ones to see, and select one (1 month)—Mrs. Davis.

Objective 3: To request volunteer assistance to escort Mrs. Davis on apartment visits and to help with the move (1 week)—service coordinator.

GOAL 2 Provide transportation

Objective 1 To complete application for K-Trans lift and CAC vans (this week)—Mrs. Davis.

Objective 2 To determine eligibility for medical escort service (2 weeks)—service coordinator.

Objective 3 ______

Figure 6.2 Client plan for Mrs. Davis

The service coordinator identified two main goals for Mrs. Davis: to find affordable housing and to secure transportation that is appropriate. These are set forth in the Client Plan (Figure 6.2).

The first objective toward the housing goal was to complete an application for a rent-controlled apartment with KCDC (the city housing authority). Due to long waiting lists, this needed to be done within the week. The next step was to determine where she preferred to live (probably close to the nursing home).

GOAL 3 ______________________________

Objective 1 ______________________________

____

____

Objective 2 ______________________________

____

____

Objective 3 ______________________________

____

____

GOAL 4 ______________________________

Objective 1 ______________________________

____

____

Objective 2 ______________________________

____

____

Objective 3 ______________________________

____

____

Service Coordinator ______________________ Client ______________

SIGNATURE SIGNATURE

Figure 6.2 *(continued)*

After the application was completed, the service coordinator arranged for a volunteer to take Mrs. Davis to look at several apartments and to meet with apartment managers to find out about waiting lists (Mrs. Davis couldn't afford to wait for a very long period of time). The service coordinator found a volunteer to help with this. Once Mrs. Davis decided on an apartment, other volunteers assisted with the move. Her son could afford to rent a moving truck and to drive the truck, although he couldn't lift or carry due to medical problems. The time allotted for these objectives was workable, and the objectives were met within a month.

The objectives for the goal of transportation were to apply for the K-Trans lift along with CAC vans. Obtaining an assessment from the Office on Aging was also an objective; that agency provided escorted transportation for medical appointments and necessary errands for people over 60. This service would be available until Mrs. Davis was accepted by the other transportation agencies.

In this case, the plan identified services and then guided the delivery of those services. The goals and objectives in the plan were developed using the guidelines suggested previously. Note that each objective clearly stated who would provide the service and gave a time frame for service delivery. The plan was implemented successfully.

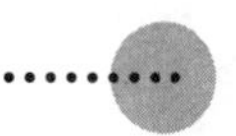

Identifying Services

Once the plan is complete and has been agreed upon by the client and the case manager, it is time to begin thinking about the delivery of services. A well-developed plan provides information about what the service is, who will provide it, what the time frame is, and who has overall responsibility for service delivery. It is the case manager's responsibility to implement the plan. What are these responsibilities? How does one begin implementation? These questions are explored next.

Identifying services has been compared to the brokering role. In both situations, the case manager is involved in the legwork and planning that is necessary for implementation. As a broker, the case manager helps clients access existing services and helps other service providers relate better to clients. This linking of clients and services also occurs as the case manager arranges for service delivery. The steps are similar.

Information and Referral Systems

One of the most helpful tools for a case manager is knowledge of what is available in the community. Who do you know? What services are available? How does one access the services? Is there a waiting list? One of the challenges facing new case managers is to establish an **information and referral system.** For case managers with experience, the challenge consists of continually developing and updating their systems. Knowing what an information and referral system is, how to set one up, and how to use it are valuable skills in case management.

Human service employers believe that the people their agencies will serve in the future will be multiproblem clients—for example, people with dual diagnoses, diverse problems, and problems of long standing (McClam, 1992). The needs of clients such as these rarely match the services available from a single agency. In these cases, the case manager finds it invaluable to have information about other available services. Many helping professionals have personal service directories to supplement existing community or agency directories.

There are four components to information and referral (Mathews & Fawcett, 1981). One component is the **social service directory,** which usually lists the kinds of problems handled and the services delivered by other agencies. In some communities, these are published by a social service agency, by a funding source such as the United Way, or (as a community service) by a business or organization. Another component is the **interagency feedback log.** Such logs provide feedback to the agencies that deliver services, to help ensure quality information and referral services. Some agencies accomplish this through referral forms that record referrals, give information on the services needed, and provide the referral agency with information on the services received. If the client takes the form to the agency providing services, it may also serve to remind the client of the appointment. A third component of information and referral systems is staff training. In these sessions, the helper may be introduced to the services of the employing agency as well as those of other agencies. Other information and referral data are that shared during staff training may include reviewing and updating referral procedures, announcing new services or ones that no longer exist, and discussing the effectiveness and efficiency of service delivery.

A final component of information and referral is the updating of social service directories. Social service directories may have two indices: one that is an alphabetical listing of agencies and one that is a categorical listing of services. Each entry in the directory lists the agency's name, address, phone number, and services. Also listed may be fees, hours of service, eligibility criteria, and sources of agency support. Here is an example of an entry:

RUNAWAY SHELTER
2535 Magnolia Avenue
Bluff City, NJ

Purpose and services: The Runaway Shelter and Homeless Youth Shelter provides shelter, counseling, and casework services to youth (ages 13–18). This is a short-term (14 days) service.

Eligibility: No eligibility requirements

Fees: None

Hours open: 24 hours

Area served: Region

Sources of support: United Way, Department of Health and Human Services

Existing directories are helpful to the case manager, but sometimes establishing one's own system is useful for filling in the gaps in published directories or for recording detailed information that may be of special interest to the individual helper.

Setting Up a System

The first step in establishing one's own information and referral system is to identify all agencies and available services. This includes listing agencies previously contacted, checking the Yellow Pages of the telephone book, and talking with other professionals. Each agency and service becomes part of a card file or a computer file that is easy to update. The file can also be expanded by talking with clients (particularly those who have been in the human service system for a period of time), meeting other professionals at meetings and workshops, and attending community meetings.

Whether on cards or in the computer, such a file is easy to use when identifying the client problem and matching it with a service. However, since a client rarely has only one problem, using the file may not be so simple. First, the client and the case manager prioritize the problems. Once this has been done, the case manager identifies which problems will be addressed by the agency and which ones need referral. These additional services can be found by checking the card file. If there is more than one resource to serve the client's particular need, the case manager works to identify the agency that can meet the client's needs in the manner responsive to the client's values and concerns.

Deborah Caudill is an 18-year-old client who needs long-term counseling to work on the anger she feels toward her father for deserting the family when she was 11. Lou Lerner, her case manager, knows that the counseling Deborah needs is beyond the scope of the services provided by the agency where she works. There are two other agencies in their community that do offer long-term counseling for adolescents. Since Deborah and Lou agree that counseling would be beneficial, they discuss these two agencies. Deborah has questions about their locations, who provides the counseling, whether it is group or individual, and how much it will cost. Lou consults her card file for the answers to these questions and provides Deborah with the information, and then they discuss the pros and cons of each option. The case manager's card file indicates that one center provides counseling services and is well known for its work with adolescents. In addition, the latest entry in the card file indicates that Jane Barkley, a previous client, had a positive experience there.

Establishing and using an information and referral system requires certain skills of the case manager. Being able to identify the client's problem, the community resources available to solve it, and the viable alternatives are all critical to the success of the system. Choosing a resource or a service requires the client's participation. The client may actually have the final say in the selection of the agency or service; the more accurate and complete the information is, the better the decision will be. Finally, good research skills are helpful as the case manager continually works to locate potential community resource alternatives and to update data on existing agencies and services.

Part of the development of a plan is identifying services to meet the client's needs. The development of an information and referral system is useful here. Throughout plan development, data gathering continues to take place.

Gathering Additional Information

Gathering additional information may be part of the planning process or part of the plan itself. To decide whether additional information is necessary, there must be a review of available information from other agencies, the referral source, employers, and others. The key to determining what is needed is relevance. Is the needed information relevant to the client and to service provision? Will it contribute to a complete array of social, medical, psychological, vocational, and educational information about the client? Once it is determined that additional information is necessary, the case manager decides how the information will be obtained. In some cases, the case manager can personally acquire the information, but it may also be necessary to consult family members, a significant other, or professionals such as psychologists, physicians, and social workers. The client also continues to be a primary source of information and is part of the decision-making process regarding the additional information needed and who can provide it. Next we introduce two data collection methods that case managers use; Chapter 7 explores what data is available from other professionals.

Data Collection Methods for the Case Manager

Two primary tools are available to the case manager for data collection: interviewing and testing. They are similar in several ways. The information is used to describe, to make predictions, or both. Each may occur in an individual or a group situation in which some type of interaction occurs. The group situation may be an interview with a family, or a test administered to more than one examinee. Both interviews and testing have a definite purpose, and the case manager assumes responsibility for conducting the interview or administering the test.

Interviewing

There are five types of interviews (Kaplan & Saccuzzo, 1993). The **assessment interview** is an interaction that provides information for the evaluation of an individual. The interview may be structured or unstructured; it uses both open-ended and closed questions. The intake interview is an example of an assessment interview in which the applicant provides information that helps in evaluating him or her and the problem in relation to the mission, resources, and eligibility criteria of the agency.

A **structured clinical interview** consists of specific questions, asked in a designated order. This type of interview is structured by guidelines to ensure that all clients are handled in the same way. The structure also makes it possible to score the responses. One advantage of this type of interview is its reliability or consistency. Flexibility is limited. Although it is a valuable source of information, the interview results should be interpreted with caution. The major limitation is its reliance on the respondent as an honest and capable interviewee who has skills for self-observation and insight.

A more comprehensive interview is the **case history interview.** This interaction includes both open-ended questions and specific questions. Topics may include a chronology of major events, the family history, work history, and medical history. Usually an interview of this type begins with an open-ended question or statement: "What was school like for you?" "Tell me about your work history." "What do you remember as the happiest times when you were growing up?" "Describe your relationship with your parents." These probes may be followed by specific questions, which may or may not be dictated by agency forms or guidelines. *"When did you quit your last job?" "What grade did you complete in school?" "Are you the oldest child?"* These questions help the case manager understand the client's background and uncover any pertinent information.

Technology is also an influence on interviewing. A computerized interview takes place via computer rather than face to face. Questions are presented and followed by a choice of responses.

Are you married? Yes No

If the answer is yes, then another question related to marriage may follow:

Is this your first marriage? Yes No

If the answer to the first question is no, then another question appears:

Did you complete high school? Yes No

The computerized interview is a good way to collect facts about a person. The limitations are that there is no nonverbal communication, and the feelings of the client are not shared. Important information may be lost as a result of these limitations.

The mental status examination (see Chapter 4) is a special type of interview used to diagnose psychosis, brain damage, and other major mental health problems. The purpose of a mental status examination is to evaluate a person thought to have problems in terms of what is known about factors related to these problems. The interview focuses on appearance, attitudes, behavior, emotions, intelligence, attention, and sensory factors. This type of interview requires the case manager to have some expertise on major mental disorders and the various forms of brain damage.

A final type of interview is the **employment interview**—an assessment procedure that aids in job selection and promotion decisions. An employment interview usually consists of both open and closed questions. You will participate in this type of interview as you pursue a career in human services. Clients may need help in preparing for such interviews. Those with past histories of difficulties such as criminal records or substance abuse may need help developing a strategy to cope with the questions.

The skillful interviewer also needs to know about **sources of error** in the interview. Awareness of sources of potential bias in the instrument itself or in the interviewer enables the case manager to compensate for any resulting distortions. A look at interview validity and reliability will help us identify potential sources of error.

It is often difficult to make accurate, logical observations and judgments, for a number of reasons. One is the **"halo effect"** (Thorndike, 1920). This occurs in an interview situation when the interviewer forms a favorable or unfavorable early impression of the other person, which then biases the remainder of the judgment process. For example, an unfavorable initial impression can make it difficult to see positive aspects of a client or a case. If a home visit to an apartment in a housing project reveals an unkempt, dirty, and very sparsely furnished living area, the case manager may find the visit unpleasant. The resulting interview with the single-parent resident is likely to be rushed and cursory, with little chance of gaining insight into any problems. They may also find it difficult to maintain eye contact with the parent, missing important nonverbal cues. Other contacts with this parent may be influenced by the memory of the physical setting.

A second cause of invalidity in an interview is "general standoutishness" (Hollingsworth, 1922). This is the tendency to judge on the basis of one outstanding characteristic, such as personal appearance. The more attractive, well-groomed individual might be rated more intelligent than a less attractive, unkempt individual. Consider a case manager who makes a home visit to investigate a child abuse report. The address is in an affluent suburb and the house is a stately two-story brick house with elaborate landscaping. The initial impression of neatness, money, and social standing may influence the investigator's interaction with the parents and the subsequent course of the investigation.

Cultural differences can also contribute to error. To take an extreme example, a case manager has been asked to visit a family that recently immigrated from India and has just moved into a rent-controlled apartment in the city. It is her last stop of the day, and she finds that she has interrupted a ceremony of *puja* (prayers of thanks for their new home). She finds family members seated on the floor around a fire. Appalled that they have started a fire on the floor, she stamps it out and begins lecturing the family on fire safety. When she finally begins to talk about the services that are available, the family does not respond.

As you can see, sources of error can prejudice interview validity. Error reduces the objectivity of the interviewer, often leading to inaccurate judgments. The more structured the interview is (see Chapter 4), the less error there will be. Because an interview does provide important information, the case manager can consider the information tentative and seek confirmation from other sources, such as more standardized procedures. Likewise, test results are more meaningful if placed in the context of a case or social history or other interview data. The two can complement each other.

The reliability of an interview is its consistency of results. In interviewing, this means that there is agreement between two or more interviewers in their conduct of the interview, the questions they ask, and the responses they make. As you might imagine, reliability varies widely. As you might expect, the reliability of structured interviews is higher. This is because they have more stringent guidelines concerning the questions and even the order of the questions. (The downside is that this structure limits what is obtained.) In general, interview data have limited reliability, because interviewers look for different things, have different interviewing styles, and ask different questions. It is important for the case manager to verify information with other sources over time.

Testing

In the previous section, testing was recommended as one way to verify the information gathered in an interview. Most people encounter tests shortly after beginning school. How we perform on tests affects our lives, and test scores have become key factors in many decisions. They influence placement in special academic classes; the assignment of labels such as *high achiever, mentally challenged,* and *average;* admission to schools and colleges; and job selection. In fact, test scores have never been as important as they are today.

Case managers encounter tests in various contexts—for example, test reports from other professionals. In some cases, the information consists of test scores and nothing more. Figure 6.3 is one example of how test results may be communicated.

In other cases, test scores are part of a written report that also gives some explanation of the scores. Box 6.1 is an excerpt from a report on a 37-year-old white male who was hospitalized for depression. He has completed two years of college and has been a personnel interviewer for ten years. To use this information, the reader of the report must have knowledge of tests and an understanding of test data.

A case manager may also encounter testing as a service offered by an agency. For example, a statewide evaluation facility located on the campus of a school offers services that include achievement testing for placement at the school and vocational testing for career development. Workers at the evaluation facility administer these tests to each client who is referred to the facility. Scores are interpreted and included in their evaluation reports, which are sent to the referring counselor. There may be other situations in which a knowledge of testing is important. For example, a case manager may be asked to select tests to be administered as part of the services required in a plan. This task requires knowing the sources of information about tests, the criteria for selecting a test,

Testing is an important means of gathering information about a client.

and eligibility for purchase and use. Such knowledge is also important when the case manager encounters a situation like the following.

A family on my caseload had trouble understanding the results of a recent assessment test that was administered at their son's elementary school. The school counselor who originally explained the results of the test used terms unfamiliar to the parents and did not answer the questions they asked. The parents feel that if they understood the results of the test, they could help their son in the areas where he was weakest. The parents have asked me to look at the test results and explain them again.

The case manager needs an appropriate level of testing knowledge in order to use tests as a resource. Because tests have assumed such importance today, particularly in decision making, case managers must think carefully about the role of testing in their work with clients. To make proper use of test results, one must understand the test being used—the purpose of the test, its development, its reliability and validity, administration and scoring procedures, the characteristics of the norm groups, and its limitations and strengths. This section presents an overview of these areas.

What is a test? A **test** is a measurement device. A **psychological test** is a device for measuring characteristics that pertain to behavior. It is a way to

Box 6.1 Test Administered and Results

Wide Range Achievement Test: A measuring device to estimate grade levels in 3 academic areas. Results from administering the WRAT show that the client is functioning at the 17.4 grade level in reading, 15.0 in spelling, and 7.1 in arithmetic. These scores are within the very superior, very superior, and average classifications, respectively, when compared with the appropriate normative age group.

evaluate individual differences by measuring present and past behavior. For example, the test your instructor will give you to measure your mastery of this material will provide an indication of what you know now. Tests also attempt to predict future behavior. You probably took the Scholastic Aptitude Test (SAT) as part of the admission requirements to college. SAT scores are usually required by higher education institutions as a predictor of success in college.

One important caution needs to be noted here: A test measures only a sample of behavior. Tests are not perfect measures of behavior; they only provide an indication. It is therefore important that case managers not make decisions based solely on test scores.

Types of tests Thousands of tests are in use today. One way to make sense out of all the tests that are available is to know how they are categorized. One classification is by type of behavior measured. Two categories are identified in this system. *Maximum performance tests* measure ability, and *typical performance tests* give an idea of what an examinee is like. These and other helpful categories are discussed in the pages that follow.

Maximum performance tests include achievement tests, aptitude tests, and intelligence tests. On these tests, examinees are asked to do their best. **Achievement tests** are used to evaluate an individual's present level of functioning, or what has previously been learned. Achievement tests that a case manager will often encounter include the Test of Adult Basic Education (TABE) and the Wide Range Achievement Test (WRAT). **Aptitude tests** provide an indication of an individual's potential for learning or acquiring a skill. Because aptitude tests imply prediction, they are useful in selecting people for jobs, scholarships, and admission to schools and colleges. The SAT is an aptitude test. In your work with clients, you will likely read about aptitude tests such as the General Aptitude Test Battery (GATB), the Differential Aptitude Test (DAT), and the Minnesota Clerical Test. When we think about how "smart" someone is, usually we mean intelligence. Tests such as the Wechsler Adult Intelligence Scale (WAIS), the Weschler Intelligence Scale for Children (WISC), the Revised Beta Examination (Beta IQ), and the Peabody Picture Vocabulary Test are **intelligence tests.** Careful consideration should be given to these tests and test

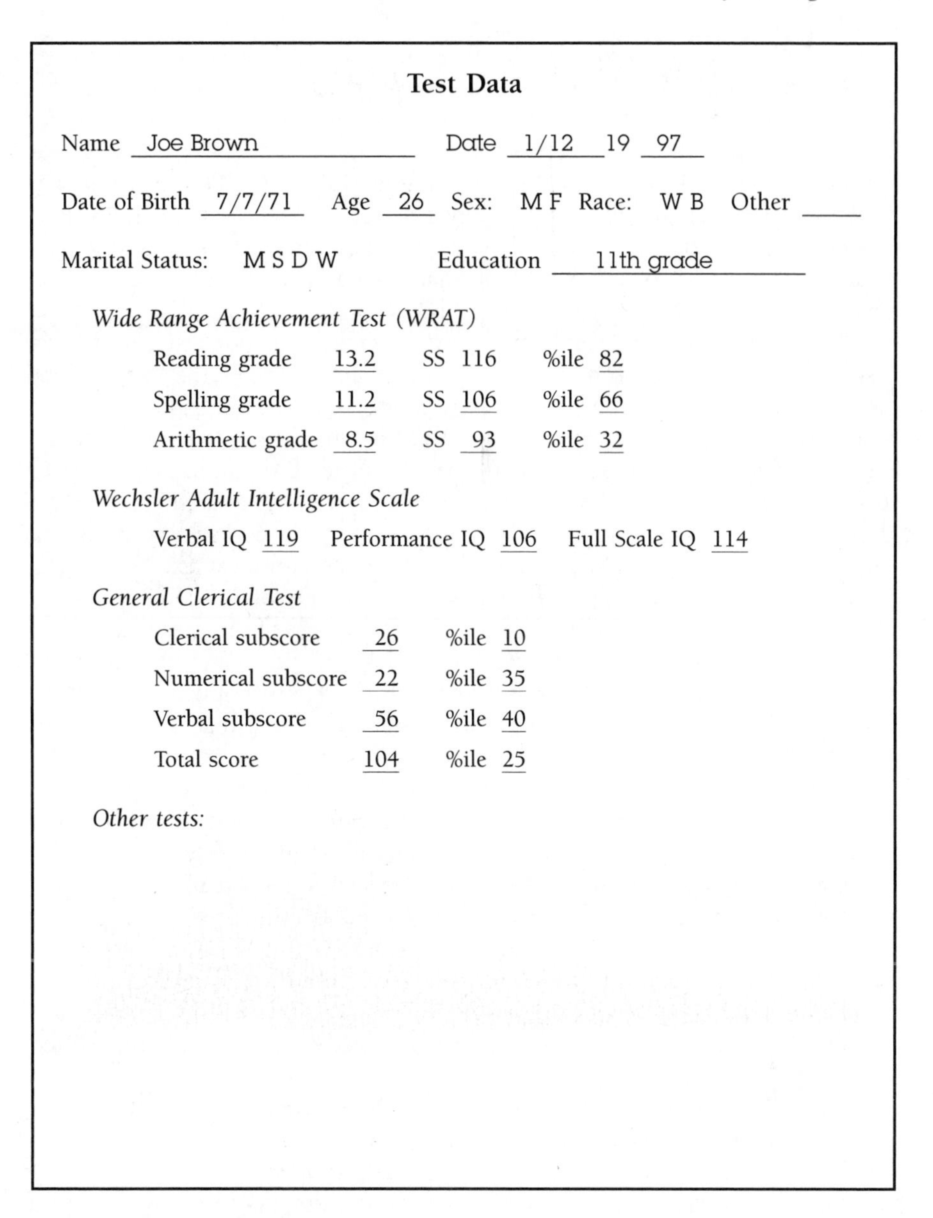

Test Data

Name Joe Brown Date 1/12 19 97

Date of Birth 7/7/71 Age 26 Sex: M F Race: W B Other ____

Marital Status: M S D W Education 11th grade

Wide Range Achievement Test (WRAT)

Reading grade	13.2	SS 116	%ile 82
Spelling grade	11.2	SS 106	%ile 66
Arithmetic grade	8.5	SS 93	%ile 32

Wechsler Adult Intelligence Scale

Verbal IQ 119 Performance IQ 106 Full Scale IQ 114

General Clerical Test

Clerical subscore	26	%ile 10
Numerical subscore	22	%ile 35
Verbal subscore	56	%ile 40
Total score	104	%ile 25

Other tests:

Figure 6.3 Test data

scores because intelligence can be defined in a number of ways. Some tests measure verbal intelligence, some nonverbal intelligence, and others problem-solving ability. The WAIS, for example, yields a Verbal IQ, a Performance IQ, and a Full Scale IQ. On the other hand, the Revised Beta Examination yields only a performance IQ score called a Beta IQ.

The other major category is the typical performance test. Such tests provide some idea of what the examinee is like—his or her typical behavior. In this category are interest inventories (California Picture Interest Inventory, Strong Interest Inventory Test, Kuder Preference Record), personality inventories (Edwards Personal Preference Schedule, Minnesota Multiphasic Personality Inventory, Sixteen Personality Factor Questionnaire), and projective techniques (Rorschach Inkblot Test, Thematic Apperception Test, Rotter Incomplete Sentences Blank). Other well-known typical performance tests are the Bender–Gestalt Test, the Vineland Social Maturity Scale, and the McDonald Vocational Capacity Scale.

There are a number of other categorization schemes for tests. Individual tests are administered to one person at a time—for example, the WAIS or the projective techniques. Group tests are administered to two or more examinees at a time. The Revised Beta Examination and the Otis Lennon school ability test are group tests, although they can also be administered on an individual basis. Tests can also be classified as standardized or informal. Standardized tests are those that have content, administration and scoring procedures, and norms all set before administration. Informal tests are developed for local use, like the test your instructor will give to you to measure your mastery of this course material. Verbal tests use words, whereas nonverbal tests consist of pictures and require no reading skills. Tests in which working quickly plays a part in determining the score are speed tests. Tests such as the Revised Beta Examination are closely timed. In contrast, power tests have no time limit, or one that is so generous that it plays no part in the score.

As you begin to explore the testing literature, you will discover that testing has a language of its own. Recognizing the categories and knowing their meanings will help you develop the vocabulary to understand testing concepts and the advantages and limitations of tests. Selecting tests requires an understanding of the terms in the following list, as well as others.

Edition: the number of times a test has been published or revised
Forms: equivalent versions of a test
Level: the group for which the test is intended (e.g., K–3 is kindergarten through third grade)
Measurement error: the part of an observed test score that is not the true score or the quality you wish to measure
Norm: the average score for some particular group
Norms table: a table with raw scores, corresponding derived scores, and a description of the group on which these scores are based
Percentile rank: the proportion of scores that fall below a particular score
Reliability: the extent to which test scores or measures are consistent or dependable—that is, free of measurement errors
Stanine: a system for assigning the numbers 1 through 9 to a test score
Test: a measurement device
Test administrator: person giving a test
Test profile: a graph that shows test results
Validity: the extent to which a test measures what it claims to measure

Selecting tests When faced with the task of choosing a test to administer, the first question must be where to find out about available tests. Perhaps the second question, which quickly follows once a case manager realizes the vast number of tests that are available, is how to select a test.

There are a number of sources of information about tests. They range from test publishers to reference books that are available in the library. These refer in turn to journal articles that provide more detailed information about a test. The general information about a test will help narrow the choices to those of interest, and it is for these tests that more specific information is gathered. Let's begin with the more general information.

Thousands of commercially available tests in English are described and critically reviewed in the many editions of the *Mental Measurements Yearbook*, published by the Buros Institute of Mental Measurements at the University of Nebraska–Lincoln. Begun in 1938 by Oscar K. Buros, the *MMY* provides comprehensive reviews of tests by almost 500 notable psychologists and education specialists (Kaplan & Saccuzzo, 1993). For each test included in the *MMY*, there is a detailed description and price data followed by references to articles and/or books about the test, along with original reviews prepared by experts. The *MMY* contains no actual tests.

Another reference that summarizes information on tests is *Tests in Print*. This volume is helpful as an index to tests, test reviews, and the literature on specific tests. Entries include the title and acronym of the test, who it was designed for, when it was developed, its subtests, the authors and publishers of the test, and cross references to *MMY*. A number of other references may be helpful to you as you narrow your selection, but *MMY* and *Tests in Print* provide the most comprehensive overviews of published tests now available.

Once the choice has narrowed, specimen sets of tests are available for purchase from test publishers. Although there are approximately 400 companies in the test industry, the top 10% are responsible for 90% of the tests used in the United States. Test publishers have catalogs that provide lists of tests and test-related items sold by that company. Companies usually offer specimen sets for sale: the test manual, a copy of the test, answer sheets, profiles, and any other appropriate material related to a particular test. The test manual, the best source of information about a particular test, provides statements about the purposes of the test, a description of the test and its development, standardization procedures, directions for administration and scoring, reliability and validity information, norms, profiles, and a bibliography. This specific information can help the case manager decide whether to use the test.

Once available information about a particular test is gathered, the case manager decides whether to select it. Then the second question, relating to the criteria for selection, surfaces. One helpful source of information is *Standards for Educational and Psychological Testing*, published by the American Psychological Association. The *Standards* is a technical guide that provides the criteria for the evaluation of tests. Among the standards that are discussed are validity, reliability, test administration, and standards for test use. Any case manager who is involved in testing should carefully review the complete standards.

Criteria for selection Many considerations go into a decision to use a test. Three primary considerations are validity, reliability, and usability. This section provides a brief overview of each consideration by defining each, identifying the types, and discussing how to evaluate. For more information, consult the *Standards for Educational and Psychological Testing* or testing textbooks that present this material in more detail.

Validity, the most important of the three considerations, is the extent to which a test measures what we actually wish to measure. Think about a time when you have had to move furniture. A yardstick was probably helpful to you as you measured the available space and the piece of furniture for a fit. Long experience has confirmed the validity of a yardstick as a tool for measuring length. Validity works the same way for a test, which is a tool for measuring behavior. Among the questions related to validity are the following. How well do math tests measure math achievement? Could we make predictions based on those scores? Does this test measure other qualities as well? These questions are answered as content, construct, and criterion-related validity are established for a test. These terms are likely to appear in the test manual's section on validity.

Validity must be evaluated in light of the intended use. There are a number of factors to consider. One is evidence of several types of validity; the more sources of evidence, the better. The quality of the evidence is also important—more important, in fact, than a quantity of questionable evidence. Included in the validity data reported should be a description of the subjects, the testing situation, and how the test was used. Finally, the evidence should relate to the purpose of the test. For example, evidence for the validity of an achievement test should be relevant to content, whereas criterion-related validity is important for tests that predict outcomes.

Reliability, the second consideration, is the degree of consistency with which a test measures whatever it is measuring—the degree to which test scores are free from errors of measurement. Such errors result in inconsistent scores from one test form to another or from one testing time to another, changes that are not attributable to such factors as the examinee's motivation, fatigue, anxiety, or guessing.

As with validity, a review of the reliability evidence of a test takes into account any errors of measurement in light of the test's intended use. For example, split-half reliability, which is often used with longer tests, yields an inflated estimate of reliability when applied to speed tests. Test-retest reliability (which involves testing the same group with the same test on two different occasions) would not be appropriate for tests in which memory plays a role from one testing situation to another. When reviewing reliability evidence, it is also important to read the descriptions of the number and characteristics of the sample, the testing situations (including the interval between administrations), and the reported reliability estimates. This information should be evaluated in relation to the test's intended use.

The final consideration is test usability, taking in all the practical factors that are part of a decision to use a particular test. The factors include economy, test administration, and interpretation and use of scores. The following questions relate to usability.

ECONOMY
Can we reuse the test booklets?
Are separate answer sheets available?
How easy is it to score the test?

TEST ADMINISTRATION
Are the instructions clear and complete?
Is close timing critical?
What is the layout of test items on the page (print size, clarity of pictures, and so on)?
Can the test be administered via computer?

INTERPRETATION AND USE OF SCORES
Are scoring keys and instructions provided?
Are norms included for appropriate reference groups?
Is there evidence that test scores are related to other variables?
Is there a clear statement about the function of the test and its development?

These questions, as well as the validity and reliability evidence, are all factors to be considered in test selection. Once a test has been selected, the test, answer sheets, scoring keys, interpretive manuals, and so on, are purchased from the test publisher.

Administration and interpretation The test manual is an important source of information about the administration of a test and the interpretation of test scores. It should identify any special qualifications in training, certification, and experience that are necessary for the people who will administer and interpret the test. Needless to say, only individuals who have the training and experience for using tests should be entrusted with them. The test manual will also describe the recording and scoring of answers. Can answers be recorded in the test booklet? Are separate answer sheets available? If the answer to both questions is yes, are the results interchangeable? Also described in the test manual are procedures to facilitate the use of the test and reduce bias. Information about the interpretation of test scores will be included; in some cases, handouts are provided. Note that it is the responsibility of the test administrator to read the manual carefully and be knowledgeable about the test and its uses.

Standardized test administration procedures are important. Many sources of error can influence test scores, and some relate directly to test administration. The testing situation itself is influenced by various factors, including the lighting, ventilation, room temperature, extent of crowding, adequate work space, and noise level. The test administrator's characteristics can also be sources of error. Giving extra help, pointing out errors after the test begins, establishing varying degrees of rapport with different examinees, or allowing additional time will all influence test scores. The administrator of a test should eliminate as many of these conditions before beginning the test and avoid deviating from the standardized procedures.

Today, computer-assisted test administration is a reality. In this testing situation, test items are presented via a computer terminal or a personal computer, and an automatic recording of test responses occurs. There are two important advantages to testing by computer: standardization of administration and scoring, and the control of bias. If a test is administered via computer, it is important to ensure that the items are legible, the screen is free from glare, and the terminals are properly positioned.

But what about the client who can't read or who reads below fifth- or sixth-grade level? Or one who has impaired vision or hearing? Or a person with epilepsy who is heavily sedated, with his or her motor processes slowed down? An initial strategy is to find individual psychological tests that can accommodate the client's problems and still measure interests, intelligence, achievement, aptitudes, or personality. The limitations of each disability need to be carefully considered in relation to the person's ability to perform on a test. For example, a client who writes slowly because of a disability should not be given a test that places a premium on speed. In some testing situations, however, special consideration cannot and should not be given to a person with a disability. If a client has an amputation below the knee and is interested in attending college, he or she needs no special considerations when taking the SAT. The limitations imposed by the disability do not interfere with the traits measured by the SAT. When adjustments are made for the person with a disability, avoid deviating from standardized procedures insofar as possible. The reason for using tests is to obtain an objective comparison of the individual with a standardized group. If the test is administered in a nonstandardized way, the norm-based scores obtained are not accurate.

When the test is given to people who are different from the groups the test was designed for and normed on, it is more difficult to get accurate results. The instructions, item content and format, methods of answering items, and many other aspects of the test have been designed to make it useful for a specific population. Knowledge of disabilities can help you decide whether modifications are needed. If deviation from established procedures or content is necessary, it is best to consult with a colleague who has psychometric expertise.

The person who takes a test has the right to receive a correct interpretation of the test scores. It is the responsibility of the test user to interpret scores accurately and meaningfully. Test scores are usually reported as raw scores, which have very little meaning. The raw scores can be converted and reported as standard scores—percentiles, stanines, or T scores—which are easier to understand and interpret. The conversion is based on the norm group or standardization group.

Once the standard scores are available, it is time for interpretation of the test. There are two essential steps in test interpretation: understanding the results and communicating them to another person, orally or in writing. The following suggestions will guide your preparation for test interpretation.

- Know the test—its purpose, development, content, administration and scoring procedures, validity and reliability, advantages and limitations.

- Avoid technical discussions of tests. Use short, clear explanations of what you are trying to communicate.
- Use the test profile as a graphic presentation of the test results. The examinee may find this easier to follow as the scores are explained.
- Explain what the score means in terms of behavior.
- Go slowly. Give the examinee time to process the information and react.

Tests are helpful tools in measuring traits common to many people. A score serves to show where a person stands in a distribution of scores of peers. How high or low a score is does not measure an individual's worth or value to family, friends, or society. Tyler (1984) suggests a guiding principle for professionals who use tests: The scores are clues to be followed. They do mean something, but in order to know *what,* we must consider each examinee as an individual, combining test evidence with everything else we know about the person. It is unsound practice for case managers to base important decisions on test scores alone. It is important to remember this in test selection, administration, and interpretation.

Summary of testing Test misuse can easily occur. Let's review some guidelines for the selection, administration, and interpretation of tests.

First, case managers should select tests that they have carefully reviewed. The validity, reliability, and usability of a test; its statement of purpose, content, norm groups, administration and scoring procedures; and its interpretation guidelines should all be evaluated in light of the intended use. The case manager should check any reviews by experts to add to his or her knowledge of the test. It is also a good idea for the case manager to take the test.

Second, case managers should use only tests they are qualified to administer and interpret. This often depends on one's ability to read the manual. Other tests require advanced coursework and supervision or practicum experiences for proper administration and interpretation. Test catalogs usually indicate how much expertise is required for the tests listed. Another helpful source of information is the *Standards for Education and Psychological Testing.*

Third, case managers who administer tests have an obligation to provide an interpretation of the test results. An understanding of raw scores and their conversion to standard scores, coupled with the ability to communicate the meaning of the scores, is necessary to do this right. In addition, it is essential to be aware of the norm groups and their applicability to the examinee. Some groups, such as Latinos, African Americans, and rural populations, may be underrepresented in the establishment of norms.

Chapter Summary

This chapter has introduced the planning phase of case management, which includes review and continuing assessment of the client's problem, the development of a plan, and the selection of services. A part of the plan may be to identify what additional information is needed and to acquire it. The case manager can gather additional information by interviewing and testing. Throughout this phase, documentation and client participation continue to be important.

Chapter Review

◆ Key Terms

goals
objectives
plan development
information and referral system
social service directory
interagency feedback logs
assessment interview
structured clinical interview
case history interview
employment interview
sources of error
halo effect
test
psychological test
maximum performance test
achievement test
aptitude test
intelligence test

◆ Reviewing the Chapter

1. Describe the two areas of concern that are addressed by revisiting the assessment phase.
2. What sources help a case manager review a client problem?
3. What role does documentation play in the review of the problem?
4. Define *plan.*
5. What activities occur before development of the plan?
6. List the characteristics of a service plan.
7. What are the benefits of establishing goals?
8. List the criteria for well-stated and reasonable goals.
9. Distinguish between a goal and an objective.
10. Identify a problem you would like to address, and develop a plan with goals and objectives.
11. List the four components of information and referral, and give an example of each.
12. Discuss the similarities between interviewing and testing.
13. Compare the five types of interviews and their roles in case management.
14. Illustrate how sources of error may affect an interview.
15. How do case managers use tests?
16. Describe the different ways to categorize a test.
17. Describe how to select a test to measure vocational interests.
18. Define *validity* and *reliability,* and describe their roles in test selection.
19. What makes the test manual the best source of information about a test?

20. Identify some sources of error in testing.
21. Under what conditions should special considerations be made for a person with a disability?
22. Identify the two essential steps in test interpretation.

◆ *Questions for Discussion*

1. Why do you think developing a plan is important?
2. If you were a new case manager, how would you begin to develop a network of available services?
3. What kinds of criteria would you use to determine whether you need to conduct a structured interview with an 8-year-old?
4. Do you believe that you will be able to determine what errors exist in the information that you gather? What problems do you expect to encounter in finding errors?

References

Boserup, D. G., & Gouge, G. (1977). *The case management model.* Athens, GA: Regional Institute of Social Welfare.

Dixon, D. N., & Glover, J. A. (1984). *Counseling: A problem solving approach.* New York: Wiley.

Hollingsworth, H. L. (1922). *Judging human character.* New York: Appleton-Century-Crofts.

Kaplan, R. M., & Saccuzzo, D. P. (1993). *Psychological testing: Principles, applications, issues.* Pacific Grove, CA: Brooks/Cole.

Mager, R. F. (1984). *Preparing instructional objectives.* Belmont, CA: Lake Publishing.

Mathews, R. M., & Fawcett, S. B. (1981). *Matching clients and services: Information and referral.* Beverly Hills: Sage.

McClam, T. (1992, September–October). Employer feedback: Input for curriculum development. *Assessment Update, 4*(5), 9–10.

McClam, T., & Woodside, M. R. (1994). The practitioner's voice: Case management for effective service delivery. *Human Service Education, 14*(1), 39–45.

Thorndike, E. L. (1920). A constant error in psychological rating. *Journal of Applied Psychology, 4,* 25–29.

Tyler, L. E. (1984). What tests don't measure. *Journal of Counseling and Development, 63,* 48–50.

Chapter Seven

Building a Case File

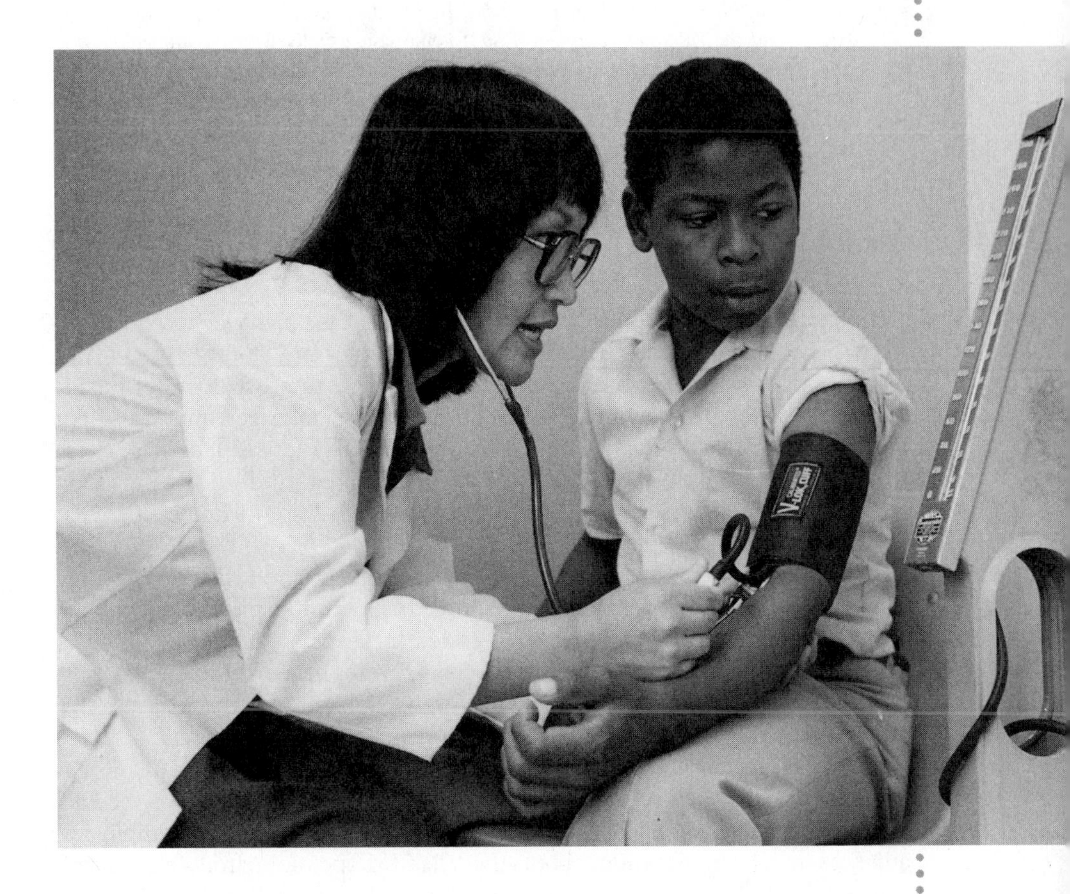

And in this particular job, it is useful to have working knowledge of medical terms, to know how to use a Physician's Desk Reference *and a pediatric manual for definitions, and to have a basic understanding of things like physical, occupational, and speech therapies.*

—Margaret Lanning, Sick Kids Need Involved People, New York City, personal communication, May 4, 1994

I have a form that I complete on each child . . . presenting problems that the child might be manifesting in school or at home; what the parents' expectations are for the child coming to school; developmental history, medical history, family history, social interaction, school history, that kind of thing.

—Linda Smith, Tennessee School for the Deaf, Knoxville, Tennessee, personal communication, October 27, 1993

We collect all the records and analyze them. I go to the schools to talk with the teacher and the guidance counselor. I do a home visit usually with the immediate family . . . and all children who come under state custody have a psychological done. They all have a medical examination. We take them to the Health Department for an EPST and D exam, which is your basic physical. And we round up the immunization records, because we have to have them for placements.

—Mark Alexander, East Tennessee Community Health Agency, Knoxville, Tennessee, personal communication, July 10, 1994

Information from other professionals comes to the case manager in two ways. When he or she receives a case file on a client from another agency or worker, the case may contain reports or evaluations from other professionals. In other situations, the plan developed by the case manager and the client may include referrals to other professionals for evaluations. In both of these scenarios, the case manager must be able to understand the information provided and (if asking for help from other professionals) to know just what to request.

The chapter-opening quotations illustrate the kinds of information that may be needed from other professionals in order to develop a plan or to provide services. The medical information, histories, or exams these three helpers mention are part of the case files of clients who have medical problems. Margaret Lanning speaks of the advantages of being familiar with medical terms and medical references when trying to decipher medical reports. Social histories and home visits offer important information to workers, such as Linda Smith, at state institutions and local service coordination teams. Agencies such as the East Tennessee Community Health Agency, where Mark Alexander is employed, gather as much information about the client as possible from all outside sources.

This chapter examines the types of information that may be found in a case file or gathered to complete one. Just which information is needed depends on the individual's case and the agency's goals, but many cases involve medical, psychological, social, educational, and vocational information. We introduce each type of information, give a rationale for gathering it, describe the kinds of data likely to be provided, and discuss what the case manager needs to know in order to make the best use of the report. For each section of the chapter, you should be able to accomplish the following objectives.

MEDICAL INFORMATION

- Tell how medical information contributes to a case.
- Decode medical terms.

PSYCHOLOGICAL EVALUATION

- List the reasons for a psychological evaluation.
- Make an appropriate referral.
- Identify the components of a psychological report.

SOCIAL HISTORY

- State the advantages and limitations of a social history.
- Name the topics included in a social history.
- List the ways social information may appear in the case file.

OTHER TYPES OF INFORMATION

- List the types of educational information that may be gathered.
- Define a vocational evaluation.

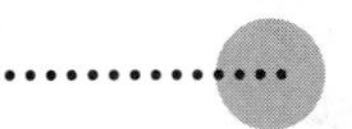

Medical Information

Knowledge of medical terminology, conditions, treatments, and limitations is important in understanding a case. Medical information may be provided on a form or in a written report. The exam and report may have been done by a general practitioner or by a specialist in a field such as neurology, orthopedics, or ophthalmology. In some cases, the case manager can interact with the medical service provider and thus be able to ask questions, request specific assistance, or offer observations. Often, however, he or she does not have this opportunity and must rely on the written report. Then the resources mentioned at the beginning of the chapter may prove particularly helpful. Many agencies have a copy of the *Physician's Desk Reference* (*PDR*) or other medical guide. Some also have a physician serving as a consultant, available when there are questions. This section introduces basic medical information to help you understand medical terminology.

Agencies approach medical information in different ways. Some require documentation of a mental or physical disability or condition in determining eligibility for services. Others use a medical examination as part of their assessment procedures. In certain situations, medical information is not gathered unless there is some indication or symptom of a disease, condition, or poor health that will affect service delivery.

Medical knowledge is particularly crucial when working with people who have disabilities. A general medical examination and specialists' reports help determine the person's functional limitations and potential for rehabilitation. It is important to set objectives that are realistic in light of the client's physical, intellectual, and emotional capacities. When a medical report covers a disability in functional terms, the description can read like the following:

> The individual has a diagnosis of schizophrenia, chronic undifferentiated type. In functional terms, this person needs work that involves simple, routine, and repetitive activity with minimal personal interaction and a structured work environment. (Brodwin, Tellez, & Brodwin, 1992, p. 7)

Often, however, the form for a general medical examination allows only a small space for the diagnosis, so the case manager reads a phrase such as "chronic back pain," "normal exam," or "emotional problems." Not very helpful, is it? Remember that the client is an important source of information; he or she can tell you about any problems. You may then need to decide whether or not a specialist's evaluation would be helpful.

Medical Exams

Generally, medical information contributes to a case in two ways. **Medical diagnosis** appraises the general health status of the individual and establishes whether a physical or mental impairment is present. For example, 10-year-old Brian Muldowny comes into state custody, abandoned by his parents. The case manager at the Assessment, Care, and Coordination Team takes Brian to the Health Department for an examination. The examination results in a diagnosis of otitis media.

Diagnostic medical services include general medical examinations, psychiatric evaluations, dental examinations, examinations by medical specialists, and laboratory tests. A medical diagnosis is helpful when the client has a medical problem or is currently receiving treatment from a physician, who may provide important information about social and psychological aspects of the case in addition to the medical aspects. When making a referral for a medical diagnosis, the case manager should help the client understand why the referral is necessary, the amount of time it will require, what they can expect to learn, and what use the agency will make of the report.

Medical consultation is used in several ways. First, the consulting physician can provide an interpretation of medical terms and information. For example, Brian Muldowny was diagnosed with otitis media. The case manager received this report, asked a colleague what the diagnosis meant, and learned that it was an ear infection. A consultation with a physician would reveal that otitis media is a severe ear infection that results when the Eustachian tubes are not properly angled. The consultation might also explain the report further and clarify possible treatments. In Brian's case, the case manager may need further information about the advantages and disadvantages of two possible treatments: insertion of tubes in the ears, and a regimen of antibiotics. A consultation with an otorhinolaryngologist (ear, nose, and throat specialist) could shed light on the medical prognosis and the extent of any hearing disability that might be expected.

The role of a medical consultant is to interpret the available medical data, determine any implications for health and employment, and recommend further medical care if needed. The case manager can make the best use of a consultant by being prepared for the meeting—perhaps specifying in writing what is needed from the consultant. This usually involves identifying problems that need to be resolved and setting forth the significant facts of the case. The case manager needs to understand medical terminology, the skills of specialists in diagnostic study and treatment programs, and the effects of disability on a client.

The medical service used most often in human services is the **physical examination,** in which a physician obtains information concerning a client's medical history and states findings. The exam data are entered into the medical record. Here we give an overview of the physical examination: the kinds of information obtained and what the case manager needs to know to make such a referral and to understand the physician's report.

Diagnosis involves obtaining a complete medical history as well as conducting a comprehensive physical exam (also called a *physical,* a *health exam,* or a *medical exam*). The results of the exam may be reported on a form provided by the referral source. Sometimes physicians use preprinted schematic drawings of various body parts or organ systems to enhance or clarify the written report. However the information is transmitted, the quality of the reporting depends on the relationship between the physician and the patient. In some cases, the patient has mixed feelings about the referral for a physical exam. He or she needs an explanation of why the referral is necessary, the amount of time the exam will take, what outcome is expected, and how the information will be used. Keep in mind that the client's socioeconomic status, language skill limitations, or cultural background may also influence how he or she feels about the referral. If it is communicated with sensitivity, and if a good relationship with the physician is established, any barriers of anxiety, depression, fear, or guilt can be overcome.

The general medical exam is done by a physician, who takes an overall look at the person's medical state. Its purpose is to evaluate the person's current state of health, focusing on two areas. First, a complete medical history records all the factual material, including what the client states and the physician's inferences from what is not said. A typical starting point is the chief complaint, as expressed by the individual. (See Figure 7.1.) If there is an illness at present, it is described in terms of onset and symptoms (including location, duration, and intensity). A family history relates significant medical events in the lives of relatives—particularly parents, grandparents, siblings, spouse, and children. Extensive information about the individual's past medical history is also collected. This may include childhood diseases, serious adult illnesses, injuries, and surgeries. A review of symptoms focuses on information about present and past disorders, which the physician elicits through questions about organs and body systems. After completing the physical exam, the physician records a diagnostic impression. The actual diagnosis is made once there is conclusive evidence—which may mean getting further studies or referring the client to a specialist for consultation.

What exactly makes up a medical exam? Techniques used during a physical exam are inspection, palpation (feeling), percussion (sounding out), and auscultation (listening). Usually, the examining physician works from the skin inward to the body, through various orifices, and from the top of the head to the toes (Felton, 1992). Special instruments are used to look, feel, and listen. More time is spent in particular areas, to ascertain whether a certain finding truly represents a change in an organ or tissue. Some parts of the exam are carried out quickly, and others require more time. More important areas may receive a

MEDICAL REPORT

CHIEF COMPLAINT: Weakness and malaise

PRESENT ILLNESS: Three weeks ago, this 40-year-old single white male had a cold, with associated mild cough and temperature elevation, which lasted two days. At that time there was a loss of appetite and a decrease in food intake.

After the cough and temperature elevation subsided, the patient noted increased weakness and general malaise. The patient reports he tires easily and is unable to sustain exercise, which was tolerated well before the onset of the cold.

FAMILY HISTORY: This patient's mother, 75, has had breast cancer and a radical mastectomy and was diagnosed with lupus ten years ago. His father, 80, had quadruple bypass surgery five years ago. There is extensive evidence of heart disease in father's family. There is a history of diabetes and tuberculosis in mother's family, although the mother has had neither.

PAST HISTORY: Patient had chickenpox at age 7. Sustained multiple fractures in a motor vehicle accident at age 16, but he is without permanent motor or neurological damage. Frequent lower respiratory infections characterized by productive cough and yellow-mucus-producing hacking cough. The patient has been exposed to tuberculosis, showing a positive PPD test but negative chest X ray five years ago, and was treated with usual course of preventative medications. Since that time patient has been free of other persisting symptoms.

SOCIAL HISTORY: Unmarried, owns home, smokes one and half packs of cigarettes a day, drinks occasionally.

OCCUPATIONAL HISTORY: Worked construction jobs following high school. Has experience with foundation work, masonry, and plumbing. Last spring he graduated from a local college with degree in accounting. Plans to start his own construction business next year.

Figure 7.1 Medical report

second, more thorough examination. The physician records the findings as soon as possible after completing the exam and shares the results with the client.

For some clients, one of the first things that occurs in the case management process is a referral to a physician for a general medical exam. As the physician conducts the exam, he or she completes a form like the one shown in Figure 7.2, which is then sent to the referring counselor. It becomes part of the client record.

REVIEW OF SYSTEMS:

HEAD, EYES, EARS, NOSE, THROAT: No frequent or severe headaches or head injuries. Wears corrective lenses for nearsightedness. Had frequent ear infections as child, but has suffered no hearing loss. Has two or three colds a year, but is without sinus pain. No problems with throat.

NECK: No significant abnormality.

RESPIRATORY TRACT: No expectoration of blood. Wheezing and shortness of breath with exertion—walking up stairs. When colds "go to chest," takes over-the-counter cough medications. Early morning nonproductive cough present. States he is not trying to quit smoking.

CARDIOVASCULAR SYSTEM: No chest pain or palpitations. No history of murmur, coronary artery disease, or hypertension. Fatigue has been increasing over the last month, making him unable to complete his morning walk, which is normally 2 $\frac{1}{2}$ miles.

GASTROINTESTINAL SYSTEM: Appetite has decreased since this cold. No indigestion. States he is on no special diet.

GENITOURINARY SYSTEM: Denies venereal disease.

NERVOUS SYSTEM: No significant abnormality.

MUSCULOSKELETAL SYSTEM: Negative.

ENDOCRINE SYSTEM: Negative.

Figure 7.1 *(continued)*

Medical Terminology

Medical reports often include medical terminology that may seem like a foreign language to a case manager who is unfamiliar with it, because physicians rely on technical words and phrases for exactness. Medical specialties also have special terminologies. Other professionals who may write reports using medical terminology are nurses, physical therapists, and occupational therapists. It can

GENERAL BASIC MEDICAL EXAMINATION RECORD

This record is CONFIDENTIAL

Section I. - (To be filled out by rehabilitation agency)

Client No. ____________

Johnson, Roy R. 7/16/60 W M S ✓ M__ D__ Sep__
(Last Name) (First Name) (Middle Name) (Date of birth) (Race) (Sex) (Martial Status)

Rt. 1 Box 68 Centerville ______ TN
(Home address: Street and number or R.F.D.) (City or town) (County) (State)

Usual occupation Plumber Description of last job Same

Last time hospitalized 5-89 Surgery ____________
(Date) (Reason) (Name and location of hospital)

Last visit to physician 18 mos. spinal cord injury Dr. Alderman
(Date) (Reason) (Name and address of physician)

Is patient now under care of physician? yes Dr. Brown
(Yes or no) (If answer is "Yes," give name and address of physician)

Patient's statement of disabilities Spinal cord injury

Signature of rehabilitation counselor Tom Chapman Date 7-20-90

Section II. PERTINENT HISTORY (To be filled out by physician.)

Back surgery; surgery twice
continued pain

Section III. PHYSICIAN EXAMINATION. (To be filled out by physician. Items checked ✓ were examined and found normal. Deviations from normal are noted. If items require additional description, please record on extra sheet.)

Height (without shoes): 5 ft. 10 in. □ Weight (without clothing) 220 pounds □ Temperature 98.4 °F

Eyes - Right ✓ LEFT ✓
(Discharge; corneal scars; strabismus; pterygium; ptosis; trachoma; fundi; cataract; intraocular tension)

Distant vision: Without glasses: R. 20/30 L. 20/70. With glasses: R. 20/20 L. 20/20
(If vision is too low to be recorded at 20 feet, indicate by recording as "less than 20/200"

Ears - Hearing: Right 20/20 feet Left 20/20 feet □ Other findings: R ✓ L ✓
(Consider denominators here indicated as normal Record as numerators greatest distance heard)
(Evidence of middle ear or mastoid disease. Drums: Normal, absent perforated, dull, retracted, discharge)

Nose ✓
(Obstruction, evidence of chronic sinus infection, polypi perforated septum, etc.)

□ Throat ✓
(Tonsils: Normal, enlarged, removed, etc.)

Mouth ✓
(Missing teeth, pyorrhea; abnormality of tongue or palate)

□ Neck ✓
(Thyroid enlargement, nodules, etc.)

Lymphatic system ✓
(Especially cervical, epitrochlear, inguinal)

□ Breasts ✓
(Abnormal discharge, nodules, tenderness, hypoplasia)

Lungs: Right ✓ □ Left ✓
(If history or physical findings indicate active or arrested tuberculosis, recommend chest X-ray, sputum examination, and consultation with chest specialist)

Figure 7.2 Medical examination form

Circulatory System: Heart ✓
(Enlargement, thrill, murmurs, rhythm)

Blood pressure {Systolic / Diastolic} 118/80 Pulse rate 76 Dyspnoea 0 Cyanosis 0 Edema 0

Evidence of arteriosclerosis None
(Type; degree; where found, as "cerebral," "brachial," etc.)

Abdomen ✓
(Scars, masses, palpable liver, palpable spleen, etc.)

Hernia ✓
(Type: Inguinal, ventral, femoral, etc. Right, left, bilateral)

Genito-urinary ✓
(Urethral discharge, varicocele, hydrocele, scars, epididymitis, enlarged or atrophic testicle)

and

Gynecological
(Prolapse, cystocele, rectocele, Cervix)

Ano-rectal ✓
(Hemorrhoids, prolapse, fissures, fistula. Prostate)

Nervous system ✓
(Paralysis, Sensation, Speech, Gait, Reflexes: Pupillary, Knee, Babinski, Romberg)

✓ (Memory, Peculiar ideas or behavior. Spirits: Elated, depressed, normal)

✓
(Neurological or psychiatric abnormalities should be described on separate sheet)

Skin ✓ (Moist, dry, clear) ☐ Feet ✓ (Weak feet, Congenial or traumatic defects) ☐ Varicose veins ✓ (Site)

Orthopedic Impairments: (*Describe*) Low back pain

Laboratory: Urinalysis: Date 8-7-90 ☐ Specific gravity 1.020 ☐ Reaction 7
☐ Albumen neg ☐ Sugar neg

DIAGNOSIS: (Indicate major and minor) Chronic lower back pain after two spinal ops for herniated disc

STATUS: of major disability: (Check appropriate terms) Permanent ✓ Temporary ____ Stable ____
Slowly progressive ____ Rapidly progressive ____ Improving ____

PROGNOSIS: Can the major disability be removed by treatment? ☐ (Yes) ☑ (No) Substantially reduced by treatment? ☑ (Yes) ☐ (No)

Physical Capacities: (Under "Physical activities" and "Working conditions" use symbols as follows:
(✓) No limitation. (X) Limitation. (0) To be avoided.

Physical activities: Walking X Standing X Stooping X Kneeling X Lifting X Reaching X Pushing X
Pulling X Other (*Specify*)

Working conditions: Outside ✓ Inside ✓ Humid ✓ Dry ✓ Duty ✓ Sudden temperature changes ✓
Other (*Specify*)

RECOMMENDATIONS:

☐ Is examination by specialist advisable? If so, specify which speciality

☐ Refraction ☐ X-ray of chest ☐ Other diagnostic procedures (*Specify*)
☐ Prosthetic appliances (*Specify*)
☐ Hospitalization (*Specify reasons and approximately duration*)
☐ Treatment (*Specify type and approximately duration*)

Remarks: Please use additional sheet for remarks and expansion of any observations.

Date 8-7-90 J. H. Jones M.D.
(Physician)
Suite 201 Physicians Office Bldg.
(Address)

Figure 7.2 *(continued)*

be a challenge for the case manager to make sense of these reports; he or she must have at least a rudimentary understanding of medical terminology.

Medical terminology follows simple rules. To analyze medical words, identify the four elements that are used to form such words: the word root, the combining form, the suffix, and the prefix. It may help to think of these elements as verbal building blocks. Let's examine each component.

Word roots The main part or stem of a word is the **word root.** In medical terminology, it is usually derived from Greek or Latin and often indicates a body part. All medical words have one or more word roots such as the following.

GREEK WORD	MEANING	WORD ROOT
kardia	heart	cardi
gastro	stomach	gastr
nephros	kidney	nephr
osteon	bone	oste

Combining forms A word root plus a vowel, usually an *o,* is the combining form, as in the following examples.

WORD ROOT		COMBINING VOWEL		COMBINING FORM	MEANING
card	+	o	=	cardio	heart
gastr	+	o	=	gastro	stomach
nephr	+	o	=	nephro	kidney
oste	+	o	=	osteo	bone

Suffixes A suffix is a word ending. In medical terminology, the suffix usually denotes a procedure, condition, or disease, as in the instances listed here.

COMBINING FORM		SUFFIX		MEDICAL WORD	MEANING
arthr (joint)	+	-centesis (puncture)	=	arthrocentesis	puncture of a joint
thoraco (chest)	+	-tomy (incision)	=	thoracotomy	incision in the chest
gastro (stomach)	+	-megaly (enlargement)	=	gastromegaly	enlargement of the stomach

Suffixes also form adjectives, express relative size, indicate surgical procedures, and express conditions or changes related to pathological processes. Examples follow.

FORM ADJECTIVES		
-al (means "pertaining to")	arterial	pertaining to an artery
-ible (indicates ability)	digestible	capable of being digested

TABLE 7.1 COMMON PREFIXES AND SUFFIXES

Prefix	*Meaning*	*Suffix*	*Meaning*
dys-	bad, painful, difficult	-itis	inflammation
macro-	large	-algia	pain
hypo-	under, below	-toxin	poison
scler-	hard	-oma	tumor
tachy-	rapid	-pathy	disease
hyper-	over, above, excessive	-osis	abnormal condition, increase

EXPRESS RELATIVE SIZE

-ole (means small)	arteriole	a small artery
-ule (means small)	granule	a small grain

INDICATE A SURGICAL PROCEDURE

-ectomy (means "removal of an organ or part") appendectomy

EXPRESS CONDITIONS OR CHANGES RELATED TO PATHOLOGICAL PROCESSES

-mania (means "excessive excitement or obsessive preoccupation") pyromania

Prefixes The word element located at the beginning of a word is the prefix. It usually denotes number, time, position, direction, or negation.

PREFIX	WORD ROOT	SUFFIX	MEDICAL WORD	MEANING
hyper + (excessive)	therm + (heat)	ia = (condition)	hyperthermia	(condition of excessive heat)
micro + (small)	card + (heart)	ia = (condition)	microcardia	(condition of a small heart)

Other common prefixes that modify word roots indicate position (e.g., *ab* means "away from," as in *abnormal*), quantitative information (e.g., *a* or *an* means "without," as in *anorexia*—defined as without appetite), qualitative information (e.g., *mal* means "bad," as in *malfunction*), and sameness or difference (e.g., *homo* or *hetero*). For other prefixes and suffixes that are common in medical terms, see Table 7.1.

There are three basic steps to working out the meaning of a medical term. First, identify the suffix and its meaning. Second, find the prefix, if any, and determine what it means. Third, identify the root words and their meanings. For example, *thermometer* consists of a suffix (*meter,* meaning "instrument for measuring") and a word root (*thermo,* meaning "heat"). Thus, a thermometer is an instrument for measuring heat. Another example is *gastroenteritis.* The suffix is *itis* (inflammation), the prefix is *gastr* (stomach), and the word root is *enter* (intestine). Gastroenteritis is an inflammation of the stomach and intestine. Remember that the vowel *o* is a combining form, linking one word root to another to form a compound word. *Osteoarthritis* is another example. The suffix *itis* means "inflammation"; word roots are *oste,* which means "bone" and *arthr,* which means "joint." The *o* is the combining vowel. *Osteoarthritis* means inflammation of bone and joint. The following list contains some common medical terms that use suffixes, prefixes, and word roots introduced in this chapter. Can you fill in the columns to work out the meaning of each term? Other examples are shown in Table 7.2.

TERM	SUFFIX/PREFIX	WORD ROOT	MEANING
tachycardia			
dysfunction			
gastritis			
nephritis			
osteopathy			
hypodermic			

It is a continuing challenge for case managers to keep current with terminology, because of ambiguities, inconsistencies, and the changing course of medical knowledge. Although most word roots have Greek or Latin origins, some occur in both but have different meanings. The root *ped,* for example, means "child" in Greek (e.g., *pediatrician*), but in Latin *ped* means "foot" (e.g., *pedicure*). Many diseases are named for individuals—for example, Alzheimer's disease and Hodgkin's disease. Some disorders are called *syndromes*—Cushing's syndrome, Horner's syndrome. *Acronyms* are also formed from the initials of lengthy names: MRI (magnetic resonance imaging) and ACTH (adrenocorticotropic hormone) are examples. In addition, medical terminology traditionally uses hundreds of abbreviations; some of the most common are listed in Table 7.3. Keeping informed about trends in medicine increases one's understanding of the meanings of terms. For example, physicians increasingly prescribe generic drugs rather than brand names (e.g., the generic diazepam rather than Valium). Keeping current with medical terminology entails awareness of chemicals, syndromes, and diseases that are newly named and sometimes given acronyms or abbreviations (e.g., AIDS, acquired immunodeficiency syndrome). It must also be remembered that words can have multiple meanings and that several names may apply to a single entity.

TABLE 7.2 SOME COMMON COMPONENTS OF MEDICAL TERMS

Component	*Meaning*	*Example*
-algia	pain	euralgia
angio-	blood vessel	angiogram
arth-	joint	arthroscopy
contra-	opposed to	contraception
derm-	skin	dermatology
-emia	condition of the blood	polycythemia
enceph-	brain	encephalitis
glyco-	sugar	glycosuria
hepat-	liver	hepatitis
hyster-	uterus	hysterectomy
leuk-	white	leukocyte
lip-	fatty	perlipemia
-oscopy	visual examination	laparoscopy
-ostomy	creation of an artificial opening	tracheostomy
-otomy	incision	craniotomy
-plasty	reparative or reconstructive surgery	rhinoplasty
pre-	before	precancerous
pyel-	pelvis	pyelogram
syn-	together	synarthrosis
tri-	three	triceps

Psychological Evaluation

The objective of a **psychological evaluation** is to contribute to the understanding of the individual who is the subject. The report writer is a consultant who makes a psychological assessment that is practical, focused, and directed toward the solution of a problem. The psychological report he or she prepares is more than a presentation of data. This section will help you determine when a

TABLE 7.3 MEDICAL ABBREVIATIONS

Abbreviation	*Meaning*	*Abbreviation*	*Meaning*
a.c.	before meals	L-1, L-2, L-3	lumbar vertebrae (by number)
b.i.d.	twice daily	L.L.Q.	left lower quadrant
B.P.	blood pressure	L.M.P.	last menstrual period
C-1, C-2, C-3	cervical vertebrae (by number)	p.c.	after meals
CBC	complete blood count	p.r.n.	as needed
C.N.S.	central nervous system	q.i.d.	four times daily
DX	diagnosis	R.L.Q.	right lower quadrant
F.H.	family history	RX	treatment
GI	gastrointestinal	S-1, S-2, S-3	sacral vertebrae (by number)
GU	genitourinary	T-1, T-2, T-3	thoracic vertebrae (by number)
HDL	high-density lipoprotein	t.i.d.	three times daily
h.s.	at bedtime	W.B.C.	white blood count
H & P	history and physical examination		

psychological evaluation is needed, how to make the referral, and how to prepare the client. The evaluation itself and the report are also discussed.

Referral

Case managers may refer clients for psychological evaluations for a number of reasons. One reason is to establish a diagnosis in order to meet criteria of eligibility for services.

Nadine is an unhappy, deeply depressed 15-year-old who is currently taking antidepressant medication. She is increasingly out of control. Yesterday, she slapped her grandmother, with whom she lives, and threatened to kill her. If she is to receive services in an inpatient treatment program, she must have a diagnosis confirming emotional disturbance.

Another reason is to provide justification for a particular service.

Jim is a 28-year-old male whose divorce will be final in a month. As the court date approaches, Jim feels more and more depressed. He is having trouble getting up in the morning, showing up for work on time, and maintaining relationships with those who are close to him. His physician has suggested counseling, but Jim's insurance company insists that he have a psychological evaluation to determine whether he needs it.

Sometimes a psychological evaluation functions as a screening or routine evaluation to obtain information about a client's personality, aptitude, interests, intelligence, and achievement.

Greg is a 35-year-old male who is the only child of elderly parents. He is mentally retarded. His parents, concerned about who will care for Greg if something happens to them, have learned of a group home where the residents live under close supervision. One requirement for acceptance into the program is a recent psychological evaluation that assesses intelligence as well as ability to function independently.

A case manager may also order a psychological evaluation to resolve contradictions or ambiguities or to add information that is missing.

Carolyn is a 10-year-old who is enrolled in public school. Her teacher is concerned about her behavior. One day she is passive, rarely interacts with her classmates, and does not participate in class. The next day, she may be loud, talkative, and disruptive. Just yesterday, she started a fight with a classmate. This has prompted her teacher to request an evaluation from the school psychologist.

Finally, a psychological evaluation may be recommended in order to answer particular questions regarding the client.

Is there brain damage? Why does the individual have trouble relating to others? How is this person adjusting to the recent amputation of her leg? Why is the client doing poorly in school?

In any of these situations, a referral for a psychological evaluation is appropriate. In each case, the case manager seeks help in order to provide the client with needed services. It is easiest to get what is needed if the consulting psychologist knows the general mission of the agency and understands the specific problem to be addressed. Having this information allows him or her to choose the most relevant and efficient approach to gathering the needed information. The referral for a psychological evaluation is usually made by a case manager, who specifies what is needed: a routine workup, testing, questions about the case, and/or a diagnosis. Thus, the psychologist is charged with a mission. It is therefore critical that the referral be more than a general request such as "psychological evaluation" or "for psychological testing." These terms communicate poorly; the referring professional has failed to express what prompted the referral. Two scenarios may result: The psychologist may ask the case manager for more specific information, or he or she may try to guess what is wanted or needed. When the reason for the referral is not clear, it is difficult for the psychologist to provide a useful report.

How does a case manager make a good psychological referral? First, it is important to be clear about the reason for referral. The case manager must clarify whether documentation of a condition or disability is needed, test scores are desired, or behavioral inconsistencies are to be explored. Specific questions also help the psychologist focus on the client's problems. The psychologist then makes recommendations to the case manager. Also, the two of them can discuss the case before the evaluation, to clear up any questions or needs. Since many referrals are made by phone or direct personal contact, such a discussion can easily take place, but it may be even more important when the referral is made in writing.

Part of making a successful referral is preparing the client for the psychological evaluation. In order to do this, the case manager needs a clear understanding of the process and the ability to explain it to the client. Some clients may be suspicious of testing or may fear that the case manager considers them crazy. Demystifying the evaluation will help dispel these attitudes.

The Process of Psychological Evaluation

The evaluation itself includes a study of past behavior, conclusions drawn from observations of current behavior, a diagnosis, and recommendations. This study requires the psychologist to assess which data are important to the client's presenting problems. In some cases, relevant information is in the client file; it is then helpful for the psychologist to have access to these documents in addition to the observations and questions from the referral source.

One of the primary ways that a psychologist observes current behavior is by testing. From the discussion of testing in the previous chapter, you know that testing gives samples of behavior. That discussion also introduced a number of tests that are useful in human services. Psychologists use many of them, notably the Wechsler Adult Intelligence Scale (WAIS) and projective tests (such as the Rorschach and Thematic Apperception Test). These tests are individually administered and scored, and psychologists have special training that enables them to use them. As a consultant, then, the psychologist decides what kinds of data to gather to carry out the assignment given by the referral source; which findings have relevance; and how these findings can be most effectively presented.

The results of the psychological evaluation are communicated to the case manager in a written report. The **psychological report** is "a document written as a means of understanding certain features about a person and his or her current life circumstances in order to make decisions and to intervene positively in a problem situation" (Tallent, 1993, p. 27). The report may appear in one of several forms, of which the most common form is a narrative (illustrated by the report that is included in this section).

Results may also be communicated as a terse listing of problems and proposed solutions. Another option is the computer-generated report, usually consisting of a sequence of statements or a profile of characteristics. Less frequently used are checklists of statements or adjectives, clinical notes, and oral reports relating impressions. Since the narrative is the form of a psychological report that is most often used in human services, let's explore it further.

Usually, the content, sources, and format of narrative psychological reports follow a similar pattern. There are three components to the content of a report. One is the orienting data, which includes the reason for the referral and pertinent background information such as age, marital status, social history, and educational record. Illustrative and analytical content is the second component; here one finds the interpretation of raw data, including test scores. The third component, the psychologist's conclusions, ends the report. This section includes a diagnosis and recommendations, which are presented with supporting evidence. The sources of the information in all three components are the interview between the psychologist and the client, test data, and observed behavior during the evaluation, as well as any available medical reports and social histories and any observations, case notes, or summaries written by other professionals involved with the case.

Among the headings that organize the report are, first, Reason for the Referral, or Identifying Data, or Clinical Behavior. This section states the reason for the assessment, identifying information, any social data, and the psychologist's observations of behavior during the evaluation. The subsequent heading, Test Results, Findings, Test Interpretation, or Evaluation, may be subdivided into Intellectual Aspects (e.g., an IQ score and what it means) and Personality (e.g., psychopathology, attitudes, conflicts, anxiety, and significant relationships). The Diagnosis section presents the main evaluative conclusions, usually expressed as a series of numbers followed by the name of a disorder or condition. The classification system used in the United States is published by the American Psychological Association in the *Diagnostic and Statistical Manual of Mental Disorders,* Fourth Edition (**DSM-IV**). It is a way of classifying all types of mental disorders. (Most agencies have the DSM-IV; in it, you can find descriptions of disorders, their prevalence, and the criteria for diagnosis.) The Diagnosis section of the report may be followed by a Prognosis section—a statement about future behavior. The Recommendations conclude the report and suggest some possible courses of action that would be beneficial in the psychologist's opinion, based on the psychological evaluation. For an example of a psychological report, see Figure 7.3.

Psychological evaluations differ according to the client's needs. The client profiled in Figure 7.3 was referred in order to *(text continues on page 191)*

CONFIDENTIAL PSYCHOLOGICAL REPORT

NAME: Scott Garrett
AGE: 7 years, 6 months
DATE OF BIRTH: 3/2/90
GRADE: 2
SCHOOL: Pineview Elementary
PARENTS: Mr. and Mrs. Scott N. Garrett
EXAMINER: Claudia Zimmerman
DATES OF ASSESSMENT: 9/30/97

Reason for referral and background information
Scott Garrett was referred by his mother, Ms. Sue Garrett, who notes that Scott doesn't enjoy reading and isn't good at it. She notes that he doesn't appear to invest in classwork, particularly seatwork. His teacher, Ms. Cole, is also concerned about Scott's progress. She notes that he is behind in reading but is on grade level in math. Ms. Cole and I discussed Scott's progress. She noted that he seems unmotivated and that his work is typically below grade level, particularly reading and language arts activities. According to the developmental history, Scott accomplished developmental milestones in a typical fashion, with the exception of speech. His first words were at about 18 months. There was no history of physiological problems except for an eye muscle imbalance problem; he is wearing corrective lens for that problem. His hobbies are typical for his age, including TV, playing ball, and riding a horse. Birth history is unremarkable, and he is the product of a full-term pregnancy. Labor and delivery were normal.

Assessment procedures
Wechsler Intelligence Scale for Children–III (WISC-III)
Woodcock–Johnson, Revised, Tests of Achievement
Brigance Comprehensive Inventory of Basic Skills

Test behavior
Scott was evaluated during two sessions. In both sessions, he was willing to work but would often ask, "Are we finished yet?" His problem-solving attack skills seemed typical for a child his age, and he displayed no signs of hyperactive or excessive distractibility in the one-to-one sessions. When

Figure 7.3 Psychological report

confronted with tasks requiring reading, he would sometimes exclaim, "Oh, no." He worked when requested to do so, and seemed pleased when praise was delivered.

In appearance, Scott is approximately average in height for his chronological age, but somewhat overweight. He is an engaging youngster, with close-cropped blond hair and round facial features; he could be described as "cute."

Test results

On the WISC-III, Scott obtained a Full Scale IQ of 106 + or − 3, a Verbal Scale IQ of 100, and a Performance IQ of 112. His Full Scale score of 106 is at the 66th percentile rank and is slightly above average for his chronological age. Chances are 2 out of 3 that the range of scores from 103 to 109 contain his true score. (A *true score* is the hypothetical average score a child would obtain upon repeated testing with the same instrument, minus the effects of practice, fatigue, etc.). Scott's individual subtest scores are as follows. (Scores can be compared to a population mean of 10, a standard deviation of 3.)

VERBAL-SCALED SCORES		PERFORMANCE-SCALED SCORES	
Information	11	Picture completion	11
Similarities	8	Coding	15
Arithmetic	12	Picture assessment	10
Vocabulary	10	Block design	11
Comprehension	9	Object assembly	12
Digit span	(13)	Symbol search	14
		Mazes	12

Scott's profile can be presented using a number of "factor scores." These factor scores have the same psychometric properties as a Full Scale score. That is, the population mean is set to 100 and standard deviation to 15 for the factor scores. The factor scores include Verbal Comprehension, 98; Perceptual Organization, 107; Freedom from Distractibility, 115; Perceptual Speed, 124. These scores reflect average to above-average performance in general, but there is some variability. For example, the Verbal Comprehension scores are considerably lower than the Perceptual Speed score and the Freedom from Distractibility score. In general, his Verbal Fluency, fund of

Figure 7.3 *(continued)*

general vocabulary words, and Verbal Reasoning and Judgment appear to be about average compared to chronological age-mates. He is able to focus attention when directed and maintain that attention and concentration to a degree significantly better than chronological age-mates. His nonverbal reasoning and synthesis/analysis scores appear to be average to slightly above average compared to age-mates. In summary, intellectual ability scores reflect average-to-better performance. This type of intellectual profile is typically predictive of average-to-better classroom performance.

On the Woodcock–Johnson, Revised, Tests of Academic Achievement, the following scores were obtained:

	GRADE EQUIVALENT	SCALE SCORE	PERCENTILE RANK
Letter–word identification	1.6	94	35
Passage comprehensive	1.7	97	43
Word attack	1.6	96	39
Reading vocabulary	1.7	95	37
Math calculation	2.8	97	43
Applied problems	1.8	97	43
Science	3.1	111	76
Basic reading skills	—	97	42
Reading comprehension	—	98	44

In general, Scott's scores on the Woodcock–Johnson ranged from slightly below average to slightly above average. His math- and science-related scores were slightly better than the reading scores. His language arts/reading abilities are approximately one grade level below his current grade level, which is somewhat consistent with teacher observations. However, he delivers unexpectedly good performance on occasions. His teacher says that he is much more capable than he demonstrates typically. His math scores were approximately grade appropriate, although applied problem-solving skills are depressed relative to straightforward calculation. Applied problems are compounded by a language component, which is not his forte. His best performance is in science, which he acknowledges as his "best subject." This suggests that Scott's relatively poor performance in language arts and reading may be motivational. Because his birthday comes late (September), he started the first grade young compared to other

Figure 7.3 *(continued)*

first-graders. Consequently, early language arts acquisition may have been particularly difficult for Scott, and there may be negative affect associated with reading skills currently. It should be mentioned that Scott's scores on the Woodcock–Johnson were compared to his chronological age-mates, not grade-mates. His scores would have been reduced by approximately 4 to 7 standard score points had age norms been applied.

Results from the Brigance reveal some specifics associated with Scott's language arts problems. That is, he missed 3 words out of 10 on the primer level, 2 out of 10 on the grade 1 level, and 5 out of 10 on the grade 2 level. Scott seemed particularly adept at calling beginning consonants but often added a sound to the consonants. For example, rather than "mmm" for M, he responded with "mu." The same was the case for the letters B, H, J, G, R, S, D, M, and F. His greatest difficulty occurred on ending sounds. For example, RIX was pronounced as RIC and LIN as LINE. Other examples of ending word problems included the following: SAT for SIB, TIDE for TID, PEN for PIN, TOX for TAX, and OX for OC. Obviously, Scott needs considerable work to master basic word lists.

Summary and recommendations

Although Scott is in the second grade, his language arts and reading skills are not consistent with that grade placement. Performance on the WISC-III indicates average to above-average intellectual ability. Results from the Woodcock–Johnson are consistent with classroom performance, which reflects relatively poor language arts and reading skills, but relatively stronger math and science skills. His medical history is normal, and there are no obvious physiological impairments. (The eye muscle imbalance apparently is being corrected by glasses, and exercise as prescribed by the physician should be continued.) Results from various tests suggest good reasoning and judgment, ability to concentrate and sustain attention, and average nonverbal reasoning and judgment. One purpose of the evaluation was to rule out the presence of a developmental reading disorder, which would be a Diagnostic and Statistical Manual IV, Axis I 315.00 diagnosis of Reading Disorder. There is not sufficient evidence to warrant this diagnosis. However, the evidence does seem to be compatible with a general developmental delay. Scott started the first grade at a disadvantage relative to other first-graders. That is, his birthday comes in September and, in general, boys develop at a slightly reduced rate relative to girls. Consequently, it is likely

Figure 7.3 *(continued)*

that his reading difficulties began because of developmental immaturity relative to age-mates, which is possibly confounded by early muscle imbalance problems of the eyes. It is possible to rule out the presence of specific learning disabilities/dyslexia, poor educational environment, low intelligence, and visual-auditory processing problems. The following recommendations are offered.

1. Scott's to-be-learned material should be individualized as much as possible to produce maximum gain and maximum motivation. He is not motivated to practice reading content. Consequently, he will need considerable support, direction, and encouragement in this area. Later, low-vocabulary/high-interest reading material can be assigned and a contract system developed to maximize reading and development automatized reading skills.

2. Because Scott's word-calling skills are poor, he should practice basic sight words, those most common in the reading content for his grade placement. A word list will be provided. These word lists can be developed into flash cards for practice.

3. Scott would profit from instruction from a tutor. To-be-learned material should be coordinated with his classroom teacher. It may be possible to obtain tutoring during the summer and after school during the next academic year. Also, there is a summer-school program available that should be good for him.

4. Retention is not an appropriate option for Scott. There are considerable negative implications, primarily social, interpersonal, and self-esteem related. In addition, the literature is controversial regarding long-term academic gains associated with retention. The evidence is not positive, especially for kids Scott's age. Tutoring is a more appropriate solution. I suspect that tutoring will be required for the next couple of years, with considerable structure necessary. Tutoring will likely be phased out during the fourth and fifth grades.

5. Scott would profit from having stories read to him. Any activity that would increase his desire to read independently is appropriate.

6. Scott is overweight, and a low-calorie diet would be beneficial. A children's weight-loss program is available at University Hospital.

Claudia Zimmerman, Ph.D.
Licensed School Psychologist

Figure 7.3 *(continued)*

to assess reading problems and to determine eligibility for special services. The tests administered and the final report would be different if the client had been referred for other reasons (e.g., behavioral problems).

Social History

For a complete case file, the client's past history and present situation must be investigated. The person's past adjustment can give indications of how he or she will adjust in the future. A **social history** also provides information about the way an individual experiences problems, past problem-solving behaviors, developmental stages, and interpersonal relationships. Some of the information in a social history may duplicate what has been gathered during the intake interview. In the social history, however, the client can relate the story in his or her own words, with guidance from the helper.

There are a number of advantages to a social history. Often the informal history taking leaves gaps, but the carefully done social history completes the picture. The case manager can then plan the appropriate integration of services and provide better information for future referrals. The social history often includes a better assessment of the client's need for services; this is especially helpful with clients who have multiple problems. A social history can also fulfill legal requirements. And finally, the process of taking a social history can help build the relationship between the case manager and the client.

There are also limitations to the social history. History taking is a preliminary activity in case management, but the client may perceive it as a phase in which solutions are put in place. Unfortunately, categorizations and judgments made at this stage may be premature. The process of taking the history can also give an inaccurate view of what will happen between the client and the case manager. Excessive questioning by the case manager may lead to a dependent role for the client, and culture-bound questions can create barriers to the development of the helping relationship. Also, an exhaustive history is not absolutely necessary to develop a plan of services. It may be helpful, but the relevant information gathered may not be for service delivery. Spending too much time on history taking can also be harmful. The client may use the process to resist significant facts. Other clients may construe it as therapy but it is not intended as such and may not even be therapeutically valuable. Despite these limitations, the social history still has the important function of completing the case file. Moreover, the case manager can use certain strategies to mitigate the limitations.

The following suggestions are adapted from McGowan and Porter's guide to history taking in rehabilitation settings (1967, p. 74). As phrased here, they apply to the gathering of information for a social history in other settings as well.

1. Do not let completion of the survey form or social history become a goal. Remember that the client is the main concern, not the paperwork.

Case managers will sometimes make a home visit in order to obtain a complete picture of the client's environment.

2. Don't divert the client from discussing some aspect of his or her history because you already have the information necessary for the form and are anxious to move on to the next topic. Important clues to understanding a client may be gained from spontaneous discussion of the aspects of the history that he or she considers significant.
3. Be sure that the client understands the reasons for the history and can see the benefits of compiling it.
4. If a client resists a part or all of the data gathering, or feels threatened by what may be perceived as an invasion of privacy, do not become defensive and punitive.
5. When additional information is deemed necessary, secure the client's permission to contact relatives, employers, friends, etc.
6. Do not let the client ramble on about all aspects of the history; maintain control of the interview.

Using these guidelines, the case manager gathers pertinent information about what appears to be the client's problem. The primary source of information is the client, who is encouraged to tell the story in his or her own way. The helper listens carefully to what is said, how it is said, and what is not said. The sequence of events, reactions, feelings, and thoughts are all taken into consideration as the client relates the history. Note taking should be kept to a minimum so that important nonverbal information is not missed.

There is no set form or procedure for taking a social history. Some agencies use forms to guide information gathering, such as the social data report shown in Figure 7.4. Others just provide guidelines for their case managers so the length and detail of social histories may vary. In all cases, the social history is prepared when a comprehensive picture of a client's situation is desired. The outline for writing it depends on what the agency wishes to emphasize, but certain topics are almost always included: identifying data, family relationships, and economic situation. Which other areas are emphasized depends on the focus of the agency and the presenting problem. For example, a social history of a couple involved in marital counseling might target such areas as family relationships and psychosocial development. For someone seeking economic assistance, important areas might be financial status, income, expenses, and work history. In general, the following areas may appear in a social history:

Identifying information: such as name, address, date and place of birth, Social Security number, military service, parents' name and address, children's names and ages.

Presenting problem: brief description of the problem.

Referral: source and reason.

Medical history: relevant hospitalizations, illnesses, treatment, and effects. Written permission is needed to obtain copies of medical records, if necessary.

Personal/family history: family life, discipline, parenting, and personal development.

Education: highest grade completed; progress; records.

Work history: training, type and length of employment, ambitions.

Present family relationships and economic situation: family members, ages, relationships, lifestyle, and income.

Personality and habits: interests, disposition, social activities, personal appearance.

The client provides most of the information for a social history, but other sources may also contribute. When the case manager has gathered material from sources other than the client, it should be inserted under the appropriate headings, with the source identified. Direct knowledge is the main source, as in the following examples.

- She did not come for her first appointment.
- The client drummed his fingers on the table throughout the interview.
- He states that his goal is to receive a high school diploma and get a job.
- The client stated that during the past week she and her husband had three fights.

SOCIAL DATA REPORT

Student's full name: ______________________ SS#: ________________

Address: ______________________________ Phone: ______________

Age: _____ D.O.B.: _____ Race: _____ Sex: _____ Hair: _____

Height: ____________ Weight: ___________ Eyes: _____________

Distinguishing marks: ____

Offense: ____

Date of offense: __________ Prior court record: ___ Yes ___ No
(If Yes, show details under Additional Information.)

School attending: _________________________ Grade: _____________

Have you been suspended from school, given detention, in-school suspension, or had truancy problems? ___ Yes ___ No (If Yes, show details under Additional Information.)

Health: __________________ Are you on any prescription medications?
___ Yes ___ No (If Yes, give type, amount, reason under Additional Information.)

Are you or have you seen a mental health counselor? ___ Yes ___ No

If Yes, where and when? _____________________________________

Have you ever used drugs or alcohol? ___ Yes ___ No

Type, amount: ______________________________

Are you employed? ___ Yes ___ No Where & hours: ____________

Figure 7.4 Social data report *(continued)*

PARENT INFORMATION

Father's name: ______________________________ Age: __________

Address: ______________________________ Phone: __________

Employment: ______________________________ Phone: __________

Educational level: 1 2 3 4 5 6 7 8 9 10 11 12 13 14 15 16 GED

Court record: ____ Yes ____ No If Yes, for what and when? __________

Use of drugs or alcohol: Type of Use: excessive, moderate, little, none

Have you had any type of counseling: ____ Yes ____ No

If Yes, when and where? ______________________________

Use of prescription medication: ____ Yes ____ No

If Yes, type and amount? ______________________________

If deceased give date and cause: ______________________________

Mother's name: ______________________________ Age: __________

Address: ______________________________ Phone: __________

Employment: ______________________________ Phone: __________

Educational level: 1 2 3 4 5 6 7 8 9 10 11 12 13 14 15 16 GED

Court record: ____ Yes ____ No If Yes, for what and when? __________

Use of drugs or alcohol: Type of Use: excessive, moderate, little, none

Have you had any type of counseling: ____ Yes ____ No

If Yes, when and where? ______________________________

Figure 7.4 *(continued)*

Use of prescription medication: ____ Yes ____ No

If Yes, type and amount: ______________________________

If deceased give date and cause: ______________________________

Parents marital status: Living together, divorced, separated, widow(er)

If separated or divorced: When? ______________________________

Discipline in home: Type: ______________________________

Does it work? ______________________________

Who provides discipline: mother, father, both, neither

Children in family:

Name	*Age*	*Lives at Home? (Yes or No)*	*Court/Arrest Record (Yes or No)*

Are you receiving any financial assistance from anyone?

Type: AFDC Amount: ______________

Food stamps Amount: ______________

Social Security Amount: ______________

Child support Amount: ______________

TennCare Amount: ______________

Other Amount: ______________

Do you have medical insurance? (Name, policy #, address, and phone)

ADDITIONAL INFORMATION: ______________________________

____

Figure 7.4 *(continued)*

The next examples are statements of information from other sources:

- Educational records indicate that the client completed the sixth grade in school.
- Her parents report that the client lived with them until her marriage two years ago.
- He was fired from his job for absenteeism.
- A psychological evaluation indicates a mildly retarded 13-year-old with a possible hearing loss.

The social history shown in Figure 7.5 combines two approaches. The Identifying Information section is a form that the case manager completes. The remaining sections are a narrative based on information compiled from several sources (listed at the end of the report). At this agency, a social history may be compiled by more than one case manager, and all who are involved in the writing of the social history sign the written report.

Another way social information appears in a case file is illustrated by the court report shown in Figure 7.6. It was prepared for juvenile court, based on social information gathered by a caseworker at the Department of Human Services. DHS caseworkers frequently prepare court reports—for example, if parental rights are being terminated or if the court requests DHS to investigate a petition for custody. All court reports have certain things in common, such as the reason for the referral to the department and the circumstances of the child, of both parents, and of the petitioner. Also included is the recommendation of the department, which the court may or may not follow. Although the format of this report is determined by the court, you will see content similarities to the social history in Figure 7.5. In this court report, a grandmother is asking for full custody of her granddaughter. A caseworker has been out to the home, completed a social history of the family, and obtained a signed release of information from the petitioner. The caseworker has also consulted with the law enforcement agencies, checked references, and obtained as much information as possible from other sources. The caseworker then writes a report, informing the court as succinctly as possible of all the relevant information gathered.

Other Types of Information

Other types of information may be relevant to the case file, depending on the agency's mission and services as well as the client's problem. Educational and vocational information, the most commonly needed, will be discussed here.

Educational information can have many parts: test scores, classroom behavior, relations with peers and authority figures, grades, suspensions, attendance records, and indications of academic progress such as repeated grades or advanced work. The sources of educational information are just as varied: school records, teachers, guidance counselors, and principals. Often, the particular information that the case manager obtains *(text continues on p. 207)*

SOCIAL HISTORY

I. *IDENTIFYING INFORMATION*

Date: 3/1/95 Name: Joe Billy Smith Date of birth: 7/14/79

SS#: 000-00-0000 TennCare/Insurance provider: Blue Cross/Blue Shield

Policy number: 000000000 Marital status: Never married

Number of children: None reported

Address: 1111 Dogwood Drive

Atlanta, GA

Telephone number: (123) 777-7777

Prevention resource: Community Health Center

Resource contact: Jim Therapist

County court: ____________

Custodial dept.: Department of Youth Development

Custodial dept. contact: Jerry Officer

Date of state custody: 2/23/95 County court: Cobb

Age: 16 years Race: Caucasian Height: 5′ 2″

Weight: 120 pounds Eye color: Blue Hair color: Blond

Unusual markings: Tatoo on left bicep

Allergies: Penicillin Current medications: None reported

Current medical problems: Ingrown toenail

Figure 7.5 Social history

Special circumstances: Ms. Smith (mother) reported that Joe was abducted from school at gunpoint and was kidnapped for five days. During the time Joe was gone, he was allegedly emotionally and physically harassed and assaulted. This took place 4/5/94 to 4/10/94.

II. *PRESENT PROBLEMS* (Current charges with dates and circumstances)

Joe appeared in Juvenile Court on February 23, 1995 on the charges of violation of a valid court order and disobeying his probation by using cocaine and alcohol. Joe plead true to these charges and was placed in the custody of the Department of Youth Development. On this same date, the court ordered an assessment of Joe and his situation.

III. *PREVIOUS PROBLEMS* (Past charges with dates, adjudications and placements)

According to Joe's court file, Joe was petitioned to court on October 1, 1994 for the charge of running away from home. Joe plead true to the charge and was placed on probation with the County Probation Service. Ms. Smith (mother) also reported that when Joe disappeared in April of 1994, she went to the Police Department to file a missing persons report. Ms. Smith stated that she was informed that a missing persons report could not be filed, but that she could sign a paper declaring Joe a runaway juvenile. This action would allow the police to search for Joe. Ms. Smith reported that she signed the form only out of concern for Joe, and that this is now in Joe's court record. No charges were ever brought against the 20-year-old male who allegedly abducted and abused Joe.

IV. *FAMILY HISTORY* (Name, social security number, current address, phone, date of birth, marital status, employment, educational level, court record, alcohol and drug problems, mental and physical health problems, possible placement resource)

A. Father: Tom Smith is Joe's biological father. His date of birth is 0/00/00, and his social security number is 000-00-0000. Ms. Smith reported that Mr. Smith lives in Dalton, Georgia, but she does not know his address. Ms. Smith stated that Mr. Smith's phone number is (123) 000-0000. Ms. Smith reported that Mr. Smith is currently remarried and is employed delivering bottled

Figure 7.5 *(continued)*

water. Ms. Smith does not know Mr. Smith's delivery route. Ms. Smith reported that Mr. Smith obtained his GED and has no prior court history, to her knowledge. Ms. Smith stated that Mr. Smith use to drink alcohol frequently—five times per week—but he did not use any drugs. Ms. Smith is unaware of any physical or mental health problems that Mr. Smith may have. Mr. Smith would not be an appropriate placement for Joe because of his sporadic interest in Joe's life. Ms. Smith stated that Mr. Smith has let Joe down many times in the past.

B. Mother: Betty Smith is Joe's biological mother. Her date of birth is 0/00/00, and her Social Security number is 000-00-0000. Ms. Smith's current address is 1111 Dogwood Drive, Atlanta, Georgia, and her phone number is (123) 777-7777. Ms. Smith is divorced and works at Kroger's in South Atlanta. Ms. Smith reported that she completed the 10th grade and then earned her GED. Per Ms. Smith, she has no alcohol or drug problems or court history. Ms. Smith stated that she deals with mild depression and seeks professional mental health services when depression sets in. At this time, Ms. Smith reports that she is taking no prescription psychotropic medication for her depression.

C. Stepparents: Not applicable.

D. Siblings:

SIBLING NAME	GENDER	DOB	FULL/HALF SIBLING
Jeff Smith	Male	0/00/00	Unknown
Ty Smith	Male	0/00/00	Full

Sibling interaction:

Ms. Smith stated that Joe gets along well with his older siblings, but he is closest to Ty. Ty's wife has a baby girl, and Joe likes to help take care of her. Joe's siblings are older than he is, and Ms. Smith reports that Joe does not get to see them as much as she would like.

E. Other: (Grandparents, relatives, boyfriends, girlfriends, etc.)

Joe's last remaining grandparent died on 2/15/92. Ms. Smith reported that Joe was very close to this grandparent.

Currently, Joe reports no girlfriend involvement.

Figure 7.5 *(continued)*

V. *FAMILY INTERACTION* (Family dynamics/relationships, current issues, financial resources, needs, risks, etc.)

Ms. Smith reported that she and Joe are very close. Ms. Smith appeared to be very protective of Joe and defended him and his actions. Ms. Smith stated that the only current issue in the home is Joe's opposition to house rules. Ms. Smith is angry about Joe's placement in the Department of Youth Development, and she is threatening to sue the state. Ms. Smith reported that she earns minimum wage and usually works at Kroger's 30–40 hours a week. Ms. Smith stated that she pays rent and utilities. Ms. Smith reported that she receives food stamps, but she is not sure how much she will be getting now that Joe has gone into custody. Ms. Smith did not feel that all of her needs were currently being met.

VI. *HOME AND NEIGHBORHOOD* (Date of home visit, type of home, adequacy of space, housekeeping standards, hazardous conditions, neighborhood description)

A home visit was made by the case manager on 0/00/00. Ms. Smith said they have been renting their two-bedroom apartment since July of 1993 and that both she and Joe have plenty of room. This case manager observed that housekeeping standards were very good. No major inappropriate housing conditions were noticed. Their apartment is in a low-level crime area of Atlanta.

VII. *CHILD*

A. Early development history (Problems with pregnancy or delivery, planned/unplanned pregnancy, parental A&D use during pregnancy, developmental milestones, serious illnesses or accidents, diagnosis of hyperactivity, etc.)

Ms. Smith reported that she had no problems with the pregnancy or delivery of Joe; she delivered Joe herself, with the help of a midwife. Ms. Smith reported that she did not use alcohol or drugs while she was pregnant, and that Joe reached all of his appropriate developmental milestones earlier than most children. Ms. Smith reported that the only serious illness Joe had as a child was pneumonia.

Figure 7.5 *(continued)*

B. Peer interaction (Relationships with peers, age of friends, activities of friends, does or does not have friends)

Ms. Smith reported that Joe does not have many age-appropriate peers. Joe's friends are usually older. Per Ms. Smith, Joe does not do much in his free time except sleep. Ms. Smith was not certain what Joe does when he goes out with his friends, but she stated that he does not go out often.

C. Education (Last school attended, grade level, major school problems; accelerated, remedial, or special ed classes, truancy history)

Joe was last enrolled at Greenbriar High School in the ninth grade. Ms. Smith stated that Joe attends regular classes. Joe has had some truancy and tardiness problems. Since the abduction from school on 4/5/94, Joe has become increasingly paranoid about attending school. It takes full cooperation with the school system to make him attend.

D. Psychological (Current and prior psychologicals, stating examiner's name, location of testing, and test dates)

Joe had a psychological evaluation, conducted at Lakeside Mental Health Institute on 4/30/94. The examiner was Dr. John Doe.

Figure 7.5 *(continued)*

VIII. *AGENCY CONTACTS AND SOURCES OF INFORMATION*

Name: Tom Casemanager Relationship: Case Manager

Address: 201 Center Park Drive, 1100
Atlanta, GA 12345

Phone: (123) 000-0000

Agency name: Lakeside Mental Health Institute

Agency contact: Dr. John Doe

Address: 5900 Lakeside Drive
Atlanta, GA 12345

Phone: (123) 000-0000

Agency name: Department of Youth Development

Agency contact: Jerry Officer

Address: 222 Jail House Drive
Atlanta, Georgia 12345

Phone: (123) 000-0000

Prepared by: ______________________________

(Name) (Title) (Date)

____

(Name) (Title) (Date)

____

(Name) (Title) (Date)

____

(Name) (Title) (Date)

Figure 7.5 *(continued)*

REPORT FOR JUVENILE COURT

Child: Lydia Maza, date of birth 8/8/85
Address: 100 Washington Pike,
Chicago, IL

Petitioner: Jorja Mitten
Address: 100 Washington Pike,
Chicago, IL

Mother: Leyla Mitten
Address: Unknown

Father: Lloyd Maza
Address: P. O. Box 18,
Hot Springs, AR

REFERRAL

The petitioner has been the primary caretaker of the child since July, 1995. The legal custodian is incarcerated in Shelby County Jail at this time. Petitioner asks that legal care and custody be given to her and her husband.

CIRCUMSTANCES OF THE CHILD

Lydia came to Chicago in July 1995, to visit her maternal grandparents. During this time Leyla Mitten was arrested for drug trafficking, possession, and dealing and sent to Shelby County Jail, where she will be eligible for parole in 1997. Jorja Mitten received a letter from her daughter asking her to care for Lydia until she is able to do so.

When asked if she would like to go back to Arkansas, Lydia stated that she would rather stay in Chicago. Lydia is a shy girl who states that she has more friends in Hot Springs, but has friends in Chicago too and enjoys playing with them. Lydia stated that she likes living with her grandparents, and she seems happy there.

Lydia's teacher at Ritta School says she is doing very well in class but seems emotionally fragile. The school records show that Lydia has missed four (4) days of school this year, all of which have been excused.

CIRCUMSTANCES OF PARENTS

Mother—Leyla, date of birth 2/26/69, was arrested in Chicago on 8/1/95 on several drug charges. She will be eligible for parole in 1997.

Figure 7.6 Court report

Reportedly, Leyla expects Lydia to be returned to her then. Leyla has alcohol and drug issues.

Father—Lloyd Maza is a spa owner in Hot Springs. He has had no contact with Lydia since July, 1995.

CIRCUMSTANCES OF THE PETITIONER

Jorja Mitten, maternal grandmother, age 58, is the petitioner in this matter. She has been married to Gus Mitten, age 59, for 30 years. The Mittens live in a three-bedroom, one-bath home, which they own and have resided in for 18 years. The Mittens have two children, Leyla and her twin brother Boyd, who lives in Red Springs.

Ms. Mitten is currently employed by a local utility. She stated that she has worked there for 19 years. She reportedly earns $3,500.00 per month.

Ms. Mitten denies any alcohol or drug abuse problems. She is a smoker, as is Mr. Mitten, and they use air filters in their home. Ms. Mitten states that she is in good health and takes Oruvail and Adalet daily under doctor's orders.

Mr. Mitten is currently employed. He stated that he has worked at Wind Industries for 22 years. He reportedly earns $4,000.00 per month.

Mr. Mitten denies any alcohol or drug abuse problems. He reports that he is in good health. Mr. Mitten takes Oruvail daily.

Mr. and Ms. Mitten are members at the YMCA, where they exercise weekly.

Ms. Mitten stated that she is concerned that Leyla is not emotionally or financially stable and is therefore in no position to care for Lydia. She stated that she is afraid that Leyla, after being released from jail, will "drag Lydia down" with her.

All references speak very highly of the Mittens and hold them in the highest regard. All stated that they have no concerns about the Mittens caring for Lydia.

A check with local law enforcement agencies revealed no prior record in Dade County on Jorja or Gus Mitten.

RECOMMENDATION

At this time, the department would recommend that custody be granted to Jorja Mitten. The legal custodian is currently unable to provide for the child,

Figure 7.6 *(continued)*

and the child states her reluctance to leave her grandparents' home. This worker knows of no reason why Lydia should not remain with Ms. Jorja Mitten.

As always, we will respect the court's wishes in this matter. We hope this information will be of assistance to the court.

Submitted by:

Illinois Department of Human Services

Tina Rachael
Caseworker 1

Figure 7.6 *(continued)*

depends on which source is contacted. Rarely is it gathered in a single report, as medical information might be. In many cases, the case manager will decide what information is needed and contact the source or sources most likely to have that information. For example, a teacher will probably be the best source of information about classroom behavior, whereas school records will provide test scores and indications of past academic performance. The contact may occur formally (in writing) or orally (by telephone or interview).

Vocational information can be important for several reasons. People seem to be happiest when their activities are satisfying and fulfill their needs. There is also the need to earn a living, and often self-support engenders self-respect. Ways of gathering vocational information range from asking the client about his or her work history to having a formal vocational evaluation done. The types of information gathered include jobs previously held, the ability to get along with co-workers, work habits (e.g., punctuality and reliability), and reasons for frequent changes in employment. How much more information is needed depends on the client's problem and the agency's mission. For example, if the client has no work experience, an exploration of vocational interests and aptitudes may be in order. For the client who has had varied employment, the focus may shift to attitudes toward work and the skills developed. The client who has a substantial record may need help in reviewing his or her experience and skills to establish a vocational objective.

Let's return to the Roy Roger Johnson's case, discussed in Chapter 1. Roy's counselor requested a period of vocational evaluation at a regional center that assesses an individual's vocational capabilities, interests, and aptitudes. Roy and the counselor, Tom Chapman, attended a staffing to hear the vocational evaluation report. Mr. Chapman later received a written report (see Figure 7.7). The report illustrates two important points. First, information about a client is integrated with other new information in order to complete the picture. In this report, you will read about work history, medical information, and test scores, as well as the results of the vocational evaluation. Second, this report is a vocational evaluation report. Vocational evaluation is a process for gathering, interpreting, analyzing, and synthesizing all data about a client that has vocational significance, and relating it to occupational requirements and opportunities.

Both vocational and educational information add other dimensions to the client record, making the case file more complete. This information rounds out our understanding of who the client is—strengths and weaknesses, abilities and aptitudes.

Chapter Summary

The information about the client that is gathered from other professionals helps the case manager see a more complete picture of the client. This information includes medical reports, psychological evaluations, social histories, and educational and vocational information. When the case *(text continues on p. 215)*

VOCATIONAL EVALUATION REPORT

ADMISSION DATA

To: Tom Chapman	Date reported: 9/24/90
From: Dan Howard	Dates attended: 9/24–25/90, 9/26–27, 9/28, 10/1
Re: Roy R. Johnson	
D.O.B.: 02/08/58	Scheduled hours: 8:00 A.M.–3:00 P.M. Monday–Friday
S.S.#: 000-11-2222	Work history: Bartender, plumber's helper
Sex: Male	Medication: None
Marital status: Single	Education: Two quarters SSCC
No. of dependents: None	Transportation: Own vehicle
Date: 11/02/90	Program manager: Jo Singletary

TESTS ADMINISTERED

Purdue Pegboard

Valpar Component Work Samples

- Size discrimination
- Upper extremity range of motion
- Simulated assembly
- Eye–hand–foot coordination
- Bennett hand tool

Reason for referral: Mr. Johnson was referred to the Vocational Training Center for vocational evaluation to assess his manual and finger dexterity. This evaluation was to consist of paper/pencil testing and situational assessment. The client completed his paper/pencil testing, but did not complete his situational assessment. This report is based on the results of the tests he took and the limited time he did spend in situational assessment.

BACKGROUND INFORMATION

The information in this section was based on statements made by the client on his evaluation interview.

Figure 7.7 Vocational evaluation report

Social/educational: Mr. Johnson was a 30-year-old white single male who lived with his mother and sister. It was disclosed by this client during his evaluation process that his mother has been severely hearing impaired since birth. Because of this, he learned sign language before he could talk. He stated that he maintained this skill.

He had recently received a settlement of $40,000 due to his injury on the job and was building a home with these funds. He did not have any other source of income. The client completed two quarters at Silver State Community College. He quit when his grandmother became ill. He also received plumber's training for 6 months at T.A.T. and attended the Georgia School of Bartending for 5 weeks. The client has held 3 jobs in the past 3 years and has not worked since April, 1989.

The client's stated vocational interest was in working as a mechanical engineer. He said he would choose this because it was within his physical capabilities, and that he had a background in mechanics.

WORK HISTORY

The client's work history was obtained from the client during his evaluation interview.

The client worked as a bartender at the Ramada Riverview for 10 months. He left to go to work at Joe B's Restaurant.

The client worked at Joe B's as a bartender for 2 months. He left this position because of a personality conflict.

The client last worked at Rock City Mechanical as a plumber's helper. He left following his injury.

MEDICAL INFORMATION

General: The client stated that he was in general good health. The client's general medical examination revealed that his only problems were related to his back injury.

Handicapping condition and functional limitations: The client was handicapped by a back injury while working on a plumbing job. It is stated in a report from Dr. Alderman that the client also had a history of previous

Figure 7.7 *(continued)*

lumbar disc disease. It is also stated that the client's recovery will be slow and that he will have a permanent impairment of 10% to the body as a whole.

The client's limitations were taken from his general medical examination, done by Doctor Jones. These limitations include walking, standing, stooping, kneeling, lifting, reaching, pushing, and pulling.

Vocational implications: The client would work best in a sedentary job, one where he can get up and move around when needed to relieve his back pain. The client was still recovering, and these limitations may change with time.

BEHAVIORAL OBSERVATIONS

General: The client was cooperative during testing and attempted all tasks asked of him. In situational assessment, the client could not work on all the contracts that we had. He wanted to wait until we had something he was able to do. When we did not get work in that was easier than what we had, he quit coming in spite of efforts to make the work easier for him. The reason stated by this client for not attempting this work was that he feared injuring his back again.

Vocational: The client did not complete the situational assessment, and the observations made were not complete.

OBSERVATIONS OF WORK BEHAVIORS

The client was observed in 13 different areas while in situational assessment. Ratings were as follows:

Rating scale for work behavior	NA	Not acceptable
	1	Not acceptable; needs improvement
	2	Satisfactory; meets criteria
	3	Excellent; employment strength

Figure 7.7 *(continued)*

Attendance	1
Punctuality	2
Co-worker relations	3
Supervisor relations	2
Work quantity	
Contract	1/3
Cleaning	N/A
Work quality	3
Work tolerance	1
Job flexibility	1
Follows work instructions	3
Use of work time	2
Works without close supervision	3
Observes all safety procedures	3
Care of tools/equipment/materials	3

Comments

Attendance: The client attended the center 43% of the time. He did not come in because he could not do the work we had (sanding and painting dumpsters).

Work quantity/contract: The client's production on the nuts and bolts contract for Smithfield Industries was 95% of industrial norms. His production on the labeling contract for A.P.S.U. was 25% of industrial norms. It is noted that this contract required 100% quality.

Work quantity/cleaning: The client could not do this.

Work tolerance: The client could work a full day but complained of pain in his feet. He said that this was because of the concrete floor.

Job flexibility: The client could not do the work on the dumpsters or the cleaning duties. He worked on the A.P.S.U. labels, tape recycling contract for Triad Corporation of Tennessee, and the nuts and bolts contract for Smithfield Industries. These were all sedentary tasks.

Figure 7.7 *(continued)*

TEST INTERPRETATION

Purdue Pegboard
Right hand—58%ile—industrial applicants
Left hand—56%ile—industrial applicants
Both hands—72%ile—industrial applicants
Right+left+both—59%ile—industrial applicants
Assembly—93%ile—industrial applicants

Valpar Component Work Sample #2—Size discrimination

	TIME	ERRORS
Assembly	115%	<100%
Disassembly	80%	100%

Performance—64%
Norm group: Seminole Community College Vocation Assessment Center

Valpar Component Work Sample #4—Upper extremity range of motion

	%ILE	MTM%
Assembly: Dominant hand	85	70
Assembly: Other hand	80	75
Disassembly	70	90

Performance: 87%ile
Norm group: Seminole Community College Vocational Assessment Center

Valpar Component Work Sample #8 —Simulate assembly
Performance—95%ile MTM—75%
Norm group: Seminole Community College Vocational Assessment Center

Valpar Component Work Sample #11—Eye-hand-foot coordination
Points Performance—25%ile MTM-50%
Time Performance—85%ile MTM-140%

Bennett Hand-Tool Dexterity Test
The client's performance on this instrument was 75%ile when compared to boys at a vocational high school.

Figure 7.7 *(continued)*

SUMMARY/RECOMMENDATIONS

Summary: Mr. Johnson was a 30-year-old man who was handicapped by a back injury, with limitations in walking, standing, stooping, kneeling, lifting, reaching, pushing, and pulling. The counselor is referred to the medical information section of this report for details.

The client's performance in testing could be considered strong, with the exception of his accuracy in performing the eye-hand-foot coordination work sample. The client's scores (except eye-hand-foot, as mentioned above) were above average when compared with the norm group.

The client's strengths in situation assessment were in co-worker relations, work quality—nuts and bolts contract, work quality, working without close supervision, observing all safety procedures, and care of tools/equipment/materials.

The client's weaknesses in situational assessment were attendance, work quantity—labeling contract, work tolerance, job flexibility.

The client had acceptable performances in punctuality, supervisor relations, and use of work time.

Recommendations: The results of the testing done here at the center to assess the client's dexterity show that he could perform a sedentary, small, hand-assembly job. He expressed no interest in pursuing this line of employment. As noted in the Medical Information section, he would need to be in a job that would allow him the freedom to get up and move around when he needed to relieve his back pain. During situational assessment, he was observed to have the need to get up and down as he worked.

As stated in the Social/Educational section of this report, the client is proficient in sign language for the hearing impaired. The client could be employed as an interpreter for the hearing impaired if he desired to pursue certification. The client said that he had never considered this, but was open to the idea. It is recommended that the counselor investigate the possibility of having the client obtain his certification.

Figure 7.7 *(continued)*

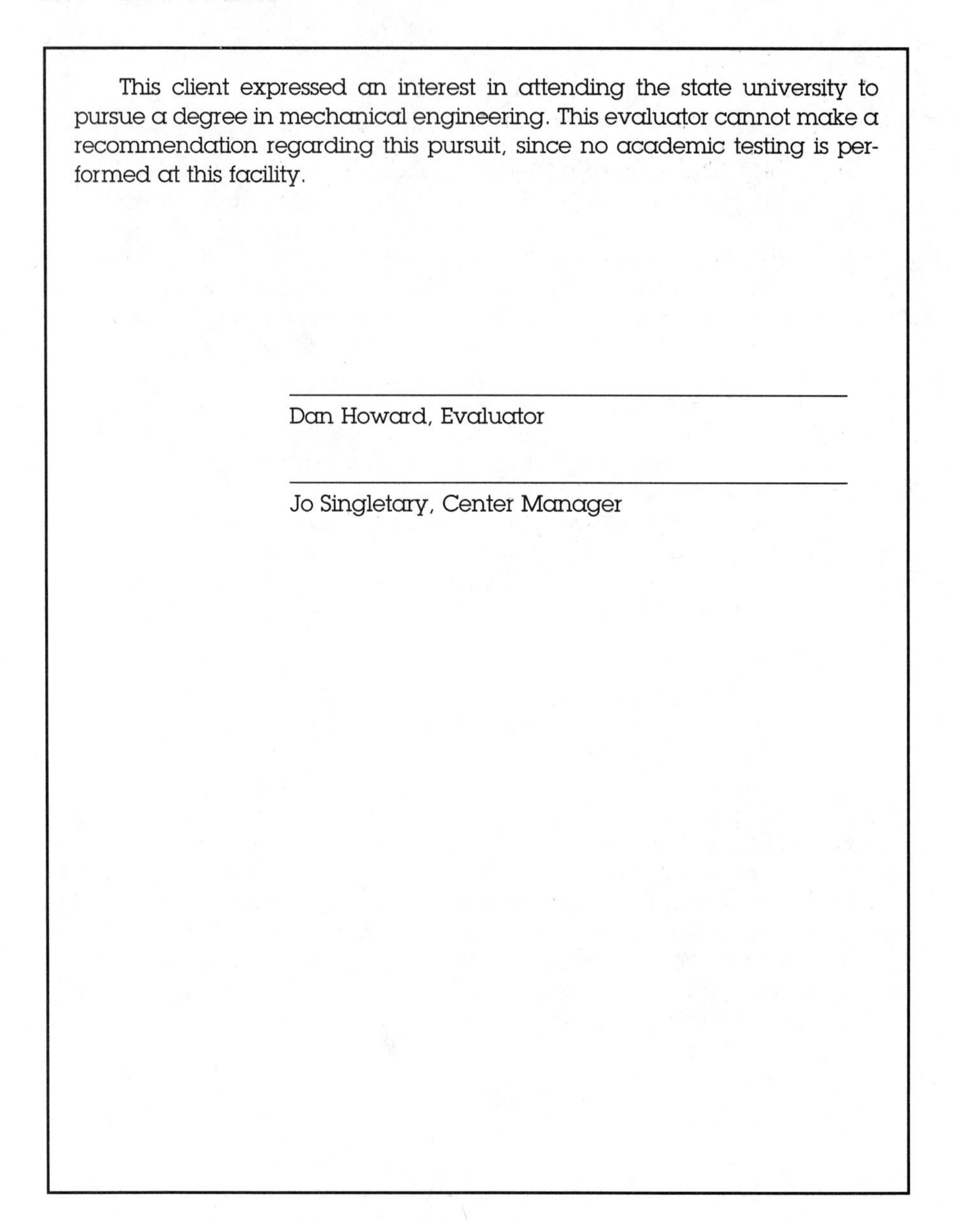

This client expressed an interest in attending the state university to pursue a degree in mechanical engineering. This evaluator cannot make a recommendation regarding this pursuit, since no academic testing is performed at this facility.

Dan Howard, Evaluator

Jo Singletary, Center Manager

Figure 7.7 *(continued)*

manager requests the information from other professionals, the goals must be clear, and it is helpful if the client's problems are identified. Once the information is received, the case manager reviews it and integrates the results with the information previously gathered.

Chapter Review

◆ *Key Terms*

Medical terminology
Medical diagnosis
Medical consultation
Physical examination
Diagnosis
Word root
Psychological evaluation
Psychological report
DSM-IV
Social history

◆ *Reviewing the Chapter*

1. Identify the resources that will help you understand medical reports.
2. How does medical information contribute to a case file?
3. In what situations would a medical consultation help you?
4. Describe a general medical examination.
5. Define each of the elements that form medical words.
6. What are the three basic steps in working out the meaning of a medical term?
7. Why is keeping current with medical terms a challenge for case managers?
8. List reasons to refer a client for a psychological evaluation.
9. How does a case manager make a good psychological referral?
10. Describe a psychological report.
11. What is a social history?
12. Describe the advantages and limitations of a social history.
13. How will the guidelines for history taking help you do a social history?
14. Complete the Social Data Report (Figure 7.4) on yourself.
15. Now write a social history on yourself, using the nine content areas of a social history.
16. Describe the three ways in which a social history may appear in a case file.
17. What do educational and vocational information add to a case file?

◆ *Questions for Discussion*

1. Why do you think medical information is important to have?
2. What difficulties do you expect to have in understanding a psychological report?

3. Develop a plan to gather information for a social history of a client who is in prison for armed robbery.
4. Do you believe that you can have too much information about a client? Why or why not?

References

Brodwin, M. G., Tellez, F., & Brodwin, S. K. (Eds.). (1992). *Medical, psychosocial, and vocational aspects of disability.* Athens, GA: Elliot & Fitzpatrick.

Felton, J. S. (1992). Medical terminology. In M. G. Brodwin, F. Tellez, & S. K. Brodwin (Eds.), *Medical, psychosocial, and vocational aspects of disability* (pp. 21–33). Athens, GA: Elliott & Fitzpatrick.

McGowan, J. F., & Porter, T. L. (1967). *An introduction to the vocational rehabilitation process.* Washington, DC: U.S. Department of Health, Education, and Welfare.

Tallent, N. (1993). *Psychological report writing* (4th ed.). Englewood Cliffs, NJ: Prentice Hall.

Chapter Eight

Service Coordination

A case manager's bible is resources. A case manager cannot begin to do any type of service to a family without resources. You cannot do it from behind your desk. You have to go out there. . . . You also have to be an advocate for both your client and your agency, so that you can get that resource, or you will have one name that will make a difference. So resources are a major tool for a case manager.

—Yolanda Vega, CSW, Director of Agency Services, Casita Maria Settlement House, Bronx, New York, personal communication, May 4, 1994

As far as doing case management in the hospital, I saw in many ways that it was crucially important to have networks in support services and know the agencies, because you certainly couldn't address all the needs of a client. I had to know who in detox I was sending my people to. How to go about getting Social Security and to help families know how to deal with the process.

—JUDITH SLATER, Kennesaw State University, Kennesaw, Georgia, personal communication, October 14, 1995

Communication with other service providers is really important. Knowledge of community resources. The case manager has to be a broker. He or she has to be able to work well with lots of other people.

—ALAN TIANO, Pima Health System, Tucson, Arizona, personal communication, October 7, 1994

One of the most important roles in case management is service coordination. Rarely can a human service agency or a single professional provide all the services a client needs. Because in-house services are limited by the agency's mission, resources, and eligibility criteria—as well by as its employees' roles, functions, and expertise—arrangements must be made to match client needs with outside resources. Case managers must have knowledge of available community resources and the skills to put them to use.

The preceding quotations reflect the knowledge and skills that a case manager uses to meet client needs. According to the case manager at Casita Maria, knowledge of resources is not enough—one must also act as an advocate so as to gain access to those resources for the client. A key to using resources, according to Judith Slater, is to have networks in place so that the case manager knows both the resource agency and the name of a contact. Perhaps the one indispensable skill in using resources is communication; Alan Tiano states that working well with other service providers depends on the case manager's communication skills.

Today's service delivery environment makes for different roles and responsibilities for the case manager. In the past, many services were provided directly by the case manager, but service delivery has become more specialized. Professionals must be careful not to provide direct services in areas in which they are not trained or lack the necessary resources. Case management has thus come to mean providing selected services, coordinating the delivery of other services, and monitoring the delivery of all of them. This shift calls for skills in advocacy, teamwork, supporting the work of other professionals, and conflict management.

This chapter explores service coordination as a critical component of modern case management. We will examine the coordination and monitoring of services as well as the skills that will help you perform these roles. For each section of the chapter, you should be able to accomplish the following objectives.

COORDINATING SERVICES

- Describe a systematic selection process for resources.
- Make an appropriate referral.

- Identify the activities involved in monitoring.
- List ways to achieve more effective communication with other professionals.

ADVOCACY

- Explain why advocacy is important.
- Name two advocacy models.
- Identify ways to be a proactive advocate.

TEAMWORK

- Describe the difference between a departmental and interdisciplinary team.
- Identify two advantages of using teams in case management.
- List the barriers to good teamwork.

SUPPORTING OTHERS

- Provide examples of ways in which case managers can support the work of others.
- Describe different types of conflict.
- List the steps in managing conflict with a win–win strategy.

Coordinating Services

In the event that a client needs services that an agency does not provide, it is the case manager's responsibility to locate such resources in the community, arrange for the client to make use of them, and support the client in using them (Cohen, Vitalo, Anthony, & Pierce, 1980). These are the three basic activities in coordinating human service delivery. In coordinating services, the case manager engages in linking, monitoring, and advocating, while building on the assessment and planning that have taken place in earlier phases of case management.

Coordinating the services of multiple professionals has a number of advantages, for both the case manager and the client. First, the client gets access to an array of services; no single agency can meet all the needs of all clients. The case manager can concentrate on providing only those services for which he or she is trained, while linking the client to the services of other professionals who are trained in other areas and have the necessary resources. Second, the case manager's knowledge and skills help the client gain access to needed services. Often, services are available in the community, but clients are unlikely to know what they are or how to get them. The success of service delivery may depend on advocacy by the case manager. Also, service coordination promotes effective and efficient service delivery. In times of shrinking resources, demands for cutbacks in social services, and stringent accountability, service provision must be cost effective and time limited. In addition, customer satisfaction is all-important. Clients have a right to receive the services they need without getting the runaround or encountering frustrating confusion among providers.

Service coordination becomes key once the client and the case manager have agreed on a plan of services and determined what services will be provided by someone other than the case manager. For services that will be provided by others, a beginning step is to review previous contacts with service providers.

What services do they provide? Is this client eligible for those services? Can the services be provided in-house? What about the individual's own resources and those of the family? Family support may be critical for the success of the plan, or the client's own problem-solving skills may be helpful. A thorough case manager will not ignore the resources of the client, the family, or significant others. The next step is referral—the connection of a client with a service provider. The final step is monitoring service delivery over time and following up to make sure the service has been delivered appropriately. These steps may vary somewhat, depending on whether the services are delivered in-house or by an outside agency, but the flow of the process is likely to be the same. Before examining these steps in detail, let's review the documentation and client participation aspects of service coordination.

Documentation is critical in this part of case management. Staff notes must accurately record meetings, services, contacts, barriers, and other important information. Also during this phase, reports from other professionals are added to the case file. Any progress that occurs in the arrangement of services must be recorded by the case manager.

Client participation is important throughout the service coordination process. This entails more than just keeping the client informed; his or her involvement should be active and ongoing. First of all, the client participates in determining the problem that calls for assistance. Then the values and preferences of the client play a key role in selecting community resources, and of course client participation is critical in following up on a referral. Clients also have the right to privacy and confidentiality—without the client's written consent, the case manager must not involve others in the case or give any outsider information about it.

Resource Selection

Once client needs and corresponding services have been identified, the client and case manager turn their attention to **resource selection**—selecting individuals, programs, or agencies that can meet the needs. Paramount in this decision is consideration of the client's values and preferences. The information and referral system the case manager has developed (see Chapter 6) is useful in this regard.

Rube Manning is a 53-year-old white male who is on parole for aggravated rape. In 1986, he had sexual relations with his 12-year-old niece; she later gave birth to his son. Both parties claim that the intercourse was consensual; the severity of the charge and conviction was due to the girl's age. The girl and the family seem to harbor no animosity toward Rube, going so far as to write a letter on his behalf to the Department of Corrections. Rube was sentenced to 23 years in prison and is now eligible for parole. Angela Clemmons is the parole officer assigned this case. She and Rube must develop a plan of services for him to pursue once he is released. Among the conditions of Rube's parole are completing a mandatory sex offender program, supporting his son, and finding employment.

There are no options for the mandatory sex offender program—there is only one available in this community. Angela senses that Rube is motivated to do everything in his power to comply with the conditions of parole. Although he does not talk much about his prison experience, he does say that he didn't like it. Angela suspects that he was abused by other inmates. Sex offenders are usually on the lower rungs of the prisoner hierarchy unless they are very strong or charismatic; Reuben is neither.

Finding employment and supporting the child are tied together. Checking her information and referral file, Angela advises Rube that there are three short-term training programs that will provide him with job skills. The first two are at the vocational school and would give him a certificate in either horticulture or industrial maintenance. The third one is on-the-job training in food services, with a modest salary until training is finished. Rube's preference is horticulture, because he grew up on a farm and thinks he would feel more comfortable outdoors. He knows that industrial maintenance is a fancy term for janitorial work, and he's not interested. The location of the food services training is not on the bus line, and Rube has no transportation of his own, but this option offers a salary immediately. Angela notices that Rube sounds interested—even a little excited—about horticulture, so she checks her card file for the phone number of her contact. (See Figure 8.1.)

In this case, resource selection has been systematic, which has advantages for both the client and the case manager. The client and the case manager proceed objectively and deliberately, taking into account Rube's values, beliefs, and desires. The rationale for the choice is articulated, and it will reinforce his motivation to follow through with the referral. Rube Manning and his parole officer have chosen the horticulture program: It is on the bus line, it builds on Rube's previous farming experience, and it is something he wants to pursue.

The selection process can also accommodate many alternatives and can tailor services to the client's unique circumstances. The conditions of Rube's parole include work, and he does want the independence, salary, and respect that come with employment. However, he is not willing to do just anything. Being a janitor doesn't appeal to him, and he does not want to work indoors. Had the parole officer ignored his feelings at this point and decided to steer him toward janitorial work, Rube would probably not be motivated to do well. At the very worst, he would do nothing, and his parole would be revoked. In addition, the relationship between Angela Clemmons and Rube Manning would not develop as a partnership. Instead, their decision to try the horticultural program takes into account Rube's wishes, along with his need for training and employment.

Being aware of the client's preferences and values is critical to the success of the selection step in service coordination. There must be a strong partnership between the participants.

Agency: Lincoln Vocational Technical School

Address: 30512 Townview Parkway

Contact Person: Lynda Johnston, Admissions
Robert Griffin, Student Services

Phone: 555-1516

Services: Short-term training programs in auto mechanics, cosmetology, horticulture, industrial maintenance, printing, and secretarial services.

Comments: Good student services and advising: Janet Evans 7/1/96

Figure 8.1 Entry in information and referral file

Making the Referral

As mentioned earlier, no helper can provide all conceivable services. Therefore, arrangements must often be made to match client needs with resources. This is done by referring the client to another helping professional or agency to obtain the needed services. Referral is the process that puts the client in touch with needed resources.

A **referral** connects the client with a resource within the agency structure or at another agency. In no way does referral imply failure on the case manager's part. Limitations on the services a case manager can personally provide are imposed by policy, rules, regulations, and structure, as well as his or her own expertise or personal values.

The case manager assumes the role of broker at this point in service coordination. The broker knows both the resources available in the community and the policies and procedures of agencies. He or she acts as a go-between for those who seek services and those who provide them. Consider the following case with regard to the referral process and the broker role.

Bethany's first client on Tuesday is Anna, a young woman who has just discovered that she is pregnant. Although the agency that employs Bethany specializes in career development services, Anna feels comfortable with Bethany and wishes to discuss her options about the pregnancy with her. On the other hand, this is a difficult subject for Bethany, because her sister had an abortion three years ago and still feels guilty and upset about her decision. In fact, the whole family is still having difficulty with it, since the sister is living at home. Bethany also knows that her training is in career development, and she has never worked with anyone dealing with an unwanted pregnancy.

The encounter illustrates a situation that is appropriate for a referral. Bethany has some personal feelings that may impair her objectivity; she recognizes that she has no experience with this problem; and her agency's purpose is career development. For these reasons, she decides it is best to make a referral to someone who can help Anna explore options related to the pregnancy. Bethany will continue to support Anna's career development efforts. In the referral process, Bethany's role is that of a broker.

Making a referral may seem like a fairly uncomplicated process, but it often results in failure. If a case manager believes that all that is necessary is being aware of client needs and making a phone call, the referral is likely to be unsuccessful. In fact, studies indicate that less than two-thirds of clients referred to other community resources actually follow through and make the initial contact (Cohen et al., 1980). Research also suggests that of those clients who do make the initial contact with a community resource, 40% fail to follow through to the first interview, and about 50% drop out before service provision is completed.

There are three reasons for the failure of a referral. The first is insensitivity to client needs on the part of the case manager. Identifying the problem but failing to grasp the client's feelings about it contributes to an unsuccessful referral. The client may not be ready for referral at this point, feeling only that he or she is being shuffled among workers or agencies. Second, if the case manager lacks knowledge about resources, the client may be referred to the wrong resource. This makes him or her feel lost in the system, think that it is all a waste of time, and believe (sometimes correctly) that the case manager is incompetent. A third reason for failure is misjudging the client's capability to follow through with the referral. Suggesting to an involuntary client that she call to make an appointment for a physical may not work, perhaps because she is new to town, is unsure who to call, doesn't have a phone, or may not even want a physical.

How can the case manager make the referral process a successful one? Two keys to success are assessing the client's capabilities and forming a clear idea of the referring professional's own role (Brill, 1995). Assessing clients' capabilities means finding out how much they can do on their own. It is good to encourage independence and self-sufficiency in clients, but some of them will prove unable to identify what they need and take the steps to obtain it. The nature of the problem, the feelings the client has about it, and the energy required for action may all contribute to feelings of being alone, an inability to act, and a lack of motivation to follow through.

In addition to assessing the client's capabilities, the referring case manager must form a clear idea of what role he or she will play in the referral process. In this, the case manager should be guided by what the client needs and what relationship the case manager has with the other professional or agency. The case manager's degree of involvement in the referral can fall anywhere on a continuum—from discussing several resources with the client, who then takes responsibility for selecting a resource and following through, to giving concrete assistance with details such as making the appointment on the client's behalf and having an agency volunteer accompany him or her to the appointment.

Bethany approached the referral process in the following way. She acknowledged Anna's concern about her situation and recognized her desire for some help. She also shared with Anna her reservations about being able to assist her, explaining that her training was in career development and she had limited knowledge about options for an unmarried pregnant woman. However, she did know of two agencies that offered just the services Anna was seeking. Anna wanted to know about these, so they discussed their services and locations. Anna was concerned about the cost of services, and Bethany was unsure about the agencies' charges. Then she checked her computer file and found that both agencies charged fees on a sliding scale. Anna didn't know what that meant, so Bethany explained that such a scale determined the fee in accordance with the individual's income. Anna was unsure how to get an appointment—who to call, how to explain the problem, and so forth. She also wondered whether she would be able to continue working with Bethany on career development. Bethany discussed all of these concerns with Anna. Together, they decided on one of the agencies, and Bethany agreed to make the initial contact. Her previous work with Anna led her to believe that once the initial anxiety of making contact was over, Anna was capable of showing up for the appointment and getting the services she needed.

BETHANY: Hello. This is Bethany Douglas at Career Development. I am working with a client who needs help identifying her options with an unplanned pregnancy. Will someone at your agency see her?

RECEPTIONIST: We do provide counseling. Let me connect you with one of our counselors.

COUNSELOR: Hello, this is Carol Fong. May I help you?

BETHANY: Yes, Bethany Douglas here. I am a career counselor at Career Development. My client has just found out she is pregnant and would like to talk with someone about her options. She is 19 and single. Could we set up an appointment for her to come see you?

COUNSELOR: Yes, I would be glad to see her. Would Monday at 11:00 o'clock be okay?

BETHANY: *(Checks with Anna, who nods)* Yes, that would be fine. Her name is Anna Swenson. She will see you at 11:00 o'clock Monday. Thank you.

Bethany used several strategies to ensure that Anna's referral was a successful one.

1. Discuss with the client the services that are provided by the resource The discussion should include why the referral is needed, how it will be helpful, and how the client feels about it.

2. Make the referral This may entail just providing the client with a telephone number and an address; or helping him or her with the initial contact, as Bethany did; or taking the initiative to contact the resource. The interaction may involve scheduling an appointment, telling what the client knows about the resource, and/or finding out what information the resource needs. Of course, before any information is released, the client's permission must be obtained.

Suppose that the referral did not go as planned. When Bethany made the call, Carol Fong might have responded differently—perhaps she couldn't possibly see Anna until next month, or her agency didn't do that kind of counseling anymore. Bethany would have two options. She could return to her card file and hope to locate another agency that provides the services Anna needs. However, suppose further that this is taking place in a small town or a rural area where there aren't any other agencies to call. Bethany's second option would be to become a **mobilizer**—one who works with other community members to get new resources for clients and communities. Bethany could try to mobilize Carol Fong and other professionals so that needed services could be made available to Anna.

3. Share the referral information with the client He or she needs to know the appointment time, the location, and the name of the person to see upon arrival (Austin, Kopp, & Smith, 1986). It is also appropriate to find out what support the client might need to follow through with the appointment.

4. Follow up on the referral The service coordinator can do this by talking with the client and the helper who received the referral. Did the client show up? What happened? Was the client satisfied with the services? With the worker? Service coordinators with thorough information and referral systems make a habit of noting such information in their files. Information from the worker who saw the client may be conveyed in a phone call, a written report, or not at all.

Bethany followed up on Anna's referral by talking with her about it the next time they met. She discovered that Anna had had no trouble finding the agency, liked the worker immediately, and felt positive about exploring her options with her. Bethany received no official report from the other agency and did not request one.

The referral process is a flexible one that can be adapted for use with any client, but client participation is vital to good service coordination. Clients participate in the decision to refer and where to refer. Their capabilities determine the extent of their involvement in the steps of the referral process—making an appointment, getting to the agency, and so forth.

The case manager's role in the referral process varies from little involvement to integral involvement, depending on the client's capabilities. The case manager's responsibilities include knowing what resources are available for the client, how to make a referral, and how to assess the client's capabilities accurately. His or her involvement does not end after the referral; the next step, monitoring services, is also the case manager's responsibility.

Monitoring Services

Once the referral is made, monitoring service delivery becomes the focus of case management. Monitoring services is more than following up on the contact; it may mean offering information, intervening in a crisis, or making another referral. The case manager continues to work in the roles of broker and mobilizer throughout this phase of service coordination. In **monitoring services,** the case manager reviews the services received by the client, any conditions that may have changed since the planning phase, and the extent of progress toward the goals and objectives stated in the plan.

REVIEW OF SERVICES

Once a case manager has made a referral, delivering the needed service becomes the responsibility of the resource—the agency or professional that has accepted the referral. However, the case manager does not relinquish the case completely. He or she remains in contact with the client to ensure that the services are being delivered, that the client is satisfied with them, and that the agreed-upon time frame is maintained. As you remember, all these are specified in the plan of services.

When checking with Rachel Vasquez after her visit to the health clinic, the case manager heard about the generous time a volunteer had spent with her in making out a balanced nutrition plan for her diabetic son. Rachel was excited about knowing what to buy, how to prepare it, and why it made for a good meal. Most of all, she was impressed by how much time the volunteer spent with her.

If there are problems with service delivery, the case manager has ultimate responsibility to intervene. Problems may be caused by the agency, the client, or both. For example, the agency may prove unable to see the client for several weeks, or may neglect to do what the client has been promised. The client, on the other hand, may fail to show up for appointments or refuse to cooperate (e.g., be reluctant to give needed information). The case manager must be aware of the situation if he or she is to know that intervention is required. The intervention in such a case involves identifying exactly what the problem is and working with the client and the resource to resolve it.

Sam Miller received a call from the VA hospital where 22-year-old Raymond Fields (who was mentally retarded) had been placed as an orderly just two weeks before. Both the supervisor and Raymond had been pleased with the match. This morning, the supervisor reported that twice in the past three days, Raymond had been seen unzipping his pants and playing with his penis in the hallways. Sam hastened over to talk with Raymond about the behavior. He told Raymond to keep his pants zipped. There was no more trouble afterwards.

CHANGING CONDITIONS

Often there is a time lag between plan development and the provision of services. During this period, the case manager seeks agency approval, if necessary, and arranges for services either within the agency or at another. It is also likely that there will be changes in the client's situation during this time. Living arrangements, relationships, income, and emotions are some of the factors that may change. Also the presenting problem may show some alteration, or additional problems may surface. Any such changes may necessitate review and revision of the plan.

Winnie Jones has been seeking employment for six months. Just as she is placed in a hotel reservations office, she discovers that she must have hip replacement surgery. Therefore, there must be changes in her plan of services and a delay in her employment.

The client's circumstances may also change during service delivery. Part of service monitoring is keeping informed of changes that occur in the client's life. Some changes may occur as a result of service delivery—for example, a client might learn more appropriate ways to express anger than hitting his spouse. Other changes may have nothing to do with service delivery yet influence it. For example, a client might decide to marry while halfway through service delivery—an action that could well affect her economic eligibility for services. Again, monitoring of services helps the case manager stay abreast and be ready to intervene if necessary.

EVALUATING PROGRESS

Monitoring services also entails continually checking progress toward the goals and objectives set forth in the plan of services. Continual evaluation may lead to modification of the plan so as to improve effectiveness or deal with new developments. In monitoring services, the case manager repeatedly asks the following questions:

- Has the identified problem changed?
- Was the referral made correctly?
- Were the desired outcomes achieved?
- Should the plan be altered?
- Should the case be closed?

Monitoring of services goes most smoothly if close contact with the client is maintained. Outcome measures focus on the client, so he or she is a key source of information about service delivery. Did the client use the resource? Was the goal of the referral attained? The case manager's responsibility continues until the client's problem is resolved. Follow-up and monitoring are performed to make sure that referrals result in the desired outcomes.

The following case illustrates how a case manager monitors services by reviewing the services received, considering any changes in conditions, and evaluating progress toward goals and objectives.

It was chilly on February 17th, but the Naylors were happy. It was Presidents' Day weekend, and they were going to have three days off. Everyone gathered in the den in front of the fireplace. Jennifer, the younger daughter, was wearing a tank top and shorts to be comfortable, since she had just come down with the chicken pox. Johanna, the older daughter, had gone to look for the kitten her grandparents had given her for Christmas. It appeared to be just another quiet evening.

Johanna came in the back door about 6:30 P.M. and said, "Mom, there's a fire in the garage!" Mrs. Naylor looked out the door that led to the garage and saw flames that were at least 10 feet tall. Calmly she said, "Everybody out," and headed for the front door. All three of them made it out safely. As the Naylors stood watching the fire consume their home, they wondered what they were going to do and where they were going to go. Would they be able to salvage anything at all?

The Naylor family lost everything. The Burn Shelter in their community immediately stepped in to provide the many services that fire victims need. Bettyjean Fleming, a case manager at the Burn Shelter, was assigned to work with the Naylor family. Once notified of the fire, Ms. Fleming went to the site of the fire to help the family with their immediate needs. Comfort, clothing, a meal, transportation to the hospital, and temporary lodging are among the services the shelter provides. Case managers work with the victims to cope with any losses they may have suffered (including family members, pets, and possessions). Once the immediate needs are met, the counselor and the victims develop a plan to meet long-term needs—which will be met by other agencies.

The Naylors had a number of needs: transportation, housing, clothing, household furnishings, and counseling. One of the most serious needs was counseling for the two girls. Not only had they lost everything they owned, but also every picture and memento of their father (who had died a year before). They desperately needed help in coping with the loss of their father, and now the loss of their home, their possessions, and their pet. Ms. Fleming learned of these circumstances from Mrs. Naylor, and they developed a plan for the long-term services. She was supportive of counseling for the girls; together, they discussed several alternatives, using the file of community services at the Burn Shelter. They decided on counseling services at the local mental health center, which is known particularly for its children's services. Mrs. Naylor agreed to call for an appointment. Unfortunately, she was told that the earliest appointment was next month; she didn't see any alternative, so she took that appointment. The following week, Ms. Fleming called to see whether Mrs. Naylor had received her insurance check. In the course of the conversation, she learned about the delay in getting counseling. She alerted Mrs. Naylor to the emergency services that were available and volunteered to call and talk with a counselor. Ms. Fleming often worked with staff at the mental health center, so she called one of her contacts, Frances Lane. Ms. Lane agreed to see the girls that very week and requested some background information, which was provided. Mrs. Naylor was grateful for the intervention since there had been no improvement in the girls' mental state.

At the next meeting with Mrs. Naylor, Ms. Fleming asked about the visit to the mental health center. Were they seen on time? How did the girls feel about Ms. Lane? Were they feeling better? What were the next steps? At the same meeting, Ms. Fleming asked again if the insurance check had arrived, how the apartment was working out, and whether Mrs. Naylor or the girls had other needs.

In this case, the case manager's responsibility did not end after the referral. Because she monitored service delivery, Ms. Fleming became aware that there were problems. She was able to intervene, using her contacts to make a difference in service delivery. She also monitored the delivery of the other services, verifying that the identified services were delivered within a reasonable time and by the professionals designated to delivery them. Not keeping abreast of developments in a case can result in delays in needed services (or even nonperformance). Then clients become frustrated and dissatisfied with both the resource and the case manager.

Working with Other Professionals

Clearly, effective service coordination depends to a certain extent on the case manager's relationship with other professionals. Both referrals and service monitoring are more easily achieved when the case manager has a relationship with the resource. The professionals on whom a case manager relies have a wide variety of cultural backgrounds, academic achievements, and job descriptions. "The most important challenge is to achieve effective mutual communication between persons of different orientation or background" (Committee on Mental Health Services, 1981, p. 161). This means providing information, being persuasive, and exercising authority when needed.

Good communication skills are critical when working with personnel from other agencies. These skills can be the deciding factor in making effective use of resources on the client's behalf. Here we present certain suggestions for enhancing communication with other helpers. First, avoid stereotyping other professionals. You may have encountered one nurse who was rude, but it is unreasonable to think that all nurses are that way. Second, don't hesitate to ask for clarification or a definition of terminology that you don't understand. It is better to ask than to pretend you know. Third, you can help others learn your own terminology by using it and explaining its meaning. Finally, be aware that other professionals may well have different styles of communication. For example, a clinical style may be more comfortable for psychiatrists. Other styles that have been identified are legal (of equal adversaries), political (of unequal adversaries), and pedagogical (teacher–student).

The Committee on Mental Health Services Group for the Advancement of Psychiatry (1981) lists 15 ways to achieve more effective communication between psychiatrists and decision makers outside the mental health field. We believe that these 15 points also work for case managers, particularly as they engage in service coordination.

BEFORE THE ENCOUNTER

1. Do your homework. Be as familiar as possible with the issues at hand and the answers you need to have.
2. Be conscious of the other person's circumstances—his values, his "language," his commitments—and respect them.
3. Be sensitive to the process of the interaction, especially the rules and practices of the specific situation in which you are involved.
4. Know your limits. Don't expect to be all things to all people. Use consultants or other resource persons who are familiar with the problem and the situation, and who can help you present your case.

DURING THE ENCOUNTER

5. Identify mutual concerns and common goals of both parties. Seek to develop an alliance for the purpose of solving the problem.
6. Define the issue and stick with it. Don't digress.
7. Listen. Keep cool, and don't argue. (At times, noncommunication may be the best communication).
8. Use simple language and concrete, familiar examples. Avoid technical jargon and difficult-to-understand abstract examples.
9. Don't bluff. Feel free to say frequently, "I don't know" and "I'll find out."
10. Keep your sense of humor but direct it at yourself, not at your questioner.
11. Keep your narcissism and self-righteousness in firm check.
12. Don't expect to win them all. At the point of impasse, back off and seek mediation.

AFTER THE ENCOUNTER

13. Remember that additional encounters probably will occur in the future. Review what happened and learn from it.
14. Do not leave unresolved issues until the next time. (Some of the most fruitful exchanges can take place at times other than when an encounter is in progress.)
15. Informed contacts and meetings with decision makers should be encouraged. (Even if issues are not on the "agenda" or resolved,

> pathways for communication can be established by getting to know each other.) (pp. 162–163)

Using these 15 points can facilitate the work of case managers with other helping professionals. Being a skillful communicator and having good working relationships with other helping professionals enhances the case manager's role as a client advocate (which is discussed next).

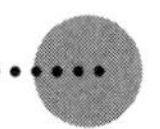

Advocacy

Advocacy is speaking on behalf of others, pleading their case or standing up for their rights. When case managers act as advocates for their clients, they are supporting, defending, or fighting for another person or group. Advocacy is also related to client empowerment and participation, as case managers support clients' involvement in decisions about their own treatment and welfare. A caseworker at the Third Avenue Family Service Center in the Bronx describes a situation faced by AIDS patients that illustrates the need for advocacy:

> It is called temporary housing, and as long as the client is living everybody is safe. If the client dies, then the family has to relocate. I have found that to be very unfair, because what happens to the children? There is no primary planning. . . . Nobody is working on this issue yet. (Carolyn Brown, personal communication, May 5, 1994)

Ellis and Hartley (1991) identify two models of advocacy: the **responsible model** and the **adversarial model.** In the responsible model, the advocate uses methods such as negotiation, compromise, and persuasion to support clients' cases. In the adversarial model, the advocate concentrates on client needs, with little concern for the context of the situation or the needs of others involved.

Advocacy is very important for case managers. Often, clients are unable to articulate what they need, nor do they understand what choices are available to them. They may not have the necessary information or the skills needed to present their positions. In some situations, what they desire is in direct conflict with people in authority. Clients may then be too intimidated to speak for themselves, or the people in power may refuse to consider client wishes.

Terry is a 14-year-old who has just been diagnosed HIV positive. She is terrified of her medical condition and will not talk to anyone about it. She just sits and stares when the caseworker, her mother, or any other member of her family tries to address the subject. Her mother wants to send her away to a residential school, but Terry will not speak of this or any related issue.

The case manager acts as an advocate for the client.

In some agencies, the people in charge may not wish to hear client complaints or grievances. There may be no way for clients to appeal decisions or discuss methods of treatment they do not support.

Clients, by definition, are not often in a position to act as advocates for themselves. Advocacy requires confidence, a feeling of control, and an understanding of the system. Most clients find themselves in the human service delivery system because they do not have these characteristics. According to Austin, Kopp, and Smith (1986) many clients are "too uninformed, too apathetic, or feel too powerless to do anything when services are denied them or when they are not being served satisfactorily" (p. 120). One caseworker at a school for the deaf tells how she remained an advocate even after her official responsibilities were completed.

> In this particular instance, we had made more than one report to Human Services, and Human Services was already involved, but then closed the case. Although our reports were not the reports that reopened the case, the case was reopened and the children were

> removed. . . . What is unique about this particular case is that usually when they are removed, I don't have any further contact with the families . . . but I just knew that the kids were going to be returned, and I felt like I needed to maintain that relationship, because I would be making recommendations to the court for return. (Personal communication, October 27, 1993)

The following are some common client problems that case managers may consider appropriate for advocacy:

- Client has been denied services, or services have been limited.
- Client has expressed interest in one method of treatment but has been given another.
- Client's family has made decisions for him or her.
- Client has little information about the assessments gathered or the decisions made.
- Client has been denied services based on factors such as race, gender, or religion.
- Client has been given treatment contrary to his or her cultural norms.
- Client has been treated with disrespect by human service professionals.
- Client is caught in the middle of a conflict between two agencies or professionals.
- Client is being given unsafe or indifferent care.
- Client does not know what his or her rights are.
- Rules and regulations do not serve the client's needs.

Case managers are well positioned to provide advocacy for clients they serve (Ellis & Hartley, 1991). In the first place, case managers know their clients. They are most familiar with their skills, values, and wishes and the treatment they have received. A measure of trust has been built up between the client and the case manager, so the case manager is likely to hear from him or her if there has been unfair treatment. Case managers are also in contact with family members, friends of the client, and other professionals with whom the client is involved. The case manager therefore has immediate access to anyone who may be involved in any unfair practice. Also, in the process of monitoring service delivery, the case manager may discover situations that warrant advocacy.

Advocacy is not an easy role to perform. The case manager must resolve certain issues before assuming the advocacy role (Austin, Kopp, & Smith, 1986). First, the case manager must gather the facts of the situation and then determine whether there is a legitimate grievance—whether making requests on behalf of the client would be justified. Then the case manager must determine whether advocacy is appropriate, rather than other methods such as problem resolution or conflict resolution. Third, the case manager must discuss the need for advocacy with the client and show willingness to speak for the client. He or she must have the client's approval before any advocacy takes place.

How to Be a Good Advocate

Many case managers do not like conflict, and they fear that, as advocates, they will not be successful in meeting the needs of their clients. Advocacy is complicated because case managers have divided loyalties to the agency, the supervisor, and the client. Personal values and beliefs of the case manager may complicate the advocacy process. The following guidelines support their work of effective advocacy. These guidelines build on the knowledge, skills, and values presented in our previous discussion of the assessment, planning, and implementation phases of the case management process.

Know the environment in which the conflict takes place. This is important when planning how to act as an advocate for the client. It is helpful to know the appropriate person to whom to make the appeal. Having a good referral network and a good relationship with many agencies is useful in understanding the environment.

Understand the needs of the client and of the other parties involved. It is helpful if the advocate can understand why the other parties are in conflict with the client or seem not to respect the client's wishes. If the advocate has this information before the appeal, strategies can be developed to meet some of the needs of all involved parties or, at the very least, articulate the common ground.

Develop a clear plan for the client. The advocate needs to be clear about what the client needs and what the other parties must give or give up to meet those needs.

Use techniques of persuasion when appropriate. Many situations are suited to persuasion—using the responsible model for advocacy. Persuasion is used to support client needs while respecting the rights of other parties as well. Persuasive techniques are similar to the ones discussed in relation to assertiveness in Chapter 11. They include stating the problem clearly, presenting critical background information and facts, explaining why the situation needs to be changed, and detailing an acceptable solution.

Once there is agreement, it needs to be stated or written, and all parties need to agree.

Use more adversarial techniques when persuasion proves ineffective. Using techniques that challenge the system directly, the case manager makes a formal appeal or takes a case through legal channels in an effort to promote change on behalf of the client. In a grievance procedure or a legal challenge, the client has an opportunity to be heard, either verbally or in writing. Usually the client's case is heard by an impartial person or group, and a ruling is made. Another adversarial technique is to use the media to take the client's case to the public.

In the case of Terry, her mother wants a solution to the problem right away. She is determined to send Terry away and has already made contact with three residential programs. She plans to move Terry somewhere as soon as possible. In reality, Terry's mother is frightened by what lies ahead for Terry and for the family. The only way she knows to cope is to distance herself from the problem. The caseworker feels strongly that Terry should have a voice in the decision; she begins her advocacy work by trying to talk with Terry about what she wants.

There does not necessarily need to be a special circumstance for case managers to assume the advocacy role. In the course of performing case management responsibilities, there are many opportunities to act as an advocate for the client. The case manager then integrates advocacy into other case management responsibilities. This is called **proactive advocacy,** and the following considerations should be observed in applying it.

- Provide quality services by involving a team of professionals and the client (or, when appropriate, a member of his or her family).
- Interact with the client to plan treatment that is congruent with his or her values and cultural orientation.
- Monitor the case and set goals and outcomes based on quality standards of professional care.
- Continually communicate with other professionals about issues that relate to client rights.
- Plan treatment that takes into consideration the client's preferences, strengths, and limitations, and provide additional support if you anticipate that he or she will have difficulty.
- Speak for the client only when he or she gives permission.
- Educate the client about the agency's policies and procedures.
- Create an environment that facilitates decision making by the client.
- Educate the client about options in treatment and about the process of making the treatment plan, and discuss the barriers that may be encountered during the implementation phase.
- Work within the system to support, modify, and create policies. Know which professionals are involved in this work and discuss your opinions with them. Volunteer for committees where policy issues are considered.
- Become involved in the political process. Get in contact with public policymakers.

Advocacy work is difficult, but being a proactive advocate makes it easier to act as an advocate when the client's rights have been violated or ignored. The case manager must also exercise good judgment in choosing when to attack barriers to the client's cause (Ellis & Hartley, 1991). At times, the need for advocacy may not be clear cut.

Martha Severn has just been asked by one of her clients not to report the $10 a week that she makes taking in laundry. Reporting this income would mean *that the client will lose some of her scholarship aid for school. Instead, the client wants Martha to try to change the eligibility rule. Although changing the rule might seem a special favor for this particular client, Martha happens to believe that this policy is a good one.*

In other instances the case manager may have to deal with competing interests—those of the agency and fellow staff members, as well as the client's.

James Dowling is a student intern at a mental hospital. He believes that one of his clients is being abused by a night technician. His client tells him of beatings during the night shift.

Ms. Wise is an elderly client who receives attendant care at home. The agency that coordinates Ms. Wise's care has a policy of keeping clients on home services for as long as possible. Cheryl Santana, a case manager at the agency, believes that clients remain much too long in home care. Cheryl makes recommendations for residential care, and the agency routinely rejects those recommendations.

Case managers must be aware of how their advocacy efforts are perceived by others. Many case managers believe that their first loyalty belongs with the institution. Others feel that they need to support the efforts of the team. Vigorous advocacy, at the expense of team camaraderie, may jeopardize the client's trust in the team. Also, it is difficult to act as an advocate for certain clients. Some people *are* dishonest, greedy, or troublemakers. With such clients, the case manager must think clearly about the legitimacy of any client demands and approach the issues with fairness in mind.

Teamwork

As mentioned earlier, the implementation phase involves working as a team. In meeting the needs of clients who have multiple problems—in child welfare, developmental disabilities, the elderly, and many other client populations—a coordinated team approach is necessary, since several professionals are involved. Sometimes referred to as the **treatment team,** this group of professionals meets to review client problems, evaluate information gathered, and make recommendations about priorities, goals, and expected outcomes. Kim Elhers, who works with adults with disabilities, describes the initial staffing for a new case.

> If there is no waiting list, then we set up a time when they can come. We do an initial staffing. We call a professional from each area. We get

> the supervisors together who are going to decide which instructor will work with this individual—the vocational instructors could have them, and/or the residential instructors could have them. (Personal communication, October 14, 1995)

There are numerous advantages of using a team to make decisions. Working as a team, professionals can share responsibility for clients as well as the emotional burdens of working with clients who have difficult problems. With a group of professionals involved, all the dimensions of the client's situation are more likely to be considered, and team members can get each other's viewpoints about the advantages and disadvantages of each decision made. In the team setting, helpers are also able to share their expertise and knowledge as they focus on each client's unique set of needs and circumstances.

Types of Teams

One type of team that is used in case management is the **departmental team,** made up of a small number of professionals who have similar job responsibilities and support each other's work. Colleagues bring their most challenging cases to the team, and their co-workers help identify client problems and generate alternative approaches to treatment. Kim Ehlers again talks about her work in staff meetings, this time with a departmental team.

> We were really excited about moving clients into their own homes, but we were not really prepared. . . . Now we have a team meeting as soon as we decide somebody is going to move, and we figure out who will assume what responsibility. (Personal communication, October 14, 1995)

Departmental teams are particularly useful when decision making is difficult or client problems create a stressful situation for case managers—for example, in child welfare work or cases of domestic violence (Wells, 1985). Within both of these settings, the departmental team shares information, offers opinions, and often makes group decisions about how to work with clients.

Thom Prassa, a case manager who works with children, describes the importance of teams: "I think the number one thing for me, why I have survived these last two and a half years, is really establishing relationships with your team, your other case managers" (personal communication, July 10, 1994).

Within the departmental team, the case manager can fill either of two roles: leader or participant. When assuming the leadership role, the case manager presents a case for review. He or she describes the client and the current status and summarizes conclusions and decisions made to date. The material is presented in such a way as to encourage feedback and dialogue with other colleagues. When the case manager is in the participant role, another colleague is presenting a case, and the case manager's role is to listen, study the situation, and provide advice and counsel.

An **interdisciplinary team** is a different way to work with other professionals to provide services to the client. As the name suggests, this team includes professionals from various disciplines, each representing a service the client might receive. Often it includes the client or a member of his or her family. Early in the treatment of a client, the interdisciplinary team gathers and shares data, establishes goals and priorities, and develops a plan. In the latter stages of treatment, the interdisciplinary team monitors the progress of the client, revises the plan, and often makes decisions about aftercare. The case manager is often the team leader, giving other team members a holistic view of the client, ensuring that the client or a member of the family is heard or representing their viewpoints, and conducting an assessment of client problems. In addition, the case manager is expected to monitor client progress between meetings, set the agenda for the meeting, give a summary of client progress, discuss next steps and any dilemmas, and help reconcile differences of opinion. At times, the team leader is placed in a difficult situation, since each helper at the table has credentials in a specialized area of professional expertise. All of them are accustomed to managing cases, being leaders, and making assessment and treatment decisions on their own.

Interdisciplinary team meetings become even more complicated when clients or family members attend (Caires & Weil, 1985). Having families and clients present leads to a more complete picture of the case, and client involvement is enhanced when he or she works with a whole team of professionals to set goals and determine priorities. On the other hand, when the client is present, many professionals are reluctant to share their knowledge and views. As team leader, the case manager has the responsibility to protect the client in these discussions while encouraging an open dialogue (Glenn Graber, personal communication, May 9, 1996).

One outcome of initial interdisciplinary team meetings is an organized, well-integrated plan designed to meet the goals that have been established to meet client needs. The Individual Family Service Plan for Bill, Linda, and Jane Smith (presented in Chapter 3) is a good example. One component of the plan was the designation of responsibility to the professionals providing each of the services listed.

Once the treatment plan is being implemented, interdisciplinary team meetings are scheduled to help monitor the client's progress. Each of the professionals involved presents a progress report, and together they make a decision about how to proceed. The case manager often meets with each of these professionals one-on-one before the meeting. In crisis situations, he or she may have to revise the plan without team approval. Interaction with other professionals can sometimes be difficult if they have not provided the services for which they are responsible or if there is some question about the quality of service delivered. The case manager is most often not their supervisor, so when such issues arise, he or she manages by persuasion and collaboration.

Leading a Team

Working on a team is an exciting experience for most case managers. They welcome the creative thinking and the support that comes from a collaborative effort. But building an effective team requires the efforts of all the members, especially the case manager. In most interdisciplinary teams, the case manager has the responsibility for leading the team and developing an atmosphere conducive to good collaboration. Truitt (1991) lists three conditions that must be met for a team to function well. First, there must be a common goal. In the case of teams involved with case management, the goal is the successful development and implementation of a plan that meets client needs. Second, each of the team members must have respect and trust for the others. Respect is important, since teams share in decisions that can radically alter clients' lives. Mutual respect is especially important in interdisciplinary teams, since each member is relied on to bring knowledge and skills in a particular area of expertise. In the case of departmental meetings, the participating colleagues continue to work side by side, so respect is important to maintain.

The third condition is an understanding and acceptance of the system of rewards, discipline, and work sharing (Truitt, 1991). This represents the context and structure in which the work is to be accomplished. There is a danger of team breakdown if team members do not view their collaborative work as a part of their job for which they are accountable. They must not think that their individual performances are of primary importance and the team efforts secondary. For a team to work effectively, all members must understand that working on the team is one part of their responsibility and that the agency expects them to treat this work seriously. In an interdisciplinary team effort, special attention needs to be paid to this aspect, since the team members are from several units within an organization or possibly from numerous organizations.

For the case manager who is involved with teams, several aspects of teamwork can directly improve the services provided and the work environment of the professionals involved. First, the clients receive services from several professionals working together. The greater the total of expertise, creativity, and problem-solving skill applied, the more effective will be the planning and delivery of services. Each professional is able to perform his or her responsibilities better because of having participated in the process of setting goals and priorities as well as planning. The professionals also have a better sense of the client as a whole person, and they have a clear conception of how their own particular treatment is integrated into the larger plan.

Teamwork also enhances members' skills in making decisions and solving problems creatively. Not only do they have opportunities to practice those skills; they can also learn from the other professionals. The environment that is fostered by teamwork is valuable to any helping professional, but especially to the case manager who is coordinating services. Better communication skills are developed in a team atmosphere; members learn to listen well and to speak to the group when appropriate (Lauer, 1996). The opportunity to share responsibility—to ask for assistance, to volunteer or give it, and to receive it—helps all team members by increasing their sense of community and reducing isolation (Lauer, 1996; Miller, 1988). At the Jewish Home and Hospital for the Aged, one case manager used team meetings to discuss difficulties, and it brought positive results.

> We would talk about difficulties, because it is very difficult to deal with some of the residents. I mean, they are not the most pleasant sometimes, and if they are sick and dying and demanding on top of that, it becomes really a tremendous burden on staff . . . so we created an open forum where people could say "I can't take it." (Personal communication, May 3, 1994)

Effective Teamwork: Barriers and Solutions

Just as there are advantages to working in teams, there are also some barriers, involving issues of trust, competence, responsibility, and communication skills (Lauer, 1996). If the case manager is functioning in a departmental or interdisciplinary team, such barriers present difficulties in providing effective services and challenges for developing a positive working environment. Let's look at each of the barriers, as well as possible approaches to overcome them.

Trust This is a condition that is basic to the group. Without trust, the members of the group have great difficulty being open with their own ideas and remaining open to the ideas of others. In the work of an interdisciplinary team, trust can be violated in a number of ways. For example, one member might discuss the case outside the team setting with professionals and friends who have no reason to know the details. Violation of trust also occurs when one member speaks badly of the work, ideas, or competencies of the group or of a member. The case manager can address the issue of trust in two ways. At the initial meeting of the team, the ground rules and expectations are set for the team, including standards with regard to confidentiality and respect for other group members. The case manager may later find that one member of the group is complaining about the work of the group or the work of another member. The case manager can handle such an incident in a straightforward manner by referring to the ground rules. Although confrontation is difficult, solving the problem in a timely manner will prevent damage to the team's future work.

Competence In interdisciplinary groups especially, a member sometimes lacks the competency (knowledge, skills, or values) to perform the work required, perhaps because of lack of training or experience. Within the framework of the interdisciplinary team, members depend on each other to fill specific roles and accept particular responsibilities. For the good of the client, any failure to do so must be addressed quickly. The case manager's position is difficult: lacking a supervisory relationship with the struggling team member, but having responsibility for the effective functioning of the team. One approach is to share with the team member observations about the quality of work thus far, pointing out strengths and weaknesses. It is helpful to give clear examples of what has not worked and how this has influenced the services provided. The case manager can also encourage improvement by offering to support the team member in writing a professional development plan.

Responsibility Unfortunately, some team members do not complete their tasks in a timely manner. Such breakdowns in responsibility occur for a variety of reasons. Helpers are often overwhelmed by their workloads, and they cannot complete all of their assigned tasks. Also, absenteeism, illness, or crises in other areas of work may prevent them from completing assignments. But each part of the treatment plan is important, and many other components of the plan are affected by any breakdown in one part. As the team leader, the case manager must help each member realize that if his or her work is not performed in the allotted time, the team must be informed (through the case manager) so that other arrangements can be made. Monitoring is important. The case manager must know where each team member stands with regard to his or her responsibilities and urge everyone to complete their work or make alternative arrangements.

Poor communication The case manager establishes the standards of communication for an interdisciplinary team. He or she also serves as a role model by listening well, speaking clearly and distinctly, demonstrating respect for the ideas of others, and ensuring that each of the group members speaks and is heard. Truitt (1991) suggests that the team leader is successful in communicating when five steps are followed:

1. Be friendly. Use a cooperative approach when opening and running the meeting.
2. Keep it simple. The agenda should be clear and concise, and the discussion should be organized. The outcome of the discussion should be clear to all who are present. The issues are likely to be complex, so the approach must be easy to follow.
3. Give a rationale. Make sure that any discussion includes not just what is decided but also the rationale for the decision. Others will ask members of the team why certain decisions were made, so each member of the team must understand the reasons.
4. Get feedback. The meeting is the right time to clear up questions, to gather suggestions, and to discuss what action each member will take following the meeting.
5. Follow up. The case manager must follow up to ensure that progress is being made. It is also helpful for all team members to receive a report of the progress made.

These guidelines will help the case manager run a smooth and effective team. Many of them are also useful when communicating with other professionals who are outside the team structure; these people can be valuable resources for facilitating the case management process. Because communication and relationships are so important, let's consider how to make and strengthen good collegial affiliations.

Supporting Others

The coordination process relies heavily on working and communicating with other professionals. The work of assessing the information gathered; finding, developing, and organizing resources; and monitoring client progress is an intensive, people-oriented process. For the success of this work, the case manager must have good relationships with other professionals. There are several ways in which case managers can support the professionals with whom they work. Such support helps establish and maintain a professional network. How exactly can the case manager provide such support?

Being a Role Model

A relationship with another professional can involve acting as a role model. Even people who have much more experience and education can learn from beginning case managers who are motivated to perform well and are always seeking ways to improve the services delivered to clients. Four characteristics contribute to a person's status as a role model for other professionals: enthusiasm, creativity, flexibility, and willingness to act (Ellis & Hartley, 1991).

Enthusiasm Showing a strong, positive attitude is important for a case manager. Among the hallmarks of enthusiasm are curiosity, willingness to learn, excitement about the responsibility, and belief in the possibility of changing clients' lives. It may be as simple as making positive remarks, such as "This can be done," "I will try to do it," or "I think we can meet the needs of this client." In dialogue with clients and professionals, it means taking positive action and focusing on solutions. Enthusiasm is contagious. When a case manager calls on another service provider with a request for assistance, enthusiasm helps create the expectation that that service provider will want to be part of the team. Of course, it is still important to be realistic and set goals that have a reasonable chance of succeeding. When hiring case managers, Suzy Bourque looks for people who will be truly caught up in their work. She just recently hired a man who "is really intuitive and insightful and respectful, and he is thrilled." She looks for someone who is "going to enjoy the work" (personal communication, October 7, 1994).

Creativity If a case manager wants to support the work of colleagues and is committed to maintaining a network, he or she must be creative—have the ability to gather relevant thoughts, facts, and ideas and fashion them into a new form. This includes generating new ideas, visualizing, and foreseeing, which can be developed over time. Innovative approaches to challenges (especially those involving limited services and multiple client needs) help everyone in the process serve clients more effectively. New ideas and alternative ways of viewing situations are constantly required of the case manager, to accommodate the many perspectives involved—those of the client, the family, the various other professionals, and the agency.

Flexibility The case management process must deal with constant change, especially during the implementation phase. The case manager needs to be flexible—able to adapt to changes in client needs, the service environment, available resources, personnel, and policies and regulations. Flexibility is necessary, for treatment plans are seldom static; change must be seen not as a barrier, but as part of the process. The case manager who regards changes as opportunities for growth will be more likely to have positive results. A case manager for a health system describes a case that required flexibility.

A few days ago, I wrote a plan that has changed some since I did the assessment, and she signed it. We have been unable to find an attendant, and she said, "By the way, I have a mother nearby who is fairly able and can assist me." . . . So now we find someone to relieve mom. (Alan Tiano, personal communication, October 7, 1994)

Willingness to act Another characteristic that motivates the work of other professionals is willingness to take action. Much of the case management process involves preliminary activities such as establishing relationships with the client and family, gathering information, and working with the client and others to form a plan. Then it is time to implement the plan. There is never a perfect time to act within this process; never has *all* the needed information been gathered or a *complete* understanding of the client's problem achieved. Being willing to put a plan in motion takes confidence on the part of the case manager; it is his or her responsibility to take the lead and set implementation in motion when appropriate.

Being a role model for others is one way for a case manager to support the work of colleagues. Positive characteristics such as enthusiasm, creativity, flexibility, and willingness to act help maintain a positive relationship with colleagues. Sometimes, however, the interaction breaks down, and conflict arises. Good conflict management is another way to support the work of others.

Managing Conflict

Even within a professional environment, conflict sometimes arises. The case manager can support others by recognizing that conflict is an inevitable part of the process and participating in its resolution (Ellis & Hartley, 1991; Smith, 1992). Cushnie (1988) suggests conflict can arise from any of four categories.

Facts There can be disagreement about data or facts. One of the best ways to resolve such conflicts is to gather more data from sources that conflicting parties agree are credible. In the case management process, this can happen during the assessment phase if there are differing opinions on the status of the client.

Methods Just *how* something is to be accomplished can be a matter of contention, since professionals often have differing standards, ways of providing services, and problem-solving approaches. One way of resolving conflicts about methods is to discuss what criteria should apply to the particular situation. This type of conflict is likely to occur within an interdisciplinary team setting.

Goals There can be a disagreement about the outcomes of the process. Professionals may have different perspectives on priorities, commitment to short-term or long-term care, the empowerment of clients, and the involvement of families. The case manager can help the team establish common ground on which to carry out the goal-setting activities.

Values Beliefs and personal philosophies differ. Conflicts of this nature are often deeply felt. To solve them, all parties must be committed to understanding each other's points of view. They must show respect for others' beliefs and find a common goal. In particular, the case manager must deal with thorny conflicts among professionals who feel that they represent the values of their particular disciplines.

According to Steve Martin, an expert in conflict resolution, several types of conflicts are common in human service settings (personal communication, August 23, 1995). The case manager may encounter any of these during the process. For example, **intrapersonal conflict** is conflict *within* an individual. In child welfare work, the case manager may experience intrapersonal conflict while deciding whether an abused child is to be returned to the home. There are conflicting feelings associated with the negative fact that the child was abused by the parents and the positive fact that the parents have since participated in counseling. **Interpersonal conflict** occurs between individuals. Such conflict can easily occur during departmental and interdisciplinary meetings because of people's differences in terms of facts, methods, goals, or values. **Intragroup** conflict is that which emerges within an established group—for example, within a family whose members are not committed to the same goals. Case managers working with elderly people often encounter families who do not agree about residential versus home care for the aging parent or relative. **Intergroup** conflict is a struggle between groups. For example, human service agencies may come into conflict as they compete for resources and for clients.

Stages of Conflict and Strategies for Resolving It

According to Truitt (1991), there are several stages of conflict development, regardless of the intensity or the subject of the conflict.

Stage One. Individuals are sure of what they want, and they set out to get it. In the process, they discover that something or someone creates a barrier to the achievement of their goal.
Stage Two. Both parties are very frustrated by the conflict.
Stage Three. The parties' dissatisfaction with the situation and inability to lessen the conflict are expressed by blaming each other.
Stage Four. The parties express their anger and frustration by saying and doing things. These reactions are based on each party's perception of what has occurred.
Stage Five. A cycle of events is created: One party acts and the other reacts. This cycle continues.
Stage Six. One of the parties, or someone outside the conflict, notices the difficulties and intervenes. This intervention is called conflict management.

There are effective and ineffective ways to manage conflict. Experts agree that the most effective conflict management incorporates a win–win strategy for all parties involved (Lauer, 1996; Smith, 1992; Truitt, 1991). Smith suggests several guidelines for using a win–win approach.

- First, begin to think and talk about the problem in terms of the needs of the conflicting parties. Many conflicts arise as individuals focus on outcomes and the results that they desire. Looking at needs instead of outcomes allows the participants to begin work on mutual problem solving.
- After there has been a discussion of needs, the participants commit to a mutual effort to solve not only their own problems but those of the other participants. It is critical that all participants be actively involved in the process, not just sitting listening to others.
- Once this much agreement has been reached, the conflict should be described in terms that are as specific as possible. Each participant can ask whatever questions and present as many facts as necessary to define the conflict clearly. In this step, individuals begin to understand that the event causing the conflict does not look the same to them as to others.
- Next, begin to identify all the differences there are between the participants. In doing this, they become able to articulate the conflict from the other participants' points of view. Mutual understanding can now take form.
- Now focus on finding solutions to the conflict. The participants brainstorm possible solutions and then evaluate each according to its potential for meeting the differing needs previously stated.
- Once an agreement is reached about how the conflict is to end, a plan is developed, with actions and responsible parties clearly stated.

The benefits of addressing conflict are numerous. Once the problem is identified and individuals are committed to working on an issue, there arises an atmosphere that promotes change. Within the human service delivery system, many improvements can be made to enhance client services. Conflict management can help resolve some of the shortcomings. Often, individual roles and responsibilities become clearer as problems are better defined. The roles of case managers can sometimes be fuzzy, and they work with a great deal of autonomy. Their work crosses department lines within their own organizations as well as in relation to external agencies. Conflict resolution represents a good opportunity for establishing clearer working relationships.

Managing conflict also improves communication. Individuals develop ways of relating to others with whom they have had difficulty, as well as better communication skills. They experience the positive outcome of facing conflict and using it as a catalyst for making a situation better. The case manager who has a positive approach to other colleagues and who is able to manage conflict always contributes to an atmosphere in which his or her colleagues will want to cooperate.

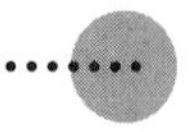

Chapter Summary

Service coordination is a key to the case management process. Knowledge of resources and a network of colleagues serve as the basis for referral and monitoring of services. Advocacy plays a part in the coordination process when the client does not receive the needed services or when his or her rights are ignored or violated. In service coordination, communication with others, especially other professionals, is all-important. This is particularly true when the case manager is working on a team to develop goals, priorities, and strategies to meet client needs. The case manager needs special skills to lead a team, overcome the barriers to teamwork, inspire confidence, and manage conflict.

Chapter Review

Key Terms

Resource selection
Referral
Broker
Mobilizer
Monitoring services
Advocacy
Responsible advocacy model
Adversarial advocacy model
Proactive advocacy
Teamwork
Treatment team
Departmental team
Interdisciplinary team
Conflict
Interpersonal conflict
Intrapersonal conflict
Intragroup conflict
Intergroup conflict

◆ *Reviewing the Chapter*

1. Name the three activities of service coordination.
2. What are the benefits of service coordination?
3. How would you use a systematic resource selection process when making a referral?
4. Under what circumstances does a case manager refer a client?
5. Discuss three reasons why referrals fail.
6. What are the steps to a successful referral?
7. Describe how the roles of broker and mobilizer apply to the monitoring of services.
8. What are the three components of monitoring services?
9. State some guidelines for working with other professionals.
10. Define *advocacy.*
11. Describe the two models of advocacy.
12. What are some guidelines for good advocacy?
13. What different types of teams are encountered in case management?
14. What conditions must be met for a team to function efficiently?
15. Illustrate how trust affects a team.
16. List barriers to teamwork.
17. What are some approaches to overcoming these barriers?
18. List guidelines for a successful team leader.
19. How do case managers support the work of others?
20. Discuss the conflicts that are most likely to occur in human services.
21. Suggest some ideas for managing conflict.

◆ *Questions for Discussion*

1. After reading this chapter, what evidence can you give that coordinating services is a critical component of the case management process?
2. Why do you think advocacy is an important responsibility for a case manager?
3. Suppose you were a leader of a case management team. How would you conduct your first meeting?
4. Do you believe that conflict should be managed? Why or why not?

References

Austin, M. J., Kopp, J., & Smith, P. L. (1986). *Delivering human services.* New York: Longman.

Brill, N. (1995). *Working with people: The helping process.* White Plains, NY: Longman.

Caires, K. B., & Weil, M. (1985). Developmentally disabled persons and their families. In N. Weil & J. Karls (Eds.), *Case management in human service practice* (pp. 119–144). San Francisco: Jossey-Bass.

Cohen, M. R., Vitalo, R. L., Anthony, W. A., & Pierce, R. M. (1980). *The skills of community service coordination.* Baltimore: University Park Press.

Committee on Mental Health Services Group for the Advancement of Psychiatry. (1981). *Interfaces: A communications case book for mental health decision makers.* San Francisco: Jossey-Bass.

Cushnie, P. (1988). Conflict: Developing resolution skills. *American Operating Room Nurses Journal, 47*(3), 732–742.

Ellis, J. R., & Hartley, C. L. (1991). *Managing and coordinating nursing care.* Philadelphia: Lippincott.

Lauer, M. (1996). *Team building: Training the wheels together.* Unpublished paper, University of Tennessee, Knoxville.

Miller, P. M. (1988). *Powerful leadership skills for women.* Shawnee Mission, KS: National Press.

Smith, S. (1992). *Communications in nursing.* St. Louis, MO: Mosby.

Truitt, M. R. (1991). *The supervisor's handbook.* Shawnee Mission, KS: National Press.

Wells, S. J. (1985). Children and the child welfare system. In N. Weil & J. Karls (Eds.), *Case management in human service practice* (pp. 119–144). San Francisco: Jossey-Bass.

Zussman, J. (1982). Think twice about becoming a patient advocate. *Nursing Life, 6,* 46–50.

Chapter Nine

Working within the Organizational Context

I saw my role in rehab very differently than I did in the hospital on the regular unit because of its team approach. . . . In rehab the leader valued social workers . . . in the hospital setting they saw us as necessary evils. . . .

—Judith Slater, Kennesaw State University, Kennesaw, Georgia, personal communication, October 14, 1995

Under the Medicaid system, our method of doing things was to identify the needs, establish a resource, and petition Medicaid to pay for it. Now, under the different MCOs, we have to make sure that the child is in an MCO that will cover the child's identified needs.

—Thom Prassa, East Tennessee Community Health Agency, Knoxville, Tennessee, personal communication, July 10, 1994

We have one behavior intervention committee . . . that is an agency-run committee . . . two people from residential, two people from vocational, a residential manager,. . . . two parents or advocates. On the human rights committee, we have people who are not from the agency. Two individuals are from our agency, and an agency person chairs it and makes all the contacts. . . . We want people who are from outside the agency to come in and look. . . . It is check and balance.

—Kim Ehlers, Mitchell Area Adjustment Training Center, Parkston, South Dakota, personal communication, October 14, 1995

The case management process takes place within the context of an agency, and client services often involve more than one organization. Effective case managers understand the organizations to which they belong. This requires mastery of three key concepts: organizational structure, agency resources, and improving services. The preceding quotations illustrate the importance of these three concepts in case management.

Organizational structure helps determine the actual job responsibilities of the case manager and the working environment. According to Judith Slater, different units within the same hospital had different expectations for her. In one unit, she was a well-respected member of a team; in another, she was tolerated and called upon only when needed.

An agency's budget constraints are also important. At East Tennessee Community Health Agency, the case managers must now deal with limitations on the resources available to help clients.

Case management work also involves trying to improve services, and one way to do this is through quality assurance programs. The Mitchell Area Training Adjustment Center has established two committees to provide oversight and quality assurance in the areas of intervention and human rights. Kim Ehlers describes her work with these two committees, which assess the merits of recommended implementation plans. Both boards operate to protect the rights of the clients.

During our discussion of organizational factors, you will meet Carlotta Sanchez, who works for the Sexual Assault Crisis Center in a city of 400,000. She has just begun her responsibilities as a case manager for the agency. Along with Carlotta, you will learn about the organizational structure of the Sexual Assault Crisis Center, the budgeting process and her role in it, the agency's informal structure, and the organizational climate within which the center's work is performed. You will also read about her participation in agency programs designed to improve the quality of services.

For each section of the chapter, you should be able to accomplish the following objectives:

UNDERSTANDING THE ORGANIZATIONAL STRUCTURE

- Name three ways that knowledge of organizational structures benefits a case manager.
- List some sources of information about an organization.
- Differentiate between formal and informal structure.

MANAGING RESOURCES

- Trace the planning and budgeting process of an organization.
- Identify other ways in which the case manager is concerned with resource allocation.
- Name the components of a budget.
- List sources of revenue.

IMPROVING SERVICES

- Identify four processes for improving the quality of services.

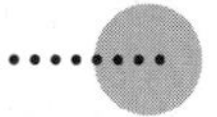

Understanding the Organizational Structure

To be an effective case manager, one must understand the organization and its structure. This knowledge is helpful in three ways. First, it gives the case manager a better understanding of his or her job responsibilities and how they relate to the goals of the agency and the specific objectives of the unit. Second, case managers can help meet client goals more easily if they use agency procedures correctly. Finally, once case managers understand the organizational climate, their work can fit appropriately into the context of the agency. They know what is expected of them, how much autonomy they have, and who can help with the difficult situations they face. Case managers face obstacles within any setting; knowledge of the particular working environment helps one identify barriers and develop strategies to cope with them.

The Organization's Plan

Several documents may shed light on the way an agency is structured. One of the first things a case manager should do is to read the agency's mission statement. Other important statements are its goals, objectives, and policies and procedures (Ellis & Hartley, 1991; Gillies, 1994; Lewis & Lewis, 1983; Morgan & Hiltner, 1992; Porter-O'Grady, 1994). As these documents are examined, it is helpful to ask the following questions:

What is the mission statement of this organization?
What are the values reflected in this mission statement?
How do these values influence the work of case management?
What are the goals and objectives of the agency?
How do these relate to the work of the case manager?
What are the written policies and procedures of the organization?
What impact do these policies and procedures have on the case manager?

A **mission statement** is a summary of the guiding principles of the agency. It usually lists the broad goals of the organization and describes those populations that will benefit from the work of the organization. It also specifies the values that guide all decisions, the agency structure, sources of funding, agency priorities, and the work of the staff (Ellis & Hartley, 1991; Lewis &

Lewis, 1983). The mission statement that follows reflects a commitment to developing human potential, retaining self-determination, and providing quality service.

> The philosophy of this agency is based on the belief that each individual should reach his or her full potential. We believe that individuals deserve respect and the right to determine their own destinies. We will work with individuals, their families, and the community to provide services, based on quality standards of good practice, that promote the satisfaction of our clients and the development of our professional staff.

Each organization has a set of policies and procedures, some of which pertain to the behavior of all employees. Such documents are often very specific, describing procedures in great detail. Many reflect a standard of practice that is determined by the agency, as well as legal requirements established by the federal, state, and local governments. Also included are standards of practice established by a professional code of ethics or by a professional accrediting body (Ellis & Hartley, 1991; Gillies, 1994; Lewis & Lewis, 1983; Morgan & Hiltner, 1992). The Mitchell Area Adjustment Training Center has a policy on the composition of the review board that approves applicants for services: It must include the executive director, a service coordinator, a nurse, and one representative from the other three units of service. At many agencies, the speed with which clients move through the system is regulated by stated policies and procedures. Pima Health System has such a time line. "Three days after they [clients] are determined eligible, they are enrolled with us. . . . Within 10 days of enrollment with us, a case manager has to do a face-to-face visit with the client and develop a care plan with the client and with the client's family members" (personal communication, October 7, 1994).

Once the case manager has a clear picture of the agency's direction, guidelines, and rules, it is important to know the **job description,** the job's relationship to the work of the department or unit, and the expectations the agency has. Case managers should ask the following questions of their supervisors or the people who are responsible for hiring new employees (Ellis & Hartley, 1991; Gillies, 1994):

> Where can I find a written description of my job as a case manager?
> What are the goals of my department or unit?
> How does case management fit with those goals and objectives?
> Does the department or unit have its own policies and procedures?
> How do they relate to my job as a case manager?

Rarely does the description of a position accurately reflect the actual work, but it does serve as a guideline. The description does define the work for which the case manager is held accountable. Job descriptions often change with reorganization, economic pressures, and changing needs of the client population. Judith Slater describes how her job as case manager has changed over time. "When I started doing case management, there was just me and my boss. . . . I covered every specialty area of the hospital. . . . It was pretty much referral by doctors. With the shift in economics and with Medicare–Medicaid guidelines, we then had to screen every Medicare patient. Initially it was referral, and then it was screening out certain types of diagnoses and discharge planning" (personal communication, October 14, 1995).

Structure of the Organization

Another way to understand the structure of the organization or agency is to find out what the relationships are among people and departments. The following questions are helpful (Ellis & Hartley, 1991; Gillies, 1994; Morgan & Hiltner, 1992):

Who has authority or control over others?
Who is accountable to those in authority?
What are the supervisory patterns within the agency?
What is the communication flow within the agency?

In applying this information to your work as case manager, consider how the relationships define authority, accountability, and communication. The following questions will be helpful:

To whom do I report?
Who evaluates my work?
Does anyone report to me?
Do I evaluate anyone?
How do I get information?
To whom do I pass information?

Two terms important in understanding the structure of the organization are *authority* and *accountability.* Ellis and Hartley (1991) define **authority** as "the ability to control resources and action" (p. 5). The lines of authority in the organization become clearer when departments are examined. **Accountability** is "being responsible to another person for your actions and use of resources" (Ellis & Hartley, 1991, p. 5). Simply stated, a major component of authority is holding individuals accountable for the jobs they perform. Each person in authority also has to be accountable for that which he or she controls.

The **chain of command** of an agency stretches from the position with the most authority to the one with the least (Ellis & Hartley, 1991; Gillies, 1994; Morgan & Hiltner, 1992). Policy information is often passed along the chain of command, from the top down.

At Helen Ross McNabb Mental Health Center, Jana Berry Morgan and Paula Hudson explain how the chain of command is used to change an implementation plan for a client.

> If the person that they are working with didn't have day treatment . . . and the case manager felt like that was a necessity for them, then the case manager can go through the client's primary clinician first. Then referrals are passed to day treatment. Then the day treatment coordinator would do an assessment of the referral and determine if the client was an appropriate candidate for day treatment. (Personal communication, April 27, 1995)

Many layers of professionals may be involved in making this decision, each person having authority for part of the decision and accountable for that part.

A document that is helpful when determining the chain of command is the **organizational chart.** (See Figure 9.1.) This is a symbolic representation of the lines of authority and accountability, as well as the information flow within an organization. An organizational chart usually consists of an arrangement of boxes that represent offices, departments, and perhaps individuals (Ellis & Hartley, 1991; Gillies, 1994; Morgan & Hiltner, 1992).

The boxes at the top of the chart represent the positions involving the most authority, supervisory experience, and control of resources. Boxes that are connected with solid lines represent departments, offices, or individuals for which there is a formal communication pattern. For example, any communication from the Executive Director would go to the four directors of the agency.

Boxes that are connected with vertical solid lines represent lines of authority. The Executive Director is responsible for the entire organization and is accountable to the Board of Directors. The Associate Director of Counseling and the Associate Director of Day Programs are supervised by the Director of Programs and Services. These two associate directors in turn supervise the counselors and the activities staff. Boxes connected by dotted lines represent departments, offices, or individuals who communicate with one another but do not have supervisory authority or control of resources. For example, all of the directors communicate with each other as they plan and implement the work of their departments.

Final authority in many human service agencies and organizations rests with the board of directors. The primary responsibility of the board is the financial health of the organization. Under normal conditions, board members do not work directly with supervisors, staff, or clients, but they do interact with the highest-ranking individual in the organization. The board focuses much of its attention on the budget (Russo, 1983). Members of the board analyze statistics such as the characteristics of clients served, the staff/client ratio, the sources of funding, and the relationship between expenditures and outcomes. The board is also involved with public relations, helping the agency establish its mission and goals and working for a positive public image (Morgan & Hiltner, 1992; Russo, 1983).

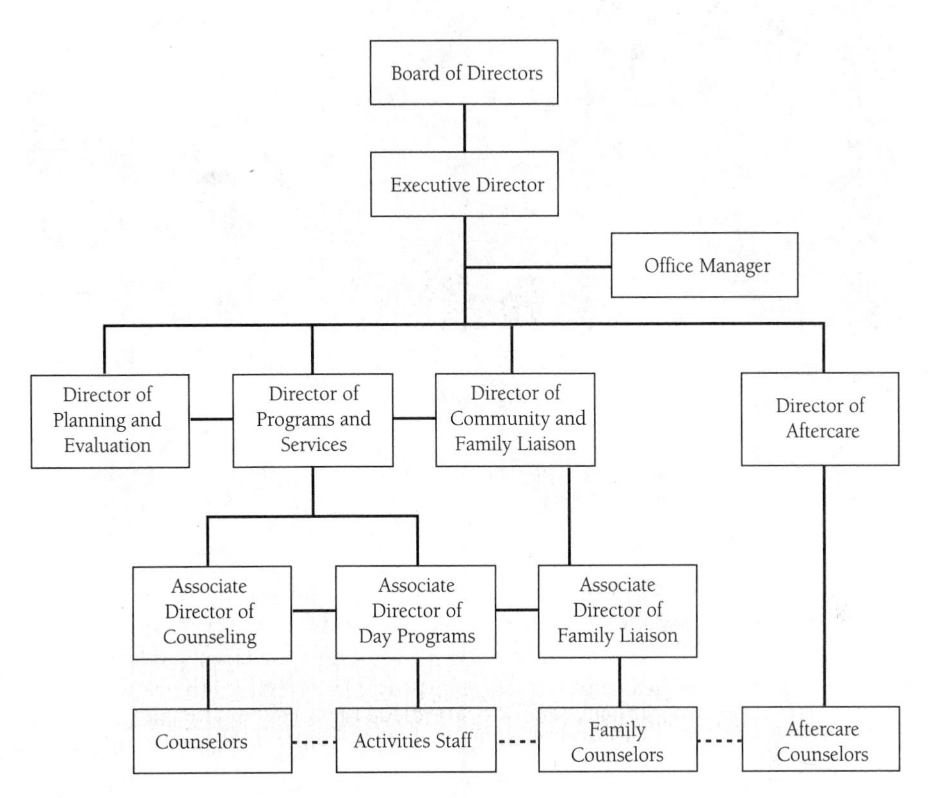

Figure 9.1 Organizational chart

A Board of Directors meeting. It is important for case managers to understand the responsibilities of their organization's Board of Directors.

Let's see how Carlotta Sanchez is adjusting to the organizational structure of her agency.

Carlotta has been in her job as a case manager for six months now. It has been quite a learning experience for her. This was her first job after graduating from the local university with a degree in human services, and she had not expected to find a job so quickly. She believes that she was hired in part because she had worked as an intern with a women's shelter during her senior year. Carlotta is not really sure how many women she has been able to help in the past six months, but she knows that she has been learning a lot. She is finally able to work with clients and families without constantly asking for help from her friend Sally, who is also a case manager. Sally has been in her job for three years and was a volunteer before that, so she understands the work of the agency well.

During the first month, Carlotta had difficulty in understanding how to prioritize her work. She was coming in an hour early and staying two hours later than the regular staff, just to complete the work that she thought she should be doing. During those first months, everyone seemed to think they knew what she should be doing, but Carlotta herself was confused about her role. Her supervisor, Ms. Ludens, who supervises the five case managers, is rarely available to talk with Carlotta about the job. Ms. Ludens is pleasant and supportive, but she is seldom accessible. At the end of that first month, Sally pulled Carlotta aside and asked if she needed any help.

Carlotta asked Sally if she could help her arrange her work schedule. Carlotta also called her college advisor to talk about the difficulties she was experiencing. Both Sally and her teacher asked several questions that she could not answer—focusing on the agency's policies and procedures. Carlotta barely remembered her orientation, which had consisted of two days of lectures on agency policy and rudimentary training in working with the target clients (primarily women who had been victims of sexual assault). During that orientation, she also got information about the legal system. At that time, everything was so new that she didn't know what questions to ask.

Carlotta asked whether she could attend orientation again the next time a new case manager was hired. She reread the mission statement and examined the goals and objectives. Some of them looked really familiar, but there were others that she did not remember, and she was not sure the agency ever addressed them in their programming. According to the mission statement and the goals and objectives, the Sexual Assault Crisis Center is dedicated to prevention and education. Carlotta had never contributed to those goals, and she did not see such activities going on in any other unit. Everyone seemed to be too busy working with clients who had already been assaulted or who were in danger.

After reading the policies and procedures, she was confused by discrepancies between what they said and what her colleagues did. She noted one particular serious violation of policy. All case managers are to follow a time line after receiving a referral. Intake is to occur within 48 hours and is followed by a staffing, but Carlotta has never been able to meet that deadline. She talked to Sally about her questions and concerns. Sally just looked at her and told her to ask the supervisor. Sally would not say another word. Carlotta thought that was a strange reaction, so she decided to ask Manuel, another case manager; he had never heard of the policy that she was questioning. She decided that Ms. Ludens must be the person to turn to.

Carlotta scheduled a meeting with Ms. Ludens. She came up with a list of questions that concerned her, including her job in relation to the mission, the time line that she had not been able to follow, and her need for more supervision. In preparing for the meeting, she found an organizational chart that showed a Director of Case Management, reporting to the Executive Director.

Carlotta felt that the meeting with Ms. Ludens went really well, although she still had unanswered questions afterwards. Ms. Ludens very carefully reviewed the job description for case manager and prioritized the responsibilities case managers have. Carlotta discovered that Ms. Ludens expected her to conduct intake interviews, coordinate a team that included all the professionals working with a client, work with the family, and provide long-term yearly follow-up with clients. She did not expect her to participate in educational or preventive activities. At times it seemed that Ms. Ludens did not understand the job of the case manager. Ms. Ludens provided a curt answer to the question about the mysterious Director of Case Management: "We no longer have that position." She gave Carlotta an equally short response when asked about the timing of intake and staffing: "That is not my policy."

Though she was still unclear about the conflict between policy and practice, Carlotta left the meeting knowing a bit more about the organization.

An organization's planning documents, the written job description for a case manager, and the chain of command (as illustrated by an organizational chart) give a picture of the formal structure. However, as in Carlotta's experience, the formal organization does not give any case manager all the necessary information about the agency. Each agency also has an informal structure that influences the work that is done and how it is accomplished.

The Informal Structure

The **informal structure** within an agency supports many of its functions. This informal structure also helps meet the needs of both supervisors and staff, as well as those of the case manager (Ellis & Hartley, 1991; Gillies, 1994; Russo, 1983). The case manager must distinguish between the formal structure and the informal structure. According to Russo (1983), the major differences between the two are as follows:

FORMAL	INFORMAL
Who supervises whom	Who advises whom on what to do
What the job description says	What the individual in the job actually does
What the formal lines of communication are	Who communicates with whom
What the policy is	What actually happens

An informal structure develops within every organization as a way of meeting the needs that are unmet by the formal structure. It enables workers to use relationships and alternative ways of getting the job done. This facilitates the work of the case manager and helps meet social and professional needs as well; co-workers celebrate successes and support each other in times of trouble. They form a network of support and cooperation based on mutual respect and long-term relationships. As discussed earlier, such a network helps the case manager locate services and monitor the progress of clients.

The informal structure of any agency has existed long before the case manager assumes responsibilities, and he or she also becomes a part of that structure. For example, through informal channels, co-workers give the new case manager advice on defining client problems, addressing issues of client motivation, finding services, and using agency policy to help meet client needs. Some individuals also provide professional advice and support when the case manager encounters difficult clients, and they rejoice together when their work goes particularly well. At times, co-workers at Jana Berry Morgan's agency team up when working with clients: "Our case managers get out in the community with clients . . . and deal with difficult situations. . . . We let our workers know . . . it is okay to take someone with you, . . . use a buddy system . . ." (personal communication, April 27, 1995). This is not a formal arrangement—just an agreement among case managers to ask for help, even when working with clients outside of the office.

The informal structure also involves an alternative pattern of communication flow (Ellis & Hartley, 1991). More individuals in the organization thus receive useful information, outside of channels. Having access to information supports the decision-making process. For example, case managers may hear about changes in policy or procedure or budget cutbacks before they happen. Advance notice gives them time to adjust their work for the coming changes or seek to modify them, serving as advocates for clients' interests.

Although the informal structure contributes to the strength and the success of an organization, it can also contribute to difficulties. For example, the informal structure tends to represent tradition, preservation of the status quo, and resistance to change (Ellis & Hartley, 1991; Gillies, 1994). This hinders case managers from advocating change to help meet the needs of their clients. Also, communication within the informal structure may be unreliable or even hurtful. Such informal information must be considered carefully but not regarded as gospel (Ellis & Hartley, 1991). Several roles, such as advocate, broker, service organizer, and expediter, depend on the case manager's status as a representative of the agency. Passing on unreliable or inaccurate information or making decisions based on such information may damage his or her professional credibility.

Since there is no official map or diagram of this informal structure, case managers must construct their own. Russo (1983) suggests that the case managers make diagrams of the interaction patterns that they observe. Roney (1965, p. 82) suggests asking the following questions about one's own case management activities, to gain insight into one's interaction with the informal structure.

> From whom do I take advice and information?
> To whom do I give advice and information?
> From whom do I take orders?
> To whom do I give orders?

Roney advises that there can be many combinations of answers to these questions. Drawing a chart based on the answers and then comparing it with the organizational chart can help illuminate the differences.

Roz Jaffe, who works in a day care setting for the elderly, describes an informal way of sharing information about clients with other staff, from the moment the clients arrive at the agency. "The drivers let us know if there is a problem. Or someone tells me they are not feeling well or there may be a problem with their home care worker. I will go to the nurse or the social worker to make sure they are aware of the problem or potential problem" (personal communication, May 3, 1994). This everyday exchange of information is very different from the formal weekly staff meetings, care plan meetings, or the charting that is completed at the end of each day.

Another helpful way of looking at the organization is to analyze the organizational climate.

The Organizational Climate

The conditions of the work environment that affect how people experience their work is the **organizational climate.** Such a climate is difficult to describe, for it is based on the values, attitudes, and feelings of people in the work setting. Case managers need to understand the climate so as to clarify their work expectations and their degree of autonomy. Ultimately, the organizational climate influences how they perform their jobs and how they relate to their clients. Two major factors influence the organizational climate: policies and procedures, and supervision (Ellis & Hartley, 1991).

As discussed earlier, each organization has a set of policies and procedures that guides the behaviors of employees. These can give one of three signals to employees (McGregor, 1960a, 1960b; Ouchi, 1981):

1. We trust you will do a good job if we supervise you closely.
2. We trust you to do a good job because you want to do a good job.
3. We trust you to do a good job because you are loyal to the agency, which is like a family.

Case managers within the first, highly supervised environment have job responsibilities that are very clear. The policies and procedures are narrowly defined and well articulated. In the second, less structured environment, policies and procedures allow more autonomous decision making, and case managers have less rigid job descriptions and responsibilities. In the third, "family" framework, an agency invests much time and energy in its case managers and expects loyalty from them in the form of long-term commitment and a focus on excellence (Ellis & Hartley, 1991; Gillies, 1994; Morgan & Hiltner, 1992).

How a case manager is supervised also contributes to the organizational climate, and the supervision is not always consistent with the agency's stated policies and procedures (Ellis & Hartley, 1991). The supervisor can establish a tone that encourages independent thinking and activity, he or she can narrow the range of responsibility and decision making. Even within an organization that has strict policies, the supervisor can foster a less rigid climate. For example, if the supervisor encourages autonomy, case managers may have the freedom to determine their own schedules, make final recommendations for client treatment, and establish agreements with other organizations, all without prior approval (Ellis & Hartley, 1991). Thom Prassa and Mark Alexander describe how the supervision structure of their agency affects their work: "We have in our office . . . divided responsibilities into a Coordinator, two Supervisors, seven Case Manager II's and 15 Case Manager I's. . . . It is really a very loosely governed office. . . . We are really given a lot of latitude to develop our own professional style, because it is a new program and because each region has its own peculiarities" (personal communication, July 10, 1994).

Let's now go back to the organization in which Carlotta finds herself.

After Carlotta met with Ms. Ludens, she decided that she really needed to work more closely with her colleagues if she was going to improve her performance and help her clients better. Sally had been very helpful, and Carlotta decided that she would also work more closely with Charlotte Jones, the legal liaison for the center. Ms. Jones had been with the agency for ten years and had a wealth of knowledge about the law, relations with the police and the community, and the welfare of clients. In staffings, everyone listened when she made recommendations. When she sought Ms. Jones's advice, Carlotta noticed that many of her colleagues were doing the same. It also seemed that Ms. Jones was willing to answer any questions that Carlotta asked.

For example, Carlotta finally broached the subject of the 48-hour turnaround from intake to staffing. Carlotta honestly felt that she was not doing her job when she had to schedule staffings after the 48-hour period. Ms. Jones smiled when she heard Carlotta's question, and she answered it straightaway: "So you have noticed. Let me tell you why we have that policy. Our Board of Directors feels that 48 hours is a reasonable time to begin to work on a case. They also believe that we have not really begun our work until the staffing takes place. We have explained many times why intake is often a longer process than 48 hours, but they refuse to change the policy. Our last director lost her job over that battle. Ms. Ludens just refuses to acknowledge the policy. We have all decided to do the best that we can. We know the policy; we try to follow it. Sometimes it is just not possible."

Carlotta also asked Ms. Jones about the Director of Case Management position that she had seen in the organization chart. Ms. Jones shook her head before she spoke. "That was another battle we lost with the board. Our budget was cut two years ago, and the board recommended that we reduce our indirect personnel costs. They wanted us to put most of our resources in direct services to clients. With the difficulty of this work, the turnover for case managers is high, and training our new case managers is critical. Your questions are evidence of that need for supervision."

Several months later, Carlotta was meeting Sally for dinner after work. Carlotta had actually left work at the conclusion of office hours that day. She told Sally that she was beginning to understand her work better, and she said, "I am even feeling more at home in the agency."

Providing case management within the context of an agency is an exciting and challenging activity for Carlotta and other case managers. Another challenge is managing resources. The next section discusses how resource allocation influences the work of the case manager.

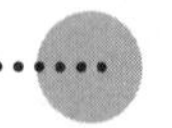

Managing Resources

Understanding the basic concepts of budgeting is important for a case manager. Much of the planning within the agency relates directly to the allocation of resources. Resource allocation determines the number of staff, the degree of operating support, the number of clients served, and the programs and services available for clients. Allocation of resources even covers the amount of time case managers spend on specific tasks.

Since costs are rising, with a simultaneous increase in the number of clients who have complex needs, there is an imperative to spend every dollar well. Allocating resources wisely requires an understanding of budgets and their applications to case management work.

What Exactly Is a Budget?

Gillies (1994) describes a **budget** as "a numerical expression of an agency's expected income and planned expenditures for a period of time" (p. 85). According to Gross and Jablonsky (1979), "A budget is a 'plan of action.' It represents the organization's blueprint for the coming months, or years, expressed in monetary terms" (p. 359). Budgeting has two purposes: planning and controlling. Sometimes the planning and controlling functions of a budget have opposing goals; planning represents the creative and expansive approach to human service delivery, whereas control is restrictive and measured. However, both are important facets of the budgeting process and have great relevance to the case management process.

Experts agree that planning and budgeting should be linked and that planning should guide the budgeting process (Gillies, 1994; Whalen, 1991). A human service agency must be clear about goals, objectives, and priorities before it begins budget preparations (Ellis & Hartley, 1991; Gillies, 1994; Lewis & Lewis, 1983; Morgan & Hiltner, 1992). A two-step process links planning and budgeting. The first step is to set up three categories: goals and objectives, planned activities, and the costs of those activities (Ellis & Hartley, 1991). For example, an initial part of the budgeting process for the Mitchell Area Training Adjustment Center was to submit a request to the state for "eight new staff to take the individuals [clients] out into the community at least once a week" (personal communication, October 4, 1994). Planners at the center believe that without eight new instructors, the parents of the clients will complain that their sons and daughters are not receiving quality care.

The second step in the budget process considers the goals and objectives and the resources available to provide the services. Revision of the budget then occurs within the parameters of the available resources. The result of the second step is a budget with revised costs to reflect the goals, objectives, programs, and services as adjusted to the funds actually available. This process allows planning to determine the budget (Gillies, 1994; Lewis & Lewis, 1983; Morgan & Hiltner, 1992). Once the Mitchell Area Training Adjustment Center receives its allocation from the state, it will complete step two of the planning and budgeting process, revising the goals for clients on the basis of the state's response to the request for eight new staff.

The case manager's work in the planning and budgeting process is determined by the agency's policies and procedures. There is now a management trend toward participatory management and budget responsibility at the unit level, so case managers may be asked to provide input into the process (Gillies, 1994; Whalen, 1991). Because case managers have knowledge of client needs, departmental goals and objectives, and new professional trends in research and service delivery, they can offer helpful recommendations for priorities and resource allocation.

Many case managers are involved directly in budgeting for their individual clients. Some have responsibility for managing an account of services for clients; they generate and monitor a budget for each one of them. This includes defining needs, setting priorities, procuring services, and paying for services rendered. Usually this begins with a fixed allocation of funds and a list of services that may be purchased for clients who have a specified set of needs. Case managers help set priorities and determine how best to use the funds. A case manager at the Pima Health System has these responsibilities: "There are expenditure limits and there are cost-effectiveness studies done for each client. The cost of home- and community-based services cannot exceed 80% of what it would cost to provide nursing home care to the client" (personal communication, October 7, 1994).

Resource allocation extends to how case managers use their time. Time is a valuable resource of the case manager, to be used for various activities such as intake, planning, coordinating resources, monitoring client progress, advocacy, providing direct counseling service, educating, interdisciplinary teamwork, building networks, and completing paperwork. Time management of each day is a resource decision, and the use of time should be linked to outcomes. (Chapter 11 will discuss time management in greater detail.)

Another purpose of a budget is to establish control of expenditures (Ellis & Hartley, 1991; Gillies, 1994; Morgan & Hiltner, 1992; Porter-O'Grady, 1994; Whalen, 1991). The budget provides a baseline from which to judge and project obligations. If an organization adheres to the budget during the year, the budget controls spending. A balanced budget, where expenses do not exceed expenditures, is important to maintain; this is one standard of good fiscal management.

The control function of the budget has a relationship to case management work and how programs are implemented and needs met. Planning takes into account the resources available and any restrictions on their use. There may be policies governing what services clients are eligible to receive (Ellis & Hartley, 1991; Gillies, 1994; Lewis & Lewis, 1983). For example, the Family Counseling Center in Tucson, Arizona, has multiple funding sources: Social Services block grants, Title III, the Older Americans Act, and the county funds. The agency's policy is to ask clients to pay for some of the services if they have funds available. Limited resources may also restrict the number of clients that can be

Table 9.1 BUDGET FORMAT FOR A HUMAN SERVICE AGENCY

Category of Expenditures	*Projected Expenditures*	*Actual Expenditures*	*Balance*
Salaries			
Supplies			
Equipment			
Travel			
Training			
Communications			
TOTAL			

served. Sometimes, planners must develop eligibility criteria that take into consideration the prevailing limitations on resources.

Features of a Budget

A budget has several components: projected expenditures, actual expenditures, and balance. (See Table 9.1.) *Projected expenditures* are cost estimates made at the beginning of the budget cycle; they designate the money that has been set aside for expenses in a given category. *Actual expenditures* represent the money that has been spent in a given category to date. At any time in the budget year, it is clear what has been spent in each category. The *balance* is the amount allocated to a given category, minus the amount already spent.

In Table 9.1, expenditure categories are listed in the left column:

Salaries are monies paid to employees, either for providing direct services to clients or for indirect services such as administration and supervision.

Supplies are necessities for maintaining the activities of the agency. Usually, such items have a short life or are used within a period of a year. Examples are paper, pens, gasoline, and essentials for repair and cleaning.

Equipment includes machines that are bought for a specific purpose, such as computers, a copy machine, and recreation equipment. These are usually one-time expenditures.

Travel represents the cost of moving personnel and/or clients. Some agencies cover clients' travel expenses to and from sites where services are delivered. At other times, case managers make home visits or transport clients to other agencies. The travel budget is also used for purposes of professional development.

Training costs often include payment to consultants who conduct staff training for professional development, as well as the cost of certification classes, books, and professional journals.

Communications covers expenditures for using telephones, sending faxes, mailing things, and any other expenditures incurred in transmitting information.

Total designates the whole amount that has been committed or spent.

Often the budget is simplified by using just three categories: salaries, operating, and capital. The *salaries* category remains the same as for the budget presented above. *Operating* includes supplies, most equipment, travel, training, and communications. The *capital* budget includes major equipment and building projects.

Sources of Revenue

Four main revenue streams fund human service programs: (1) federal, state, and local government-sponsored programs; (2) grants and contracts; (3) fees; and (4) private giving (Ellis & Hartley, 1991; Gillies, 1994; Lewis & Lewis, 1983).

Federal, state, and local government funds These are available for programs and services. As in the case of Medicare and Medicaid, they may be direct reimbursements for client care. In the past, this was considered to be stable long-term funding. In today's climate of rising costs, balanced government budgets, and diminishing role of government in human services, such funding is less secure. There is a movement to allocate financial support to local levels of government in order to increase accountability and to develop programs closer to where the problems occur.

Grants and contracts Many agencies write proposals for grants and try to develop contracts to secure funding from both government agencies and private corporations. Sometimes, staff members are assigned to write proposals to acquire grants and contracts. Sponsoring agencies often establish priorities and set restrictions on resource allocation. If human services agencies and their target populations are to benefit from such external funding, the priorities of the granting or contracting body must be similar to those of the agency (Ellis & Hartley, 1991; Lewis & Lewis, 1983).

Fees Such revenue is generated in two ways. First, many clients have insurance that covers the costs of the services they receive. The insurance is managed through a third party, and the payment is made according to the policies of the insurance agreement. In the managed care environment, there may be restrictions on the range or quantity of services available (Hicks, Stallmeyer, & Coleman, 1993). Fees for services are also collected from individuals who choose not to use their insurance, who have exceeded the amount the insurance will pay, or who are ineligible for government or insurance support. In such cases, fees are often assessed on a sliding scale: the higher the person's income, the higher the fees; the lower the income, the lower the fees.

Private giving This is becoming an important source of revenue. Human service agencies solicit private individuals, businesses, and corporations for donations of money, equipment, and professional expertise. The current trend is to establish relationships with such donors, in the hope that funding can be stabilized by a long-term partnership.

How an agency is funded influences the case management process. Government funding and insurance reimbursements have direct effects on the services provided. Any change in funding has the potential to expand, restrict, or alter services. Many programs have changed over time as funding patterns have shifted. For example, Angela LaRue, a parole officer, describes changes in an adjustment counseling program for parolees. "We used to have an adjustment counseling therapist. The therapist did this for free for a while. Then we got funding. . . . Then he did it for free when we lost the funding" (personal communication, October 18, 1995).

Case managers may be involved in grant writing or may serve as consultants to those who are making applications, and working on grant-writing projects provides an opportunity to speak for the needs of clients. In addition, case managers may be involved in fundraising. For many years, private fundraising was the sole responsibility of the board of directors and the executive director. As private giving becomes a more important part of agency revenue, case managers can expect to increase their participation in fundraising activities. The case managers at Casita Maria solicit private donations to enhance the services they provide. In one Casita Maria program, clients receive an allowance to buy a bed and a kitchen set for a new apartment. With private donations, caseworkers are able to get their clients additional furniture. Yolanda Vega, Director of Agency Services, shared one client's appreciation: "She was more than thrilled, because she has got her apartment in order" (personal communication, May 4, 1994).

Carlotta Sanchez, the new case manager at the Sexual Assault Crisis Center, has little experience with budgets. Let's see how she begins to learn about the resource allocation process at the agency.

To Carlotta, it seemed that the more she talked with her colleagues, the more she discovered the importance of the Board of Directors. She knew that Ms. Ludens spent quite a lot of time with board members, and during the week of a board meeting, the whole staff was busy guessing what changes the meeting might bring for the agency's functioning. Ms. Jones had also told Carlotta that funding was very tight for the coming year. Carlotta had discovered that although the budget year for the agency began January 1 and ended December 31, budget preparations began very early in August.

Carlotta still did not have a supervisor, and she had little way of getting key information about the agency unless Ms. Ludens spoke at the large staff meeting or sent an "all staff" memo. Sometimes, Carlotta heard news from the other case managers or Ms. Jones. By the middle of the summer, several of her colleagues were expressing worry about the budget for the coming year. Several funding sources were cutting back, and the staff had no way of guessing how the agency would be affected. The state government had experienced a shortfall for the second year in a row and was talking about reducing support to social services. United Way had increased the list of agencies that it supports and might decrease its funding to the center. Donations from the public sector were also down from the previous year.

In September, Ms. Ludens called the entire staff together and asked them to work the budget for the coming year. They were to consider three scenarios: a budget increase of 2%; a budget decrease of 2%; a budget decrease of 4%. She established a budget team and asked them to describe what impact the increase scenario and the two decrease scenarios would have on programming and staffing. She also asked the budget team to consult with everyone in the agency. Carlotta was chosen to represent the case managers on the team.

Carlotta learned a lot during the two months she worked with the budget team. They talked about priorities, the cost of programs, the target population, and how to set and measure expected outcomes. Carlotta was amazed at how her own responsibilities would change under each scenario. It was a stressful period, and involving the entire staff in the budget planning took a lot of time and effort. The group had the three scenarios ready for Ms. Ludens by the deadline, and she prepared to present them to the board at its meeting the following week. The whole staff was holding its breath; the actual budget total would not be available until final word about state funding arrived.

One other effort related to organization influences the roles and responsibilities of case managers in the delivery of human services. The next section discusses how agencies involve their case managers in the process of improving services.

Improving Services

In recent years, there has been an increasing emphasis on providing quality services. This has resulted in the use of four processes: linking outcomes to cost; conducting utilization review; planning quality assurance programs; and promoting continuous improvement. Each of these addresses the issue of quality in a different way.

What Is Quality?

Quality is a term that is very difficult to define. Persig (1974, p. 163) expresses the dilemma this way:

> Quality . . . you know what it is, yet you don't know what it is. But that's self-contradictory. But some things are better than others, that is they have more quality. But when you try to say what quality is, apart from the things that have it, it all goes poof! There's nothing to talk about. But if you can't say what quality is, how do you know what it is, or how do you know it even exists?

In the fields of health and mental health care, *quality* is understood to mean that the delivery system meets its objectives (with a particular emphasis on client needs) in relation to four variables: the limitations of the methodology used; the client's status; the environment; and the resources available (Brook & Lohr, 1985; Donabedian, 1983; Savitz, 1992; Steffen, 1988). Defining quality and measuring it are only one component in providing quality services. There must then be an identification of any problems and a corresponding change in service delivery (Savitz, 1992). This section presents two approaches to defining and measuring quality: linking objectives to outcomes and conducting utilization reviews. We will also discuss two approaches to improving quality: conducting quality assurance programs and implementing continuous improvement programs.

One way to identify the quality of services provided is to evaluate the outcomes of programs in relation to the resources spent (Gillies, 1994). To do this, programs and the necessary resources for implementation must be weighed against expenditures. Outcomes such as the number of clients served, the programs delivered, and client progress are matched to actual expenditures. To perform such an evaluation, Feldman (1973, p. 44) suggests providing answers to the following questions:

> What are the objectives of this organization?
> What programs are available to move toward these objectives?
> What will each of these programs cost in human and financial terms?

The results of this analysis are helpful to a case manager, for one of the most frustrating elements of case management is the difficulty of determining one's own effectiveness and efficiency (Gillies, 1994; Lewis & Lewis, 1983; Morgan & Hiltner, 1992). If case managers can begin to link their activities to the objectives of the organization (progress made by the client) and to the cost of the service delivered (combining the costs of direct service to the client with any support services), they will begin to see the relationship between their work and the benefits to their clients.

One type of quality analysis has emerged from the managed care environment: utilization review, which has been developed to oversee service provision and monitor costs. It is a response to accusations that managed care substitutes cost-effectiveness for quality care. In the area of mental health, managed care has also been criticized for underfunding necessary treatments (Knesper, Belcher, & Cross, 1987; Savitz, 1992; Sharfstein, Krizay, & Muszynski, 1988; Wyszewianski, Wheeler, & Donabedian, 1982). Let's examine utilization review and its impact on the case management process.

Conducting Utilization Review

A utilization review is conducted by peers within the human service delivery system. Two approaches are used during utilization review: pretreatment review and second-opinion mandates (Corcoran & Gingerich, 1994; Tischler, 1990). In **pretreatment review**, a managed care professional must review and approve a treatment plan before service delivery. **Second-opinion mandates** apply to certain diagnoses; there must be consultation with a second professional before a treatment plan can be approved. Gillies (1994) states that utilization review is designed to determine the proper treatment for each of the client's problems or conditions, the appropriate course of the treatment process as a whole, and how the plan submitted must be revised.

Utilization review is often applied because the treatment is provided on a prepayment basis. In other words, since limited funds are allotted for each client, it is critical to assess client needs and the cost of relevant services before providing such services. The responsibility of people who perform utilization review is to provide for the treatment measures that represent the best use of resources in light of the anticipated outcomes.

In many managed care PPOs, health care and mental health care are provided to employees of a corporation or to members of a designated group at discounted rates, in return for a high volume of clients. To manage a PPO cost-effectively, the diagnosis and treatment regimen must be shortened (Feldman & Fitzpatrick, 1992; Gillies, 1994; Mullahy, 1995). The ultimate goal is to spend the least amount of time and provide the lowest-cost treatment that restores the client to good health or mental health (Feldman & Fitzpatrick, 1992; Gillies, 1994; Mullahy, 1995). Another goal is to keep professionals out of legal difficulties. To meet these goals, another process has developed—one that emphasizes quality care. The process of quality assurance documents that the care provided has met quality standards (Porter-O'Grady, 1994).

Planning Quality Assurance Programs

The focus of **quality assurance** programs in the human services is on developing standards of client care (Gardner, 1992; Mullahy, 1995; Porter-O'Grady, 1994; Savitz, 1992). Relevant to the work of the case manager are two quality assurance measures: the standard treatment plan and the client satisfaction survey.

An agency develops a standard treatment plan to ensure common practices in intake, assessment, goal setting, planning, implementation, and evaluation of outcomes. Agency practice must establish standards of care for common assessments within the scope of the agency's activities. The work of the case manager and others must then be compared with the standards thus established. Thus, each case manager provides the same set of services for each client. When services do differ, reasons must be provided. For the quality assurance process to be of real value, the case manager must be able to link assessment with implementation and with positive outcomes.

Client satisfaction is another component of quality assurance programs. As a dominant criterion for evaluating their work, it is viewed both positively and negatively by professionals. Some would question the client's competence to judge the quality of services (Savitz, 1992). On the other hand, there is evidence that clients who are pleased with their treatment are more likely to follow the plan and to be straightforward in reporting whether their goals have been met (Steffen, 1988). Since case management aims to empower clients and increase the extent of their responsibility for their own lives, client satisfaction can be considered a key element in providing quality service.

Figure 9.2 shows how client satisfaction is measured by one agency that uses service coordinators to manage cases. Investigating satisfaction is difficult; it requires an atmosphere where clients believe that someone wants to hear what they have to say. It is also helpful to hear from clients who terminate services prematurely or whose treatment proves unsuccessful. Their identification of problems in service delivery is beneficial to agencies that value quality assurance, for this feedback can serve as the basis for needed improvements. Unfortunately, such clients are sometimes difficult to reach.

The final approach to improving the quality of services is **continuous improvement** (also known as total quality management). This movement first began in industry but is fast becoming an accepted part of quality assurance in the human service delivery system (Feldman & Fitzpatrick, 1992; Hicks, Stallmeyer & Coleman, 1993). Both supervisors and case managers have a role in the continuous improvement process. Supervisors are charged with providing leadership that encourages all staff to participate in improving service delivery; case managers are expected to provide quality work and to collaborate with others in a spirit of cooperation and teamwork.

Edward Deming (1986), a leader in the total quality movement, suggests several guidelines to be observed if work toward quality in an organization is to succeed. First, note that *all* processes and procedures can be improved, so each component of the process is subject to scrutiny and possible change. Another guideline of Deming's is that customer needs (in this case, those of clients) must be considered. A third component of continuous improvement is the requirement of teamwork across units and departments to solve problems and improve outcomes. Clearly, this goal is consistent with the commitment of the case manager to work effectively with colleagues.

If an agency is participating in a total quality program, the case manager may be asked to conduct peer reviews or join a quality circle. In a *peer review* system, professionals establish a process for evaluating one another. Another feature of continuous improvement is the *quality circle,* in which those who serve on case management teams work together to solve problems.

1. Date contacted by case manager ______________________

2. Number of meetings with case manager ______________________

3. How would you rate the frequency of contact with the case manager?

   Adequate ________ Too many ________ Too few ________

4. Were the case manager's goals and intervention consistent with your needs?

   Definitely ________ Somewhat ________ Not at all ________

5. Please describe your diagnosis. ______________________

   ____

6. Has the status of your difficulties changed since you had your first contact with our agency?

   ____

   ____

   Please describe: ______________________________

   ____

   ____

7. Check the appropriate column for the following.

	EXTREMELY HELPFUL	SOMEWHAT HELPFUL	HELP IN THIS AREA WASN'T NECESSARY
Effectiveness of our services in dealing with your situation	________	________	________
Understanding your problems and intervention	________	________	________

Figure 9.2 Quality assurance questionnaire

Obtaining information from your case manager to share with you (e.g., treatment alternatives, instructions to be followed)	_______	_______	_______
Assisting with arrangements for necessary services	_______	_______	_______
Offering support, encouragement, or assistance in coping with your difficulties or those of a family member	_______	_______	_______

8. Were there other ways in which the case manager was helpful? Please describe:

____

____

9. How would you rate the work of your case manager?

   Excellent _______ Good _______ Fair _______ Poor _______

10. In terms of effectiveness, our agency intervention came

   At the right time _______ Too early _______ Too late _______

11. Were there other areas of assistance that could have been helpful? Please list:

____

____

12. How would you like to see our services changed? _______________

____

Figure 9.2 *(continued)*

13. Would you recommend our services to others in a similar situation?

 Yes __________ No __________

Signature: ______________________________ Date: ________________
 (Optional)

Figure 9.2 *(continued)*

Source: Adapted from *The Case Manager's Handbook,* by C. Mullahy, pp. 260–262. Copyright © 1995 Aspen Publishers, Inc. Adapted with permission.

Another area of focus for the continuous improvement approach is the management of time (mentioned previously in the section on managing resources). Managing time well means knowing how time spent is related to priorities and expected outcomes. Good time management techniques for the case manager will be discussed in Chapter 11.

Let's revisit Carlotta as she becomes involved in her agency's efforts to institute continuous improvement.

Once she had participated in the budget team, Carlotta's confidence increased, and she felt that she was able to add something of value to the agency. She began to look around for other committees that she might join. She felt that she was too new to the organization to join the Personnel Committee, and she was not very interested in fundraising and development activities. (Dealing with large crowds and planning events were not much fun for her.) However, she was interested in the Total Quality Management (TQM) program that was beginning to be applied to the agency's volunteer training. The TQM group included two volunteers, the volunteer coordinator, a board member, a client, and a member of the Mental Health Association. The group wanted a member who worked directly with clients, and they invited Carlotta.

In the beginning, Carlotta thought that she had never been to so many meetings. They had training on six consecutive Saturdays just to become familiar with the principles of TQM. Carlotta's father worked for a chemical company in a neighboring state, and he had told her about his terrible experiences with quality teams. He thought that they were a waste of time and gave workers false hopes that things would get better. Carlotta was more optimistic, and she really enjoyed meeting and working with people she normally had little contact with. In the first sessions, the group learned about the quality movement and the progress it had fostered in other organizations. By the third Saturday, the group had started focusing on the volunteer program.

First they had to outline the mission, goals, priorities, and stakeholders and customers of the agency. Carlotta thought that the agency only had one customer—the client—but she learned differently. After much discussion, it was decided that the agency had many customers, among them the board, the citizens of the county, the local government, and the police in the city and the county. The list was much longer, but the point for Carlotta was that more people depended on her agency than she had ever realized.

The first six Saturdays were only the beginning of the meetings, and the group spent a lot of time outlining what happens in the agency, exactly how the work flows, and the role of volunteers in that work. They learned about the training and supervision of volunteers, their actual work experiences, how clients responded to working with them, and how the volunteers themselves felt about their work. They sent out a survey and then interviewed clients and volunteers about their experiences, to determine what they valued about their experiences with the agency and how satisfied they were. Carlotta was not sure exactly how this work would turn out, but it gave her a different perspective, and she enjoyed working with these new colleagues. She came to be confident that all her work would help provide better conditions for volunteers and also for clients.

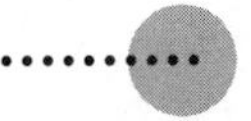

Chapter Summary

Case managers work within an organizational context and must understand the essentials of agency structure. This enables them to use the organization more effectively to meet the needs of clients. They can then see more clearly how their work is integrated into the mission of the agency. Many case managers are also expected to manage resources. A thorough understanding of the budget process and its effects on the resources available for clients helps case managers serve as effective advocates for their clients' interests. Because they are naturally committed to improvement in the services they deliver, case managers can enthusiastically embrace several methods of quality assurance, such as utilization review and continuous improvement. These processes are also designed to contain costs.

Chapter Review

◆ *Key Terms*

Mission statement
Job description
Authority
Accountability
Chain of command
Organizational chart
Informal structure
Organizational climate
Budget
Contracts
Fees
Pretreatment review
Second-opinion mandates
Quality assurance
Continuous improvement

◆ *Reviewing the Chapter*

1. Why is it important for a case manager to understand organizational structure?
2. List sources of information about an organization and describe the type of information each provides.
3. How can a case manager find out about relationships within an organization?
4. Describe the relationship between authority and accountability.
5. What is the purpose of the informal structure? How does it operate, and how is it different from the formal organizational structure?

6. How does the organizational climate affect the work of the case manager?
7. Name the two steps that link planning and budgeting.
8. What role do resources play in preparing a budget?
9. Describe how budgeting for clients takes place.
10. In what other ways does resource allocation affect case management?
11. Name the components of a budget.
12. Design a budget for yourself for one week, using the categories discussed in this chapter.
13. How does an agency get money to support its work?
14. Describe the two ways in which fees are generated.
15. How does the agency's funding affect case management?
16. How do linking objectives to outcomes and conducting utilization reviews relate to defining and measuring quality?
17. How do quality assurance programs and continuous improvement programs contribute to better services?
18. Describe pretreatment reviews and second-opinion mandates and their roles in utilization review.
19. How does the use of standard treatment plans affect case management?
20. Discuss client satisfaction as an element in quality case management.

◆ *Questions for Discussion*

1. Why do you think it is important to understand the informal structure?
2. Can you provide a rationale for the statement, "Case managers should understand the process of managing resources"?
3. What do you think will happen as case managers learn more about how to improve services?
4. Do you think it is a good idea to involve clients in the process of improving services? Why or why not?

References

Brook, R. H., & Lohr, K. (1985). Efficacy, effectiveness, variation and quality: Boundary crossing research. *Medical Care, 23,* 710–722.

Corcoran, K., & Gingerich, W. J. (1994). Practice evaluation in the context of managed care: Case-recording methods for quality assurance reviews. *Research on Social Work Practice,* 4(3), 326–337.

Deming, W. E. (1986). *Out of crisis.* Cambridge, MA: MIT Press.

Donabedian, A. (1983). The quality of care in a health maintenance organization: A personal view. *Inquiry, 20,* 218–222.

Ellis, J., & Hartley, C. (1991). *Managing and coordinating nursing care.* Philadelphia: Lippincott.

Feldman, J. L., & Fitzpatrick, R. J. (Eds.). (1982). *Managed mental health care.* Washington, DC: American Psychiatric Press.

Feldman, S. (1973). Budgeting and behavior. In S. Feldman (Ed.), *The administration of mental health services.* Springfield, IL: Charles C Thomas.

Gardner, D. (1992). Measures of quality. In M. Johnson (Ed.), *The delivery of quality health care* (pp. 42–58). Chicago: Mosby.

Gillies, D. (1994). *Nursing management.* Philadelphia: Saunders.

Gross, M. J., & Jablonsky, S. F. (1979). *Principles of accounting and financial reporting for nonprofit organizations.* New York: Wiley.

Hicks, L., Stallmeyer, J., & Coleman, J. R. (1993). *Role of the nurse in managed care.* Washington, DC: American Nurses Publishing.

Knesper, D. J., Belcher, B. E., & Cross, J. G. (1987). Preliminary production functions describing change in status. *Medical Care, 25,* 222–237.

Lewis, J. A., & Lewis, M. D. (1983). *Management of human service programs.* Pacific Grove, CA: Brooks/Cole.

McGregor, D. (1960a). Theory X: The traditional view of direction and control. In *The human side of enterprise* (pp. 33–43). New York: McGraw-Hill.

McGregor, D. (1960b). Theory Y: The integration of individual and organizational goals. In *The human side of enterprise* (pp. 45–57). New York: McGraw-Hill.

Morgan, E. E., & Hiltner, J. (1992). *Managing aging and human service agencies.* New York: Springer.

Mullahy, C. M. (1995). *The case manager's handbook.* Huntington, NY: Aspen.

Ouchi, W. G. (1981). *Theory Z: How American business can meet the Japanese challenge.* Reading, MA: Addison-Wesley.

Persig, R. (1974). *Zen and the art of motorcycle maintenance.* New York: Bantam Books.

Porter-O'Grady, T. (Ed.). (1994). *The nurse manager's problem solver.* St. Louis, MO: Mosby.

Roney, J. (1965). A case study of administrative structure in a health department. *Human Organization, 24*(4), 345–357.

Russo, J. R. (1983). *Serving and surviving as a human service worker.* Pacific Grove, CA: Brooks/Cole.

Savitz, S. A. (1992). Measuring quality of care and quality maintenance. In J. L. Feldman & R. J. Fitzpatrick (Eds.), *Managed mental health care* (pp. 143–158). Washington, DC: American Psychiatric Press.

Sharfstein, S. S., Krizay, J., & Muszynski, I. L. (1988). Defining and pricing psychiatric care "products." *Hospital Community Psychiatry, 39,* 372–375.

Steffen, G. E. (1988). Quality medical care. *Journal of the American Medical Association, 260,* 56–61.

Tischler, G. L. (1990). Utilization management of mental health services by private third parties. *American Journal of Psychiatry, 147,* 967–973.

Whalen, E. L. (1991). *Responsibility center budgeting.* Bloomington: University of Indiana Press.

Wyszewianski, L., Wheeler, J. R., & Donabedian, A. (1982). Market-oriented cost containment strategies and quality of care. *Health and Society, 60,* 518–550.

Chapter Ten

Ethical and Legal Issues

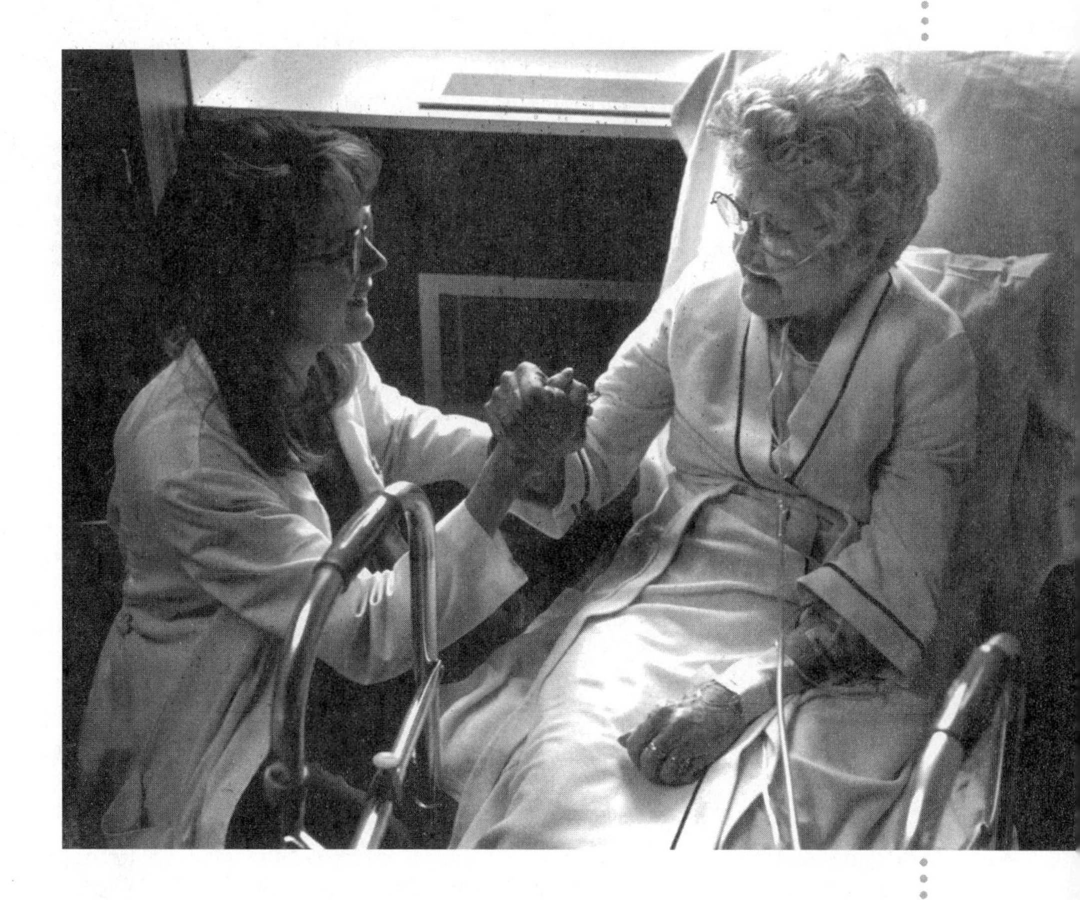

Case managers wear lots of hats. Advocating for the client when needed . . . and one of the things in our system is the gatekeeper role. . . . But the case manager really works for the client and for the HMO too . . . so you can't just be a client advocate. You have to determine what services are truly needed and appropriate for the client.

—ALAN TIANO, Pima Health System, Tucson, Arizona, personal communication, October 7, 1994

We case manage our clients to the best of our ability. The outcome is something else and if Ms. S. decides she wants to remain in a dangerous situation—as long as we have given her what information and resources we have—then it is out of our hands. People have the right to make decisions that may be stupid in our opinion. People have the right to poor judgment. Who is to decide what is poor? And I think one of the rules of case management is to support people's right to make their own decisions.

—Suzy Bourque, Family Counseling Agency, Tucson, Arizona, personal communication, October 7, 1994

Now legally the home attendant is not supposed to give medication to a person. The person can say, "Gee, it's time, and here it is," but they are not legally supposed to do it. So she doesn't give him medicine at night, sometimes he could use something for sleeping and I didn't blame her for not doing it, and you know there is nobody else to do it.

—Rosalyn Baum, Jewish Home and Hospital for the Aged, Bronx, New York, personal communication, May 3, 1994

I have a client whose family thinks that she should not be on the medicine. Her family tells her that we are bad for her and . . . she listens to her family . . . and then she sees what they are telling her may not be in her best interest.

—Jana Berry Morgan, Helen Ross McNabb Center, Inc., Knoxville, Tennessee, personal communication, April 27, 1995

Effective human service delivery often requires a delicate balance of commitment to the client, the agency for which the case manager works, laws and regulations, court rulings, and professional codes of ethics. These conflicting interests can create crises that require the case manager to make difficult choices (Kelly, 1992). The chapter-opening quotations reflect some of the tensions that case managers face. Alan Tiano describes the conflict resulting from case managers' dual responsibility—to their clients and to the agency. Advocacy means working for the client's interest, and gatekeeping entails upholding the rules, regulations, and restrictions of the agency, which are not always best for the client.

Suzy Bourque speaks of a different type of dilemma, involving the mandate to grant client autonomy whenever possible. Case managers may see clients choosing alternatives that are not in their best interests, and families are not always supportive of the client or of the case management process. The wishes of the client, those of the family, and the needs of the client must be weighed against professional guidelines, agency policy, and government regulations.

In the situations such as those described above, as well as many others, finding the appropriate resolution is difficult and challenging. Case managers must continually ask themselves certain questions: What is in the client's best interest? What is the right choice ethically? Am I operating within the guidelines of the agency that employs me? Case managers look to codes of ethics, the law, and agency policies and procedures to guide their practice. Mediating disagreements among family members and clients, working with potentially violent individuals, honoring client preferences, maintaining confidentiality, and upholding complicated rules and regulations are among the thorny issues with which case managers grapple. The following pressures increase the challenge:

- Individuals are becoming more aware of their right to make decisions about their own care.
- Families are becoming more involved in relatives' care.
- New technological, psychological, and economic interventions are continually being developed.
- Dealing with finite resources, case managers must control costs and allocate resources equitably.

For each section of the chapter, you should be able to accomplish the objectives listed.

FAMILY DISAGREEMENTS
- Define the "daughter from California" syndrome.
- List guidelines to follow to encourage positive participation of families.

WORKING WITH POTENTIALLY VIOLENT CLIENTS
- Describe why violence is becoming more prevalent in modern society.
- Apply the steps in addressing issues of violence in the workplace to a specific case management situation.

CONFIDENTIALITY
- List reasons why the issue of confidentiality is so difficult.
- Describe guidelines to follow when discussing confidentiality with clients.

DUTY TO WARN
- Define *duty to warn.*
- Show how the case manager works with a team on issues involving the duty to warn.

AUTONOMY
- Describe the difficulties that arise with regard to granting client preferences.
- Explain how guidelines can help a case manager who faces issues of autonomy.

BREAKING THE RULES
- List the sources of rules and regulations.
- Tell how to decide when to break a rule.

Family Disagreements

One of the most difficult ethics issues for the case manager is the "daughter from California" syndrome—family disagreements over the care of an incompetent client. This syndrome is named for the case of Mrs. M., an elderly woman whose two daughters disagreed vehemently about her care (Molloy, Clarnette, Braun, Eisemann, & Sneiderman, 1991).

Mrs. M. was a woman 83 years of age with a five-year history of Alzheimer's disease. She had been cared for by her 60-year-old daughter and placed in a hospital when the daughter could no longer handle her. Because of medical complications, the mother was unable to walk, incontinent, and demented. The daughter recommended a "do not resuscitate" (DNR) order. She insisted that her mother be given medical treatment only for comfort and pain. The daughter was sure that this measure coincided with her mother's wishes, although there was no written directive from the mother.

Complications arose when a second daughter, who had not seen her mother in years, arrived from California. This daughter was outraged with the DNR order and demanded that her mother receive treatment to continue her life for as long as possible. The daughter from California was angry at her sibling and denied that her mother was in a serious state of decline. She dominated treatment team discussions and threatened to sue the entire team. The team worked with this daughter to help her with her denial, guilt, and anger. The sister who had previously provided the primary care gave in to the California sister's wishes, and aggressive medical treatment for the mother resumed. The California sister returned to her home, and the DNR (with treatment only for comfort and pain) was put into effect once more. The mother died two weeks later, and the daughter from California did not return.

The legal question in this situation is: Who can make the decision to terminate life-prolonging measures when the patient is incompetent and has left no clear directions? In 1990 the United States Supreme Court, in a 5–4 decision, upheld a ruling from the Missouri Supreme Court that "it is constitutionally permissible for a state to require that life-prolonging measures for an incompetent patient remain in place unless the patient had left behind a clear directive excluding such measures in the event of a future state of incompetency" (Molloy et al., 1991, p. 397). The state rule is not mandatory, but when an agreement cannot be reached, the court becomes the advocate for the patient and makes the decision.

This case is one example of the complex decision making that must occur when the wishes of the client, the family, and helping professionals are not in agreement. The situation becomes more complicated when the goals of the organization or the ethical standards of the professionals are also being challenged. Molloy, Clarnette, Braun, Eisemann, and Sneiderman (1991) propose the following hierarchy of decision making to help professionals proceed when the patient is incompetent and the family cannot agree on the treatment.

1. If the patient while competent had signed an advance directive, then his/her wishes must prevail, even over the contrary views of family members.
2. If the patient has left no directive, it is necessary to turn to the family. An in-family discussion of treatment options is called for, although, as a general guideline, actual decision-making authority should be allotted in accordance with the following prioritized list: spouse, adult children, siblings, other family members. (The term *spouse* would include the patient's significant other.)

3. If there is unresolvable in-family conflict regarding case management of the patient, the court should be petitioned to appoint a guardian to act on the patient's behalf.
4. In the event that there is no one to speak for the patient, then the patient's health care providers should assume that role. (Molloy et al., 1991)

Interdisciplinary teams need strategies for working with difficult families, especially when the client is incompetent (Callaghan, 1987; Libow, 1981; Lynn, 1988; Molloy et al., 1991). Here are some guidelines:

- Schedule a time for all family members to attend a team meeting. As many family members as possible should take part, as well as all professionals involved.
- Prepare a summary of the client status, in consistent and clear language. Do not use jargon or complex medical terminology.
- Describe the possible points of decision making with regard to treatment.
- State the dilemma clearly and ask the family to make the decision. The family must come to a consensus.
- Ask the family to designate a spokesperson to speak with the treatment team. Communicate any changes in client status or prognosis through the spokesperson.
- Give the family a deadline to make the decision.
- If the family cannot reach consensus, work with them. If this is not effective, ask one member to begin to make the decision. Begin with the spouse if possible.

The case manager has a specific role in this process. He or she must ensure that the mission and goals of the organization are upheld, that professional ethics are respected, that the team is kept informed, that the family has support, that the family understands its rights and responsibilities, and that the decision is made in a timely and fair fashion.

Working with Potentially Violent Clients

Cases of violence are increasingly common in the human service delivery system. **Violence** is defined as "any physical or verbally assaultive behaviors, including hitting, biting, punching, choking, pinching, scratching, throwing, pushing, cursing, threatening, striking, or injuring with a weapon or item. . . . Violence is also physical harm to one's self or others, or the physical destruction or damaging of property" (Distasio, 1994, p. 14). Individuals who have the potential for violence are often human service clients. Since they have complicated and possibly long-term problems, their services are usually coordinated by a case manager. Distasio (1994, pp. 1–2) gives the following example of what can happen.

> Harriet is a psychiatric social worker who was employed on the psychiatric unit of a large teaching hospital. She had been concerned about the anger her presence elicited from Mark, a 30-year-old chronic paranoid schizophrenic. He was very verbal in expressing his anger toward her. Mark had been granted weekend passes so that the staff could evaluate his ability to live in the community. After his last weekend pass, Harriet saw Mark standing in the hallway, conversing with another patient. Harriet continued walking down the hall, and as she neared him she caught a glimpse of something long and bright. In the next few seconds, Mark moved quickly, closed the distance between them, and stabbed Harriet in the abdomen. Harriet spent two weeks in intensive care, was on sick leave for the next two months, and needed extended psychotherapy to help her cope with the trauma of what occurred. . . . Harriet wonders if she will ever again be able to successfully confront what has become for her a personal demon—her fears and anxieties about working with violent or potentially violent patients.

Case managers have a responsibility—to themselves and everyone else involved—to look for violent tendencies in their clients. Since it is the case manager who gathers much of the information about the client and monitors client treatment and progress, he or she is in a position to know whether there has been any history of violence or warning signs during treatment. The case manager then warns the interdisciplinary team of the possibility. If any other members of the team report danger signals, the case manager passes that information to all involved. This may also require changes in the treatment plan or the addition of other professionals to the team.

Our society is more violent today than ever before. Stresses are exacerbated by personal pressures, including divorce, illness, hopelessness, and lack of skills for coping with difficulties. In addition, the policies of deinstitutionalization and treatment within the least restrictive environment have released into the community many individuals who do not have resources to care for themselves. The problem is compounded by a lack of social support for many individuals, changing values, easy access to weapons, and a tendency in our culture to condone violence.

Some human service settings are more susceptible to violence than others (Blair & New, 1991; Blank & Mascitti-Maur, 1991). The most vulnerable seem to be psychiatric settings, nursing homes, emergency departments, and outpatient settings (Haller & Deluty, 1988). In these settings, there is often insufficient support for staff who work with potentially violent clients.

Many agencies do not recognize the potential for violence in their clients. Service providers are not educated about this potential, and protocols for working with violent clients have seldom been established. Many agencies just expect their clients to be violent. It is "part of their culture," and staff members are simply expected to deal with it. Some agencies reward staff for keeping things quiet—putting the emphasis on controlling clients rather than working with them in a therapeutic way. In such an environment, professionals may hesitate to report any signs of violence, fearing that they will be accused of not doing their jobs. Also, administrative factors may contribute to poor management of violent clients: understaffing and insufficient supervision (Distasio, 1994).

The case manager can support the work of the team by acknowledging the potential for violence and preparing for manifestations of it. What exactly can the case manager do to address the issue of violence? The following eight steps, adapted from Distasio (1994), can be used as guidelines.

1. Help the members of the interdisciplinary team understand violence as a process: how to identify a preassaultive stage, what patient characteristics are predictors of violence, what helping situations increase the likelihood of violent behavior, and how staff behaviors and characteristics relate to violence.
2. Define aggression and the factors that are associated with it, such as lack of control, lack of freedom, and confusion.
3. Agree on intervention goals, such as calming the client, defusing the situation, and ensuring the safety of the client and others.
4. Teach intervention strategies to meet the goal of reducing violence or previolent behavior.
5. Provide guidelines for professional behavior.
6. Develop team responses to violent behavior.
7. Develop ways to document any violent events.
8. Encourage the agency to develop protocols for violent behavior. This may include a statement of institutional policy, a process of documentation, employee follow-up, and counseling for affected employees.

Confidentiality

In the helping professions, the obligation of **confidentiality** is fundamental to developing a relationship between the helper and the client. When the client is assured that information disclosed during the helping process will be kept in confidence, he or she feels free to share concerns and issues. The fuller the disclosure, the greater the opportunity for the case manager to gather valuable information about the client and his or her situation, which facilitates assessment and treatment planning. Trust between the helper and the client is a prerequisite to the success of their relationship. Case managers are in a unique position with respect to confidentiality, for they work with the family and friends of the client as well as with professional colleagues. As discussed in Chapter 4, this requires a special sensitivity to confidentiality issues.

One of the first points of discussion between case manager and client must be confidentiality and its meaning within the case management process. Four standards for confidentiality must be stated:

1. The case manager will keep the confidences of the client except in instances of information about harm intended to self or to others.

2. The case manager will share basic information with colleagues on the team; he or she will inform the client who those colleagues are and what that information is.
3. With the consent of the client, the case manager will give family and friends certain information (Kane, 1993).
4. In most states, communication between case managers and clients is not legally privileged. That is, the case manager can be compelled to testify in court as to what the client may say.

Even with these guidelines articulated, dilemmas often arise concerning confidentiality. Consider the following situations.

An 18-year-old has just been diagnosed as pregnant. There is no legal obligation to inform the parents of the young woman or the father of the child, but the care coordinator always encourages clients to disclose this information. This young woman refuses. The discussion is closed and the matter remains confidential, even though the young woman's mother is the care coordinator's best friend.

An elderly man is furious with his social service coordinator because she told his daughter that he was dying of pancreatic cancer. The service provider knew that the daughter's husband had just asked for a divorce. The daughter was devastated, because she had been counting on the father's support.

A counselor has been asked by his minister for information about a client who is a member of the congregation.

Having sensitive information about clients often causes dilemmas for case managers. To avoid problems, some decide to limit what information they seek to gather about a client: The less information they have, the fewer confidentiality problems they will encounter. Certainly, case managers must carefully choose what information to seek, but strict limits on information gathering are not always recommended. A complete history of the client enables the case manager to develop a treatment plan that will meet the needs of that client. *Relevant* information must be gathered, and it helps the case manager understand how to work with family members and friends. Since the case manager is also coordinating care and monitoring progress, abundant information also helps him or her give better guidance to other professionals on the team.

The case manager should address certain matters before collecting any information, to anticipate any confidentiality problems (Kane, 1993, p. 150).

1. What does the case manager need to know to do the job, and why?
2. What should become part of the permanent record?

3. What understandings should exist between case manager and client about why the information is being sought, how it will be used, and the client's right to refuse to answer?
4. Under what circumstances should information be shared?

At the beginning of the process, the case manager determines what information is needed to determine eligibility, conducts a comprehensive assessment, sets priorities, and develops a treatment plan. During the course of any case, there will emerge information that the case manager needs to know but would not have thought to ask about at the outset. A good outline of the information to be gathered keeps him or her focused on appropriate and relevant areas to probe. The client often asks the case manager why certain information is necessary; at other times, he or she discloses much more than is needed for the process to proceed.

Another dilemma is how much information to put in the permanent record. The case manager must record all information that documents the work with the client and describes his or her history. But not everything the client reveals needs to be recorded, especially if it is not relevant to the presenting problem. Unfortunately, the confidentiality of written information is not always secure, regardless of whether the record is on paper or computerized.

Once the case manager has decided what information to gather and what to record, he or she explains why each piece of information is important. Then he or she decides what parts of the information gathered will be shared with family, friends, and other professionals, and discusses this with the client. However, professionals and family members are bound to ask questions beyond what the client and case manager have agreed on. The client and the case manager negotiate about each such piece of information, and the client must give permission for disclosure. He or she also has a right to refuse to answer any question; if the information is necessary for determining eligibility, the case manager explains this. He or she can ask the client whether there is another way to obtain the necessary information.

Even following these guidelines, the case manager is bound to face issues that warrant further consideration. Three examples are the short cases presented earlier. Let's consider how these situations might be resolved. In the first case, the care coordinator gathers confidential information about the 18-year-old's pregnancy. The information remains confidential unless the agency has a policy mandating disclosure to parents or to the father of the child. If this is so, the coordinator should have informed the young woman at the time of intake. The care coordinator's obligation to inform supersedes the confidentiality guarantee. Such a policy is less likely to apply here, since the young woman is of legal age. The situation becomes complicated if the young woman's mental competence is in question, or if her health or that of the fetus is threatened in any way. The fact that the care coordinator is a close friend of the young woman's mother does not have any bearing on the issue of confidentiality. If there is no legal or policy mandate, the coordinator must not inform her best friend, even though she is the potential grandmother in this case.

In the second situation the social service coordinator is torn between her allegiance to the dying father and her responsibility to his daughter. The difficulty here hinges on the coordinator's definition of who the client is. She has chosen to behave counter to the wishes of the father by breaking confidentiality with regard to his physical condition. Before doing so, she should ask herself the following questions: Is the father competent to request that his daughter not be told? Did he give the coordinator other information indicating that the daughter should be told, in spite of his reaction after the fact? Does she believe that it is in the father's best interest for the daughter to have this information? Does the coordinator see the daughter as the primary client? If so, why? The coordinator should not violate the father's request for confidentiality unless the answers to these questions provide sufficient justification.

In the third case, the counselor is asked for information by a professional who is not involved with the client's case, at least within the established service delivery system. The counselor is under no obligation to give the information unless there is an established need to know and the counselor gains the client's consent to share the information. It would be a different matter if the counselor had reason to believe that the client might harm himself or others. There would then arise a *duty to warn,* changing the counselor's obligation from confidentiality to a duty to share information.

Duty to Warn

A case manager works in a nursing home for the elderly. One client, Mrs. Eddy, constantly expresses anger toward her husband. He lives at home and comes to visit her twice a day. He appears devoted to her and is her only contact with the world outside the nursing home. Today Mrs. Eddy says that she has a gun and plans to shoot her husband. The case manager is unsure whether to believe her. All the rooms are cleaned three times a week, and it is doubtful that a gun would remain unnoticed. Also, Mrs. Eddy is bedridden and has severe arthritis in her hands and fingers.

The **duty to warn** arises when a helping professional must violate the confidentiality that has been promised a client in order to warn others that the client is "a threat to self or to others" (Costa & Altekruse, 1994). The *Tarasoff* court decisions in 1974 and 1976 establish that mental health practitioners have a duty to warn not only others working with the client and the police, but also the intended victim. The *Tarasoff* rulings govern the law of most states (Fulero, 1988). It is a difficult judgment for a case manager to break confidentiality and invoke the decision to warn, since confidentiality is such a fundamental responsibility. Violation of confidentiality is otherwise considered unacceptable practice in most instances.

The law and professional codes of ethics provide guidance in this matter. Further court rulings have clarified the duty to warn foreseeable victims, as first prescribed in the *Tarasoff* decision. However, state laws on professional confidentiality may conflict with the duty to warn (Costa & Altekruse, 1994). For example, a Texas statute requires helping professionals to keep confidences that seemingly violate the *Tarasoff* rule. Thus, the helping professional may face situations where there are two conflicting responsibilities, both of them legal. Since professionals are personally liable when they break confidences, 13 states have passed statutes limiting the professionals' liability when they report such threats (Hulteng, 1989; Lake, 1994).

Professional codes of ethics do not always provide clear guidelines in this matter. Since case managers do not have their own professional code, they use the code of whatever discipline they hold credentials for. For instance, the Code of Professional Ethics for Rehabilitation Counselors states:

> Rehabilitation counselors will take reasonable personal action, or inform responsible authorities, or inform those persons at risk, when the conditions or actions of clients indicate that there is clear and imminent danger to clients or others after advising clients that this must be done. Consultation with other professionals may be used where appropriate. The assumption of responsibility for clients must be taken only after careful deliberation, and clients must be involved in the resumption of responsibility as quickly as possible. (Woodside & McClam, 1994, p. 261)

Members of the helping professions are encouraged to use codes as initial guidelines, but ethical codes are not written to cover every possible situation (Gladdings, 1991). They are established as principles, which professionals should use to guide their behavior. Such principles sometimes conflict and often do not address new issues that emerge.

The duty to warn is especially difficult for the case manager. He or she must consider the legal implications should someone be hurt. In addition, there are complications in dealing with this issue in the context of the team.

Since the case manager sometimes provides direct services and also coordinates the treatment plan, the following questions may arise:

When do I give information that is related to duty-to-warn issues to other professionals who work with the client?

Do I tell individual professionals on the team on a "need to know" basis, or do I provide the information to the entire team?

What standards do these other professionals have concerning the duty to warn and confidentiality?

How do I involve the client in determining the answers to the preceding questions?

What guidelines for confidentiality and duty to warn are established by my agency? Do these conflict with my professional code of ethics?

What are the guidelines of the other professional organizations represented on the interdisciplinary team? Are these in agreement with those of the professional organizations I belong to?

What legal standards are applied to the duty to warn and confidentiality in my state?

Because this issue is so complicated, Costa and Altekruse (1994) make the following suggestions:

Get informed consent. Many case managers make it standard practice to discuss the concept of confidentiality with clients at the outset. Because of the legal ramifications, many case managers now ask the client to sign an agreement stating that the case manager will report any information suggesting that the client may harm self or others (Margolin, 1982).

Plan ahead using consultation. The case manager must research this issue thoroughly before working with clients and discuss it with the supervisor and other agency officials familiar with relevant policy and procedure (Corey, Corey, & Callanan, 1988). This issue should also be discussed with the team, to raise everyone's level of awareness and develop approaches for addressing these situations.

Develop contingency plans. The case manager must think through exactly how he or she will handle a client who threatens harm to another and who claims to have a weapon available (in the home, office, or car).

Obtain professional liability insurance. Many helping professionals either have their own professional liability insurance or are covered through the agency's liability policy. It is important to know if the agency has such a policy; consult with supervisors about this.

Involve the client. There are advantages and disadvantages to involving the client if the case manager believes that the situation warrants a duty to warn. One way to involve the client is to first remind him or her of your duty to violate confidentiality when there is a threat to harm self or others. The case manager can ask the threatening client to warn the intended victim. Although involving the client in this way can serve to maintain the trust of the relationship, it may also enrage the client and put the helping professional in danger. The client may feel cornered and become even more determined to carry out the destructive act (Snider, 1985).

Obtain a detailed history. The case manager's development of a detailed client history should continue throughout the process. As information is acquired, special attention should be paid to any indications of past acts of violence and evidence of impulsivity or anger. Look for other clues or warnings of danger, such as threats, a plan, or a weapon.

Document in writing. The case manager should document any indications of violence, threats, or expressed anger. Such notes are to be written as objectively as possible.

Costa & Altekruse (1994, p. 349) give guidelines for case managers when dealing with clients who represent a threat to themselves or others.

- Remind the client of your ethical and legal obligation to warn.
- Invite the client to participate in the process if possible.
- If possible, develop a plan with the client to surrender any weapons.
- Inform your supervisor, your attorney, law enforcement personnel, the local psychiatric hospital, and the intended victims.
- Keep the client committed, or have him or her committed for treatment.
- Keep careful records of all actions taken.

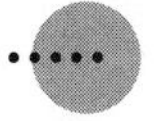

Autonomy

Client autonomy is a fundamental value in the case management process. As discussed in Chapter 1, case managers must be committed to the principles of client empowerment and participation, as well as to the ultimate goal of the process—that clients become able to manage themselves. Clients are to be involved in the process as much as possible, and treatment plans must support client choice and promote self-sufficiency. At the same time, the case manager has the obligation to provide the client with quality services and to act in his or her best interests. Sometimes, the case manager may believe that what the client prefers is not in his or her interest; this situation can entail an ethical conflict.

In an effort to encourage client participation, case managers can regard **autonomy** in a broader sense than just giving the individual the right to make choices (Hofland, 1993). Positive autonomy means that the case manager works to broaden and strengthen the autonomy that the client has. If client preferences will result in danger to the client or others, the case manager must find a way to make those preferences more appropriate.

There are several instances in which client autonomy is not an absolute priority—for example, when client preferences interfere with other clients or other helping professionals, when the client is not competent to make decisions, and when clients need protection from their own decisions (Hofland, 1993). In such situations, the case manager must have a clear conception of the reasons why autonomy should be restricted. The following case exemplifies the issue.

Mrs. Zeno is married and has three children. She has been a client of Child and Family Services for the past two years. She is now going to school, working for an associate degree. She intends to work full-time after she completes her education. According to Mrs. Zeno, her husband takes all the money that she receives from Child and Family Services and from welfare. He gives her a small allowance, from which she buys groceries and pays for the rent and utilities, but there is never enough money for clothes for her children or educational supplies for herself. Mrs. Zeno is reluctant to challenge her husband. She does not want to disrupt her home, and she needs her husband's support if she is to finish school. In her last meeting with her case manager, Mrs. Zeno firmly stated that she would do nothing about her husband's behavior. The case manager believes that Mrs. Zeno should, at the very least, talk with her husband about the problem.

In dealing with issues of autonomy and self-determination, asking the following questions can help the case manager and guard the client's autonomy.

What are the facts of the case? Before any autonomy issues can be thoroughly understood, the case manager must gather relevant information.

Can you understand the history of the case and the current situation, and view it from the client's perspective? It is important to understand the client's perspective, which determines his or her preferences and choices.

Is the client competent? It is difficult to determine competence. Clients may be able to make good decisions at certain times but not at other times. Also, they may demonstrate competence in one area of decision making but not in other areas. Only with great caution should a client be declared incompetent.

Counseling at a women's center

What does the client want? Learn exactly what clients desire. Inform them fully of all the alternatives available, and help them understand the consequences (positive and negative) for each alternative.

What are the barriers to what the client wishes? For many reasons, clients may not be able to get what they want. Sometimes, policies and regulations restrict the case manager's ability to support client choices. Often there are resource problems: The services are not available, or funds are insufficient to pay for them.

Can these obstacles be overcome? It is the responsibility of the case manager to break down barriers to client autonomy. When making any decision that violates client autonomy, the case manager must inform him or her of the reasons for making that decision.

What are the risks involved in allowing client autonomy? In determining whether to violate client autonomy, one must calculate the risks to the client and others. If a client's decision would cause harm to self or others, the case manager must overrule him or her.

What is the advice of the treatment team? The opinions of the treatment team must be sought and considered; these professionals have much at stake in their own work with the client. If any of the professionals involved do not regard client autonomy as a high priority, this issue must be addressed early in the case management process.

How can the case manager guard the values of the client? The values of the client should guide the planning of services. If the client has conflicting values, the case manager works with him or her to sort through the priorities.

Client Preferences: One Component of Autonomy

To develop the care plan, case managers take into account client preferences. In every part of the plan, client preferences receive priority, and everyone involved must respect those preferences. Sometimes this creates problems, as illustrated in the following examples.

In developing a plan for Ms. Toomey, the case manager has limited funds, but part of the money available is designated for health and recreation activities, three days a week. Ms. Toomey hates any mention of good health; she especially dislikes exercise and fresh air. Ms. Toomey really loves to go to the movies; she does not watch them at home. The case manager is trying to decide whether she should use the resources available to pay for an attendant to take Ms. Toomey to the movies three days a week (which is Ms. Toomey's preference) or to have the attendant walk with her in the park at least once a week. The case manager is not sure that she can justify movies three times a week to the funding source.

Mr. Krutch needs child care for his baby while he works during the evening at a factory. The available child care workers are almost all Latino and speak little English, but they work well with children. Mr. Krutch refuses to have a Latino person in his home or caring for his child.

The case manager should consider the following guidelines when faced with issues involving client preferences.

Ask what the client needs. The case manager must have a clear idea of the client's needs.

Ask why the client has a particular need. For each of the needs listed, articulate clearly why the client needs the proposed services or support. The client and the case manager then understand why each of the needs is important. During this process, the cultural background of the client must be considered (Williams, 1993). Too often, case managers do not realize the preferences of their clients, particularly when they come from different cultures.

Can each of the needs be addressed? It is possible that not all the client's needs can be addressed. Sometimes resources are limited, but the client usually has some choices available. The resources available may determine which needs can be met. The case manager must tell the client what is available, and he or she must also promote client interests by advocating the introduction of services that are not currently available.

How are client preferences met? Respect for client autonomy means that the case manager works to meet the preferences of the client. Creative advocacy and problem solving can help in finding ways of meeting client needs. By advocating institutional adjustments to client preferences, the case manager builds trust and strengthens the relationship. Within a good relationship, clients are more willing to make their preferences known.

When is a preference not legitimate? The case manager has a responsibility to follow the agency's policies—one client is not entitled to an undue share of the case manager's time or of any other resource at the agency's disposal.

Let's look at how to apply these guidelines to the cases that were presented earlier. In the case of Ms. Toomey, the case manager faces a dilemma. Ms. Toomey does not want to take part in any activities involving exercise or good health. She has made it clear that she only wants to go to the movies. The funding agency insists that its money not be used to pay for three movies a week. The agency will only fund one movie per week; its criteria for funding eligibility require that there be a variety of activities and that they be related to physical health. In this case, the treatment team agrees with the funding agency. It suggests that an exercise program and an activity that will challenge her intellect will better support the goals set for Ms. Toomey. With this information, the case manager can discuss options with Ms. Toomey. The case manager is confident that she and Ms. Toomey will find a solution. She also knows that she has limited time to spend on the matter, since she has responsibility for 35 other cases.

The service provider working with Mr. Krutch to find child care is in a difficult situation. When Mr. Krutch expressed his disdain for individuals of Latino origin, the care coordinator listened carefully. Two issues emerged. First, most of the available child care workers are Latino. They are very able and have excellent recommendations. Second, the agency has a carefully stated commitment not to discriminate in hiring based on race, gender, religion, or national origin. On the other hand, the care coordinator working with Mr. Krutch realizes that it is important for him to have a good relationship and a feeling of trust for whoever cares for his child. In this case, the care coordinator decides to try to find a non-Latino child care worker, but she also gives Mr. Krutch a realistic picture of what candidates are likely to be available. She explores Mr. Krutch's reasons for his reluctance to hire a Latino worker, and she prepares him for the possibility that he may have to give such a worker a try. One criterion she used when deciding to seek a non-Latino worker was to ask what harm this might do to the client. Mr. Krutch was vehemently against such a hire, and going against his wishes would not have been helpful to the goals of the care process.

Breaking the Rules

One of the most difficult dilemmas that a case manager can face is whether to comply with laws, regulations, and rules of practice when these do not appear to meet client needs. Two roles of the case manager collide—representing the agency versus serving as the client's advocate.

Ms. Dimatto is a case manager at an AIDS center in a small community. One of her clients, Mr. Sams, is dying. Mr. Sams is living with his partner, and his family resides in a nearby town. He has a good deal of family support. Mr. Sams is bedridden now, and his partner works full-time and cares for him at night. His family shares the responsibility for his care during the day. His mother stays with him three days a week, and his father three days a week. There is one day when Mr. Sams is by himself. During the other days, his parents read to their son, talk with him, give medications, and feed and bathe him. Mr. Sams' partner takes part in a self-help partners' group sponsored by the AIDS center. The center also works with the family, helping with any crises that arise and finding needed resources. In addition, the agency gives the parents a small stipend for the care that they give their son during the day.

The family is facing a problem. Next month, the mother is having bypass surgery, and she will need her husband to help her during her recovery. Neither she nor her husband will be available to provide care to their son.

Mr. Sams knows that he needs daily care, so he has consented to hire an aide to stay with him and provide him the help that he needs. However, according to agency policy, his parents cannot receive any agency support for any care they give him for six months after the aide is hired.

To protect the client, the agency has established a rule that members of the family who receive in-kind or monetary support for client care must be providing such care on a continuous basis. If the care is interrupted for more than 14 days, the family cannot be supported by the agency as caregivers for six months. This rule works for stability of care and discourages families from providing help only when it is convenient for them. Another policy states, "The assistance must be provided by an individual whose sole responsibility is the care of the client." Thus, in this case, the father may not be reimbursed if he cares for both his wife after her surgery and for his son.

Ms. Dimatto, the case worker, has a conflict to resolve. On the one hand, in accordance with the rules, she could hire an attendant to stay with Mr. Sams until his death. His parents could visit him as they wish, but they would not receive any compensation for the care they give their son, nor could the father receive a stipend for caring for his wife. Ms. Dimatto could also suggest that the parents return after 14 days, asking the father to care for both his wife and his son. This solution would violate another policy—that the care of Mr. Sams be the father's sole responsibility—since the father would also be taking care of the mother during the postoperative period.

This is an example of the tension between the case manager's responsibilities as an employee of an agency and as an advocate for client interests. One question to ask (Kane, 1993) is, "Is it right or responsible to break rules for the welfare of the client?" Regulations established by governments and agencies are made to apply to everyone and every situation, but the reality is that few rules cover all situations. The obligation to follow regulations must be considered on a case-by-case basis. Kane (1993) proposes some guidelines for such consideration.

What is the purpose of the rule, and how does it help the client? It is assumed that the rules have been developed to keep the client from harm. In Mr. Sams' case, the rule that encourages continuity of family care and penalizes any break in that service is written to discourage families from supporting the client only when it is convenient. The rule supports the importance of stability in care, especially for those who are ill and dying. The regulation that requires the attendant to have this client's care as his or her sole responsibility is likewise formulated for good reason—to ensure that the client receives maximum attention.

Who made the rule? What are the consequences of violating it? Rules originate from many sources: the federal, state, and local governments; local associations; and individual agencies. Each carries authority and imposes consequences for violating its rules. The case manager must know the source of the rule and the consequences of violating it. In Mr. Sams' case, the rule was instituted by the agency two years ago, because families were taking agency stipends but were not providing good care for their relatives—who were the agency's clients. The agency finally decided that it was more important that clients receive quality care than that the care be given by family members.

What does the client lose and what does the client gain if the rule is followed? The case manager carefully articulates the client's gains and the losses if a rule may be broken. The weighing of the advantages and disadvantages helps the case manager think through the issue and place client welfare at the center of consideration.

Is this a life-or-death situation for the client? The answer to this question often gives perspective to consideration of the rule's effect on client welfare. If the answer to this question is yes, the case manager has an argument worth considering for bending or breaking the rule.

Asking for help. In breaking or ignoring rules for the client's benefit, the case manager must keep in mind three considerations: good decision making; use of supervision; and asking for exceptions (Hornyak, 1993). He or she must ask the questions posed above to have a firm basis for decision. When using supervision or asking for an exception, violation of the rules may be only one of several concerns. When asked for help, the supervisor may respond in one of several ways. First, he or she may tell the case manager not to violate the rule. At that point the case manager must decide whether to obey the rule and the supervisor or to break the rule and ignore instructions; there are negative consequences for both actions.

Second, the supervisor may give the case manager permission to violate the rule. If the case manager does so, both the case manager and the supervisor are liable for the consequences. The supervisor may also ask the case manager not to act until an appeal for an exception has been filed.

Appealing for an exception. Appealing for an exception to a rule is a very different action from just bending a rule or violating a policy. The issue becomes more public. There is open dialogue about the rule and its purpose, the possible precedent of granting an exception, and the actual issues of the case for which the exception is requested. An appeal involves a foray into the realm of political activity; it is advocacy at a different level.

Petitioning for an exception can provide an impetus for change. It helps those who make the rules see that rigid enforcement may violate the very intent for which the rule was made in the first place. In some situations, there is no way exceptions can be made. It can be useful to appeal for an exception even when there is no clear channel for such an appeal. Authorities may then understand the need to establish appeal procedures, since rules are not appropriate in every situation that arises.

Chapter Summary

Many ethical and legal issues confront case managers. They are committed to respecting the rights of clients and their families, but they must represent the agency and work within limitations of resources. Family disagreement, confidentiality, and autonomy are three of the issues that challenge case managers, and there are guidelines for approaching each dilemma.

Chapter Review

Key Terms

Violence
Confidentiality
Duty to warn
Autonomy

Reviewing the Chapter

1. What pressures make it difficult for case managers to confront ethical issues?
2. Describe the main issue in the "daughter from California" syndrome.
3. List ways that case managers can include family members in the case management process.
4. Why is violence so prevalent in human services today?
5. List eight steps case managers can follow to confront potential violence.
6. Explain the importance of confidentiality in the case management process.
7. How does confidentiality affect recordkeeping?
8. How does the duty to warn relate to confidentiality?
9. Describe how the case manager can experience conflict between the duty to warn and the requirement of confidentiality.
10. List the considerations for the case manager when giving duty-to-warn information to the interdisciplinary team.
11. Why can it be difficult to grant client preferences?
12. Relate the case of Ms. Toomey to the guidelines for thinking about issues of autonomy.
13. How does the case manager determine when it is okay to break the rules?

Questions for Discussion

1. Why do ethical issues arise from the involvement of families in the case management process?
2. Confidentiality involves some special difficulties. Discuss them. What conclusions can you draw?

3. Do you think that it is a good idea always to grant client preferences? When is this not appropriate?
4. What advice would you give a new case manager about breaking a rule to meet a client's need?

References

Blair, D., & New, S. (1991). Assaultive behavior: Know the risks. *Journal of Psychosocial Nursing, 29*(11), 25–29.

Blank, C., & Mascitti-Maur, J. (1991). Violence in Philadelphia emergency departments reflects national trends. *Journal of Emergency Nursing, 17*(5), 31–321.

Callaghan, D. (1987). *Setting limits. Medical goals in an aging society.* New York: Simon & Schuster.

Corey, G., Corey, M. S., & Callanan, P. (1988). *Issues and ethics in the helping professions.* Pacific Grove, CA: Brooks/Cole.

Costa, L., & Altekruse, M. (1994). Duty to warn guidelines for mental health counselors. *Journal of Counseling and Development, 72*(4), 346–350.

Distasio, C. A. (1994). Violence in health care: Institutional strategies to cope with the phenomenon. *Health Care Supervision, 12*(4), 1–34.

Fulero, S. M. (1988). Tarasoff: Ten years later. *Professional Psychology, 19,* 84–90.

Gladdings, S. (1991). *Group work: A counseling specialty.* New York: Macmillan.

Haller, R., & Deluty, R. (1988). Assaults on staff by psychiatric in-patients: A critical review. *British Journal of Psychiatry, 152,* 174–179.

Hofland, B. F. (1993). Use of facts to resolve conflicts between beneficence and autonomy. In R. A. Kane & A. L. Caplan (Eds.), *Ethical conflicts in the management of home care: The care manager's dilemma* (pp. 36–45). New York: Springer.

Hornyak, R. (1993). Perspective from Indiana. In R. A. Kane & A. L. Caplan (Eds.), *Ethical conflicts in the management of home care* (pp. 179–180). New York: Springer.

Hulteng, R. J. (1989). Commentary—The duty to warn or hospitalize: The new scope of Tarasoff. *Liability in Michigan, 67,* 1–28.

Kane, R. A. (1990). What is case management anyway? In R. A. Kane, C. King, & Urv-Wong (Eds.), *Case management: What is it anyway?* Minneapolis: Long-Term Care Resource Center, University of Minnesota.

Kane, R. A. (1993). Uses and abuses of confidentiality. In R. A. Kane & A. L. Caplan (Eds.), *Ethical conflicts in the management of home care* (pp. 147–158). New York: Springer.

Kelly, L. Y. (1992). *The nursing experience.* New York: McGraw-Hill.

Lake, P. F. (1994). Revisiting *Tarasoff. Albany Law Review,* 97–173.

Libow, L. S. (1981). The interface of clinical and ethical decisions in the care of the elderly. *Mount Sinai Medicine, 48,* 480.

Lynn, J. (1988). Conflicts of interest in medical decision-making. *Journal of the American Geriatric Society, 36,* 945.

Margolin, G. (1982). Ethical and legal consideration in marriage and family therapy. *American Psychologist, 7,* 788–801.

Molloy, D. W., Clarnette, R. M., Braun, E. A., Eisemann, M. R., & Sneiderman, B. (1991). Decision making in the incompetent elderly: "The daughter from California syndrome." *Journal of the American Geriatrics Society, 39*(4), 396–399.

Snider, P. D. (1985). The duty to warn: A potential issue of litigation for the counseling supervisor. *Counselor Education and Supervision, 25,* 66–73.

Williams, O. J. (1993). When is being equal unfair? In R. A. Kane & A. L. Caplan (Eds.), *Ethical conflicts in the management of home care: The case manager's dilemma* (pp. 206–208). New York: Springer.

Woodside, M., & McClam, T. (1998). *An introduction to human services* (3rd ed.). Pacific Grove, CA: Brooks/Cole.

Chapter Eleven

Professional Development

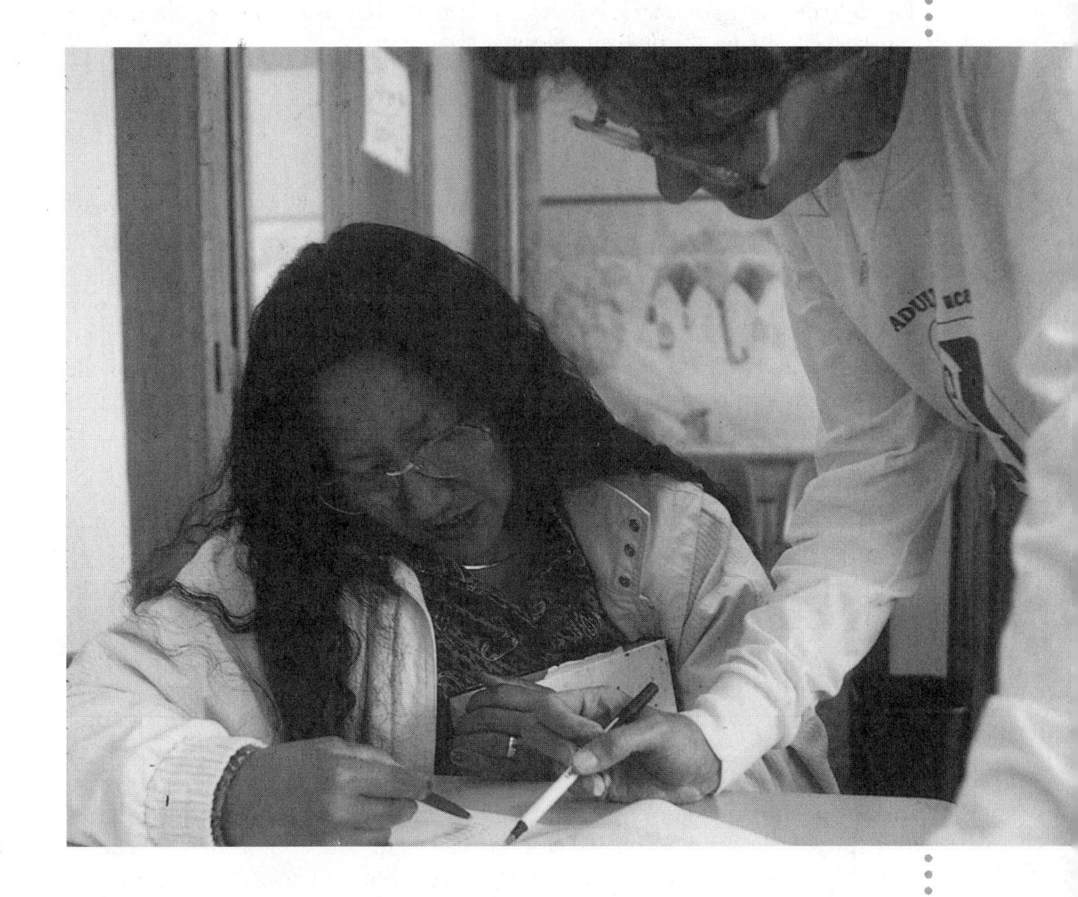

It's not easy. At the end of the day you are tired. And you go home—you know what I am talking about. I have had times I have had to go into the restroom and cry for a few minutes because it was a real hard case.

—Cathy Lowe, manager, Private Industry Council, Knoxville, Tennessee, personal communication, August 30, 1993

I think that in order to succeed at anything you need to check the reason, the real reasons why you are choosing this work. . . . You have to have a love for the field. You need to have a love for the families or the population that you are serving.

—Sandra Flecha, caseworker, Casita Maria Settlement House, Bronx, New York, personal communication, May 4, 1994

s a case manager, how much time do I have to deal with this family? How much bonding am I going to do? If I know that I have a year, how much am I going to disclose and how close am I going to get?

—ZULMA RESTO, caseworker, Casita Maria Settlement House, Bronx, New York, personal communication, May 4, 1994

The hardest thing that I do sometimes is to try to work with people who . . . have difficulty looking at other people's ideas . . . people who are not willing to compromise or work together to come up with a plan.

—LESLIE BADAINES, Tennessee School for the Deaf, Knoxville, Tennessee, personal communication, October 27, 1993

One of the most important commitments that human service organizations make to their case managers is to help them develop as professionals. As discussed in previous chapters, case management is a challenging responsibility. Providing quality services takes a long-term commitment to the development of a solid knowledge base, effective skills, and the values that support positive client growth. The agencies that employ case managers do have an obligation to provide opportunities for professional development, but case managers also need to assume responsibility for improving their own professional performance.

This chapter suggests ways in which case managers can facilitate their own growth. We begin by discussing burnout, a response to intense work-related stress—a topic that is particularly relevant for case managers. In the first quotation at the beginning of this chapter, a case manager at the Private Industry Council describes how she sometimes needs a brief emotional outlet before proceeding with her stressful work. This is one way case managers may respond to the difficulties they face.

Later in the chapter, we discuss three areas of professional development that enhance the case management process and also help reduce stress and burnout. These three areas of growth and learning are developing a case management philosophy, managing time, and developing assertiveness skills. The other three chapter-opening quotations illustrate each area, respectively. Sandra Flecha's words present her philosophy of case management: she approaches her clients through a serious commitment to human services and real empathy for the people with whom she comes in contact.

Time management is always a challenge for case managers. Also at the start of the chapter, Zulma Resto, another caseworker at Casita Maria Settlement House, lists the questions that she asks herself before getting involved with a client. An important variable for her is how much time she has to spend with the client; that time restriction has a determining effect on the kind of relationship that she can establish. The third challenge, acting assertively, is illustrated in the final quotation. Leslie Badaines describes her difficulty in working with helpers who have already formulated their opinions and made their plans without undertaking consultation. Her work with staff often calls for assertive behavior that helps move others into a more flexible frame of mind.

Throughout this chapter, we will follow the experiences of Ms. Song, a case manager at a department of human services. Her case illustrates the difficult facets of professional development. In her work of assisting young mothers with parent education and vocational training, she is committed to professional development, both for her own growth and for the improvement of client services.

For each section of the chapter, you should be able to accomplish the objectives listed.

UNDERSTANDING BURNOUT
- Name the factors that contribute to burnout.
- Recognize the symptoms of burnout.
- Apply strategies to alleviate burnout.

DEVELOPING A PERSONAL PHILOSOPHY
- Identify the components of a personal philosophy of case management.
- Write your own philosophy.

MANAGING TIME
- Analyze how your time is spent.
- Use techniques to manage time effectively.

ACTING ASSERTIVELY
- Speak assertively in order to deal more effectively with clients and colleagues.

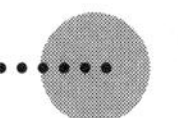

Understanding Burnout

Burnout was first defined by Christine Maslach (1978) as a phenomenon experienced by those who work in the human service arena. Those most vulnerable to burnout are "people whose work in one way or another involves direct contact with different kinds of recipients—welfare clients, patients, children, prisoners . . ." (Maslach, 1978, p. 56). According to Maslach, **burnout** is the "emotional exhaustion resulting from the stress of interpersonal contact" (p. 56). Drew Priddy is a social worker in the San Francisco Bay area. She writes about burnout in connection with working with children and families affected by crack and cocaine.

> I began to cry as I thought about these three small boys. I cried a little at first, then a flood of tears overcame me. . . . Somehow my denial quit working that night. I was now facing the horror head-on. . . . About six months later I took a leave from my job because of "severe stress anxiety." . . . I returned to my job after two months . . . the blinders would not come back. I could not stop talking and thinking about crack and the children affected by it. . . . One colleague called her supervisor one morning and said she could never come back. I understand completely. (Priddy, 1990, pp. 197, 198, 199)

Burnout and Case Management

Professionals with case management responsibilities are particularly susceptible to burnout, because many of the factors that contribute to burnout are integral to their work. Among those factors are the nature of the clients, the stresses of dealing with bureaucracy, and a personal tendency to react negatively in stressful situations.

THE NATURE OF CLIENTS

Much of the work of case management involves clients who have very complex and long-term difficulties: children who are at risk, adults who have disabilities or are elderly, and people with medical problems such as AIDS or cancer. In fact, modern-day case management emerged as a methodology designed to handle the overwhelming stress of serving clients with multiple problems. Case managers daily come face to face with clients who have little hope and who often confront long-term illness and severe social difficulties (Eagan, 1993; Halfon, Berkowitz, & Klee, 1993; Hiscott & Connop, 1990; Priddy, 1990; Savicki & Cooley, 1994).

Competency is a key issue facing case managers in providing services to such clients. Many of them have problems for which the helper needs extensive training. Working with many of these populations requires a degree of expertise that is achieved only after years of study and experience. Children with special physical disabilites, the elderly, and substance abusers are among the target populations that require specialized training (Cole, Pearl, & Welsch, 1989; Halfon, Berkowitz, & Klee, 1993; Ratliff, 1988).

THE BUREAUCRACY

Case managers also struggle with the bureaucracy in which they work. One pervasive difficulty is the size of caseloads. Case managers often lack sufficient time to give each client quality treatment; they must settle for services that are adequate at best. In addition, the programs offered and the resources available often do not match clients' needs. Case managers find themselves in the middle—understanding all too well clients' needs, yet knowing the limitations of the system.

Poor or limited supervision is also linked to burnout. Without enough orientation, communication, training, or support from good supervisors, case managers are left on their own to solve problems and acquire education and training as best they can (Arches, 1991; Eagan, 1993; Ratliff, 1988; Savicki & Cooley, 1994).

Lack of control and autonomy within the bureaucracy is another factor that contributes to burnout in case managers. According to Arches (1991), lack of control is manifested in three areas: ideological, productive, and technical. Many case managers find they do not have any input into the philosophy of the agency, its organization, or the policy decisions that guide their work with clients. Too often, whoever controls the funding also makes these key decisions.

At present, many agencies depend on multiple funding sources, each of which wishes to dictate the goals and priorities of services delivered (Arches, 1991; Savicki & Cooley, 1994).

Lack of control over the working environment also creates frustration for case managers. They are not able to serve their clients well when they have little authority to determine the size of their caseloads, the time and resources they can spend on clients, the timing of assessment, and the recommendations for and delivery of services. At some agencies, autonomy is further limited when administrators dictate the details of case management. Such micromanagement can extend to the smallest tasks of case managers, leaving no room for them to use their own skills to shape the process to match the needs of individual clients (Arches, 1991; Hiscott & Connop, 1990; Savicki & Cooley, 1994).

Limited resources is another aspect of the bureaucracy that contributes to burnout in case managers. This is very frustrating when financial considerations seem to drive decisions about policy, procedures, and available services. In dealing with insurance reimbursements and managed care utilization review, the work of case managers is often compromised. Although time that is spent for counseling, diagnosis, and evaluation is reimbursed, much of the "management" time is not. Activities such as finding placements, meeting with clients or colleagues, advocacy, and direct service do not count as billable services, so the importance of the service coordination role is discounted (Arches, 1991; Ratliff, 1988). Also stressful is the emerging trend to serve more clients with fewer resources. More and more, *accountability* refers to finances, not necessarily measuring outcomes in terms of client needs.

PERSONALITY TRAITS

People who choose to work as case managers have certain personality traits—sensitivity to the suffering of others, the ability to show empathy, a willingness to help others—that make them vulnerable to burnout (Hamilton, 1992; Ratliff, 1988; Wilson & Kniesl, 1988). Working with clients in trouble day after day can be emotionally wrenching and physically draining. A worker at the Third Avenue Family Service Center in the Bronx describes her work with domestic violence. "I get very uptight sometimes doing casework with a man who battered his wife. . . . It pinches. . . . Before coming here I worked with a coalition on domestic violence. . . . I saw the atrocities that were committed against children and the wife. . . . I took a leave in order to get out of the burnout" (Teresa Rodriquez, personal communication, May 5, 1994). Many case managers, after working so long and so hard with difficult clients, feel that they have little to show for their efforts. They become discouraged and feel ineffective because of their clients' slow progress.

Case managers sometimes have unrealistic expectations for themselves and for their clients. When expectations are high, progress is barely observable, and job satisfaction suffers. In addition, case managers receive mixed messages from the public. There is often little support for programs that help people in need, yet human service workers are nonetheless expected to make superhuman efforts to overcome impossible barriers (Priddy, 1990; Ratliff, 1988).

Recognizing and Preventing Burnout

Certain observable symptoms, related to work habits, psychological and physical well-being, and relationships with clients, can indicate that case managers are reacting negatively to stress. Absenteeism, tardiness, and high job turnover are several indicators of widespread burnout. A case manager might also have symptoms such as gastrointestinal problems, substance abuse, exhaustion, sleep disturbance, lowered self-concept, inability to concentrate, and difficulties in personal relationships. Their attitudes and working patterns may change: They become disorganized, have difficulty working with colleagues, blame clients for their own problems, and depersonalize clients by referring to them with disparaging names (Arches, 1991; Maslach, 1978; Ratliff, 1988). These changes soon affect job performance, and case managers feel increasingly more frustrated and worn down by their responsibilities. Burnout is a devastating problem for the agency, the case managers affected, and the clients served.

However, there are ways for case managers to change their environment and work schedules to reduce the stress. Among them are finding time to plan and reflect, redesigning the job or parts of it, collaborating with administrators to change the culture, requesting a temporary change of assignment, finding someone trustworthy with whom to share thoughts about work, rethinking priorities, and delegating.

Other suggestions for combatting stress and burnout involve personal strategies such as keeping thoughts of work at bay during nonwork hours, scheduling time to wind down between work and home, scheduling time for fun and for hobbies, exercising regularly, and keeping a journal. One psychologically healthy approach to stress is to begin to accept personal limitations, set realistic goals, and focus on success and small steps of progress. Linda Smith, at the Tennessee School for the Deaf, reflects on her efforts to establish realistic goals for herself. "You start out idealistic. But you need to temper the expectations so that they are realistic for the client, but also that they are realistic for you. Because if my expectations are more realistic, then I am not going to feel so discouraged about the result of the work that I put in" (personal communication, October 27, 1993). Also, case managers must understand that not everyone can be helped, and they must distinguish between what can be controlled and what cannot (Hamilton, 1992; Kerfoot, 1994; Lachman, 1994; Maslach, 1978; Ratliff, 1988).

Many of the problems case managers encounter can be addressed through professional development efforts, such as developing a personal philosophy of case management. Such a philosophy can improve stability and focus for case managers who are caught in the unrelenting demands of their work.

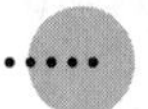

Developing a Personal Philosophy

A **philosophy** is "a system of thought that reflects beliefs, values, and knowledge, which form the basis for understanding specific phenomena or concepts" (Grant, 1994, p. 44). A philosophy of case management is a set of statements reflecting the beliefs that guide case management activities. Principles, ethical standards, and professional values guide case management practice. An individual can combine these elements in a unique way. Mark Alexander, a service coordinator in youth services, expresses his philosophy as follows. "I would consider myself a child advocate. . . . What I try to do is social change. . . . I think integrity is real important. . . . Our philosophy is teamwork" (personal communication, July 10, 1994).

Three steps are involved in developing a personal philosophy.

Step One: The Principles of Case Management

The first step in developing a personal philosophy is to define the relevant values of case management. Reviewing the guiding principles of case management from Chapter 1 will help you understand these values.

Integration of services is an effective way to meet the multiple needs of clients by coordinating the work of the various professionals involved. Service integration facilitates good priority setting and positive interaction and reduces the fragmentation and duplication of services.

Advocacy is an important focus of case management, because many clients are not able to access the system for the help they need.

Equal access means providing the opportunity for all clients to receive assistance. Affordability ensures better access to services; eligibility criteria should not exclude people who lack traditional economic, social, and political access.

Quality of service is measured against standards or criteria for effectiveness and efficiency. Providing services that meet professional standards of excellence demonstrates respect for clients.

Clients are full participants in case management activities. In each phase, case managers include the client and seek ways to expand his or her role in the process.

Evaluation is key in case management. The effects of the process must be monitored, as well as the particular services provided for the client. Clients participate in the evaluation process.

In discussing the principles that guide the practice of case management, it is also important to examine the goals of that process. Three goals were presented in Chapter 2.

Continuity of care means addressing client needs from the time of initial contact with the agency all the way to termination of services.

The whole person receives attention during case management. Many human dimensions must be considered as the case manager and the client address the problems.

Client empowerment stems from the longstanding belief that all individuals, regardless of their needs, have integrity and worth. This belief leads the case manager to give the client a central role in the process.

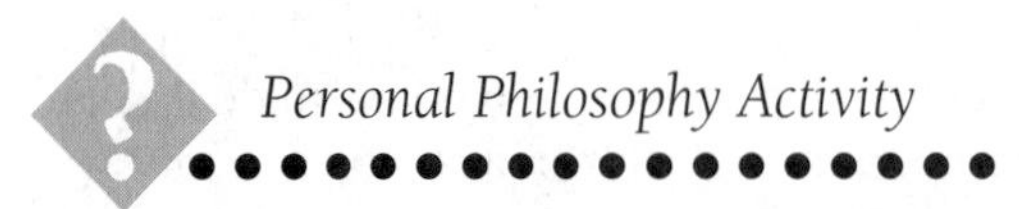

Personal Philosophy Activity

List the three principles that you most support. Explain in several sentences why they are important to you.

Step Two: The Ethics of Case Management

A second step in preparing a philosophy of case management is to think about the ethics that guide the process. Chapter 10 discussed ethical considerations in greater detail. For present purposes, three are central to case management activities.

Confidentiality is the promise the case manager gives to the client to share information only with those authorized to have such information. The case manager is the clearinghouse for information about the client. He or she must constantly weigh what information to share with other professionals and family members so that they may give the client necessary support and services.

Self-determination is the client's right to participate in decision making, particularly when the decisions relate to goals and services. Self-determination supports the goal of client empowerment that is central to the process.

Self-worth of the client must be fostered by the case manager and others who provide support and services. Unless the client develops a belief in himself or herself, the process is likely to produce only minimal temporary results.

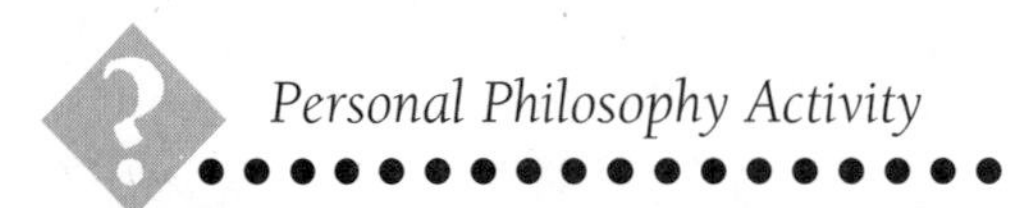

Personal Philosophy Activity

List the two ethical principles to which you are most committed. Explain how these will influence your work as a case manager.

Step Three: Personal Values of Case Management

The final step in shaping a personal philosophy of case management is to identify the values that support it. Sometimes it is difficult to pin down what values actually shape one's ideas, opinions, and actions, but one way is to examine past choices about work (Rokeach, 1973). Let's look at some of these choices.

One such choice is the decision to become a helping professional. People choose this profession for several reasons, such as a desire to help others or a sense of gratitude for help they have received. After the decision to work in human services, other choices present themselves, such as working with a specific population or concentrating on case management as a career.

Values also guide how individuals function as case managers. They are reflected in one's personal definitions of helping and case management, views of the rights and responsibilities of a case manager, standards of personal competence, conception of professional development, and expectations of how professionals should work together. Thom Prassa stresses the importance of personal values as the foundation for case management. "Your personal life is very valuable in this setting. . . . Your life goals, your values, your upbringing . . . are very important to get into this kind of work" (personal communication, July 10, 1994). Values also apply to the treatment of clients—one's personal view of human potential, attitude toward clients' rights and responsibilities, and view of who should be eligible for assistance.

Personal Philosophy Activity

Answer the following questions and then list the values that support your answers.

Name two criteria you would use to determine whether individuals need help.
What help do you think clients have a right to expect?
Name three rights that clients have.
List three responsibilities that clients have.
Name three rights case managers have.
List three responsibilities case managers have.

To finalize your personal philosophy, review what you have written and summarize it in a one-page statement. ◆

Philosophies of case management can be written in different formats. One way is to list the most important considerations in the work. Gillies (1994) offers the following list of principles.

1. I believe that individuals have the right to have a life in which they can work to fulfill their highest potential.
2. I believe each individual is unique.
3. I believe that each individual has intrinsic self-worth.
4. I believe that individuals have the right to solve their own problems.
5. I believe that at times individuals will have difficulties that they are unable to solve on their own.

6. I believe that it is important for professionals to be able to assist during those times.
7. I believe that professional assistance must be based on standards of excellence.
8. I believe that professionals must work together.
9. I believe that professionals must have autonomy to use their skills.
10. I believe that both clients and professionals have a right to be satisfied with their work.

Sujat Gandhi, a social services student, developed his own philosophy during his field experience as a case manager in community mental health.

My name is Sujat Gandhi, and I am working as a field student at the local mental health center. Before I began my field work, I studied human services and I learned about the helping process and case management. My professors asked me to write a philosophy of case management. I based it on what I had discovered about my own philosophy of helping and what I had learned about case management.

For me, the first commitment is to the uniqueness of each individual. I am a person of color and a minority in this country. I have felt discrimination, and I also know that I have discriminated. I know that it is easy to stereotype individuals when you work in the human services. When I am a case manager, I may be tempted to put clients into categories because it will make my work easier. I will try not to do that.

My second commitment is to allow each client to determine the problem and to work with that client to solve the problem. I am not sure if there is a fancy professional name for this, but I do know that empowerment is important. I want my clients to solve their own problems with my support, so they can solve other problems later when I am not there.

I also believe that since clients have many problems, it takes many professionals to solve them. I am committed to working with other professionals. In human services, we call this the interdisciplinary approach. *In case management, this is especially important. One of the major functions of the case manager is to get professionals to work together to help the client. I want to do this.*

Of course, Sujat's philosophy of case management was written before he had worked as a case manager, so it will change through the years as he gains experience and knowledge. Importantly, however, Sujat did have a philosophy when he began his field experience, and he used it to guide his professional efforts.

One challenge that is common for case managers is managing time. The next section describes why this presents such difficulty for them and explains ways to use time more effectively.

Managing Time

Time management is particularly important for case managers, but the nature of their work makes it especially difficult (Cassell & Mulkey, 1985; Holbrook, 1983; Sheridan, 1988). There are so many types of activities required of them: interviewing clients; writing reports; meetings with clients, team members, and others; arranging services; transporting clients; and monitoring services. Each of these takes specific skills, a special orientation, and considerable time. In addition, the various activities require interaction with many different individuals, including colleagues, clients, and families. Each contact takes time—to make the initial contact and to monitor the work and maintain relationships with those involved (Cassell & Mulkey, 1985; Ellis & Hartley, 1991; Gillies, 1994). Another facet of case management that makes time management difficult is the uniqueness of each case. At the beginning of work with a client, it is hard to predict the full scope of the services needed or the intensity with which the client will need support (Gillies, 1994). An average caseload for case managers in home health care may be 25 clients, but rarely do they have "average" clients.

Another difficulty is the lack of consistent demands on time. As a result, it is very difficult to establish a regular schedule and to plan and then perform certain tasks at appointed times. When crises interrupt this complicated work, the case manager must constantly weigh and shift priorities. Even the best time manager must set aside the plan for the day to assume additional responsibilities that emerge and demand attention.

In thinking about managing time, Cassell and Mulkey (1985) suggest asking four questions:

- What personal characteristics, both strengths and weaknesses, influence how case managers manage time?
- Exactly how is time spent?
- What techniques are used to control time?
- What is the time management system used? What principles are followed?

Personal Characteristics

Ultimately, case managers must develop systems of time management that fit their own styles and work situations. Gillies (1994) suggests that personal styles are reflected in the approach used to manage time. For example, case managers who are goal-centered structure their time to meet their goals and priorities. They regularly review their work as it relates to their goals and priorities. Other case managers might be plan-oriented, spending much of their work time following a plan they had previously established. They make deadlines and know the status of their projects, even their long-term ones. Another style is completion-focused. Case managers who have this style are able to clarify how to bring a project or task to its completion, and they are good with details. These are just three examples of time management styles. Sometimes it is

	I can	I cannot
1. Describe my job	______	______
2. Perform skills required	______	______
3. Understand the target population	______	______
4. Locate relevant resources	______	______
5. Develop a good network for referrals	______	______
6. Gain consensus	______	______
7. Develop relationships with others	______	______
8. Maintain relationships based on honesty	______	______
9. Concentrate on a task	______	______
10. Work quickly	______	______
11. Learn easily	______	______
12. Motivate others	______	______
13. Problem-solve quickly	______	______
14. Maintain good physical health	______	______
15. Maintain good mental health	______	______
16. Set priorities and make a plan	______	______
17. Follow a plan	______	______
18. Make decisions	______	______
19. Organize my work	______	______
20. Say "no"	______	______
21. Overcome fear of new situations	______	______
22. Be creative and help others be creative	______	______
23. Operate with a team	______	______
24. Try to improve my skills and abilities	______	______
25. Be confident about my professional future	______	______
26. Be confident about my personal goals	______	______
27. I am satisfied with my present goals at work	______	______
28. I feel comfortable and optimistic about achieving my goals	______	______

Figure 11.1 Characteristics related to managing time

helpful to do an assessment to identify personal characteristics that relate to managing time. Figure 11.1 is an assessment instrument case managers can use to focus on their time management skills.

Once this instrument has been completed, responses to the assessment questions are analyzed for strengths and weaknesses. The form shown in Figure 11.2 is helpful in organizing this information. After personal characteristics have been determined, it is important to find out how case managers use their time.

Characteristic	*Strength*	*Weakness*

Figure 11.2 Elaboration of strengths and weaknesses related to time management

How Time Is Spent

Before case managers can manage time effectively, they must assess how they actually use time during the workday. A time record can be used to establish time spent and outcomes achieved. A time log records activity every 15 minutes, describes and summarizes interactions, and covers at least two weeks, so as to capture variations in workflow (Cassell & Mulkey, 1985; Ellis & Hartley, 1991; Fritz, 1991). Once such a log is completed, the data are analyzed to reveal how time has been allocated.

There are several ways to analyze the data. One is to examine the distribution of time in various case management activities, such as intake interviewing, reading reports, team meetings, and making community visits (Cassell & Mulkey, 1985). It is also helpful to know how much time is spent managing cases and providing services. A form like Figure 11.3 might be used for such an analysis.

Time Management Techniques

The time analysis of the tasks of one day, considered in light of the extent of control the case manager has, provides the basis for managing time. The third component of time management is the planning and implementation of more effective use of time in several steps. It begins with setting goals and priorities and ends by assigning particular activities to times when they can best be completed.

Laekin (1976) recommends that desired goals be categorized into three time spans: the next five years, the next three months, and this day. The next step is to assign priorities to the goals. For example, an ABC scale can be used, where A is the most important, B is moderately important, and C is least important. The goals can then be sorted into categories: short-term, intermediate, and long-range. How do case managers use this information in planning for the day? They focus on the most important activities and choose tasks that are ranked A (most important) from each of the categories—short-term, intermediate, and

	Number of Hours/Week	%
Managing		
Intake interviewing		
Report writing		
Team meetings		
Community visits		
Making referrals		
Monitoring the plan		
Community outreach		
Subtotal		
Direct service		
Direct service to clients		
Direct service to families		
Subtotal		
TOTAL		

Figure 11.3 Time allocated to case management activities (managing and direct service)

long-range. In Figure 11.4, Ms. Song has completed such a list of priorities using two categories: (1) meeting the needs of her clients and other responsibilities and (2) tasks that must be accomplished.

The next step is to develop a list of tasks that will help meet established goals, then decide how to structure each day so as to use work time most efficiently.

Although use of time is partly determined by the rhythm of the agency and the personal preferences of case managers, the following guidelines will help structure time (Cassell & Mulkey, 1985; Clark & Clark, 1992; Ellis & Hartley, 1991; Fritz, 1991; Gillies, 1994; Truitt, 1991).

WORKFLOW

- Pick the two most important goals and do them first.
- Handle the toughest jobs first.
- Alternate difficult and easy tasks.
- Group similar tasks.

Misha White	
Set up intake interview	A
Read chart	B
Talk with the foster care unit	C
Call Dr. Rogers	A
Susan Minniti	
Follow up family visit	A
Set up appointment with pediatrician	A
Write case notes for last week	B
Talk to Susan about ideas for schooling	B
Ask Susan about her physical	B
Becca Clebo	
Check employment status	B
Jo Morales (and David)	
Prepare for staff meeting	A
Other	
Get coverage during vacation	A
Check in-service time	B
Fill out quality assurance forms	A
Talk to Bruce about new sources for food	B
Find eye exam resource	A

Figure 11.4 Prioritized list of tasks for Ms. Song

ASSIGNING TIME

- Plan some uninterrupted time.
- Plan time for the unexpected or for crises.
- Allot time around deadlines.
- Assign time to plan.
- Assign time for paperwork.
- Have a list of quick tasks to use as filler.

BEING CLEAR WITH OTHERS

- Understand assignments.
- Delegate tasks that others can or should do.
- Develop a system for monitoring the work of others.
- Say "no" when the assignment is inappropriate or there is not enough time to complete the task.

Three issues of time management relate particularly to case management: people versus paperwork, large caseloads, and efficiency and effectiveness.

Managing paperwork This has always been a challenge for human service professionals. As early as 1981, Karin Eriksen discussed the dilemma of how to maintain complete records when that takes time away from contact with clients and their families. As accountability has come to be more strongly emphasized and cases more complex, paperwork responsibilities are critical to the success of case management.

Case managers can use several techniques to complete their paperwork in a timely manner: finding privacy to write and reflect, scheduling time regularly to complete assignments, and choosing to work on paperwork early in the day if procrastination is a problem. Case managers can also use partners for support and computers to facilitate their work, in addition to customized systems of sorting and filing. (Chapter 12 will have more information on paperwork and computers.)

Kim Ehlers, at the Mitchell Training Center, does paperwork during hours when the agency is closed. "I try not to do as much paperwork. . . . My door is never closed. . . . We can take two days a month. We can also take a day off during the week and come in and work on a Saturday when there is no one there, to get the paperwork done. That is wonderful!" (personal communication, October 14, 1995). Each case manager has to develop his or her own system to make the use of time more effective.

Managing a heavy caseload Many case managers find themselves having too many clients to serve them all well. Managing time then becomes critical in meeting client needs. One complicating factor is the difficulty of estimating the care needed for the "average" client. Client needs differ, resources may be difficult to obtain, families may need particular attention, and client goals may shift in the middle of the process. Certain warning signs appear when the caseload is too heavy or time is not being managed well. Among them are phone calls not being made or returned; missed deadlines for providing a service, making a home visit, following up, or maintaining progress; incomplete paperwork; and hours of overtime work, despite which the helper is unable to catch up (Cassell & Mulkey, 1985; Fritz, 1991; Sheridan, 1988).

Time management techniques can help deal with heavy caseloads. One technique is to prioritize on a monthly, weekly, and daily basis; this helps maintain focus on what work must be completed. Another is to map out each day spent, making clear the outcomes and tasks that are not completed. Finding more efficient ways to gather information, write reports, and monitor client progress are also helpful. It is possible to differentiate among clients in terms of time allocation; by reviewing regularly, one may discover that too much time and effort have been allotted to particular clients. Case managers can also get in the habit of telling clients what services they can expect, to avoid backtracking and rehashing later.

Efficiency and effectiveness It is important to make this distinction. *Efficiency* relates to the amount of energy spent and time saved on a task. Being more efficient does not necessarily mean providing better or more effective services. There are two ways to improve efficiency and also increase *effectiveness.* First, the most efficient way of completing a task may actually improve the quality of the service by eliminating delays in service delivery. Second, the time and energy saved can be spent in other ways that benefit the client directly. At the Family Counseling Agency in Tucson, Arizona, helpers have been able to manage time more efficiently (to the direct benefit of clients) by screening referrals with a telephone interview. "We ask about her daily living activities and her independent living skills and income in order to prioritize. . . . We spend a lot of time on these calls. . . . I have my most experienced staff doing intake, because sometimes if you spend 45 minutes on the phone, you can really give them some good resources. Maybe they don't need to be in our system at all, and we can steer them where they need to be" (Suzy Bourque, personal communication, October 7, 1994).

When Time Is Managed Poorly

When case managers do not manage time wisely, there are many undesirable outcomes. According to Fritz (1991), the following negatives can be associated with poor use of time:

- *Excessive tension.* Tension is exhausting and results in poor judgment and ineffective relationships.
- *Reliance on excuses.* Defensiveness about ineffective performance may make others hesitate to depend on the person.
- *Indecisiveness.* This leads to more ineffective time management—shifting from task to task and procrastinating.
- *Perfectionism.* To demand perfection often means not bringing projects to completion.
- *Negative emotions.* Inefficient work habits generate guilt, worry, and hostility—emotions that hinder appropriate goal setting and planning.
- *Insecurity.* This often results in valuable time wasted in trying to impress others.
- *Workaholism.* This is an indication that time management has broken down.

When such negative outcomes result, clients are not served well, and case managers find little job satisfaction. These factors have a strong relation to excessive stress and burnout. Learning how to manage time is a worthy goal for case managers. Let's see how Ms. Song manages her time.

The following scenario traces the beginning of a day with Ms. Song. At its conclusion are questions that will help you use the principles of time management to assess her use of time and provide suggestions for improvement.

While driving to work (already late), Ms. Song listened to the radio. She heard an announcement about last night's state legislative session; representatives had voted to withdraw the $3 million subsidy for the jobs program for young mothers. This decision will have a major impact on young mothers and children—the people with whom Ms. Song works as a case manager at the Department of Human Services. She stopped at a convenience store to *buy a paper so that she could read more about the decision. She read the lead story on the front page, then flipped to page 8, where she perused an article about a local doll show and then another about the dangers of motorcycles. Her son wanted one for his birthday, and Ms. Song was concerned. By the time she looked up, she was already late for work. Ms. Song scurried to her car and headed for the office.*

As Ms. Song entered the human services building, she reflected on how much she loved just being there. She began a ritual series of visits with colleagues. On a given day, it could take anywhere from 15 to 30 minutes just to reach her office. Today was no exception; the legislature's action was a major topic of conversation. Ms. Song's late start made planning for the day difficult to accomplish. As on every other day, she felt that she was starting off behind schedule.

She hurried into action, because she needed to make her phone calls early. Seven new clients had been referred to her last week, and she needed all seven assessments completed by the end of this week. This particular deadline always causes Ms. Song stress. She usually has to talk to the young mother, visit the parents or other family members, and talk to a former teacher and at least one other professional who is involved with the client—just to have enough data to make the initial assessment.

Ms. Song made eight phone calls. She arranged four family visits, and three of them were for evenings this week. She had a personal rule that she would work only two evenings each week, but that was already broken, and it was only Monday. She then played phone tag with an obstetrician and another colleague in her unit.

All of this was interrupted by three crises. One involved child care for a client who had disappeared, another a client who had contracted mononucleosis and could not go to class, and the third a colleague who needed her help supporting a distraught client during an intake interview. Right before lunch, Ms. Song decided that she had better make a list of everything she must accomplish before leaving for a visit to the local high school. She also had appointments to interview two clients today in the office; she had to find time to read the records of the four whose homes she would visit; and one client needed her help preparing for a job interview. About this time every day, Ms. Song doubts that she will ever be on top of her job responsibilities.

The following questions will help you understand how Ms. Song manages time and how she can improve her time management skills.

1. What personal characteristics help Ms. Song manage time? What personal characteristics keep her from being a good time manager?
2. Exactly how does she spend her day? What tasks does she perform? Is she managing? Providing direct service? Who controls her time?
3. How would you help Ms. Song prioritize?
4. What principles of managing time would be most relevant to Ms. Song?

5. What suggestions would you give to Ms. Song if she were to ask for your help?

Time management is one skill that can improve the performance of case managers. Another is learning how to act assertively.

Acting Assertively

Communication is "a process of exchanging ideas and feelings. It occurs when meaning is conveyed from one person to another, verbally and nonverbally" (Hamilton, 1992, p. 338). Important communication between individuals occurs even when there is silence. Communicating effectively is always a challenge, and one's ability to communicate can always be improved. Specifically, the work of case management can be facilitated by **assertiveness.** To communicate assertively is to express oneself with confidence and conviction (Alberti & Emmons, 1986; Jakubowski & Lange, 1978; Lee & Crockett, 1994; Williams & Long, 1991). Communicating assertively is also positively related to "building self-concept, increasing self-confidence, and decreasing stress and anxiety" (Lee & Crockett, 1994, p. 422).

Defining Assertiveness

There are several elements to communicating assertively. The first is recognizing the equal rights of the speaker and the listener. A second is maintaining a positive attitude about the communication and wanting it to be fair. An assertive communicator is also comfortable with negative feelings such as anxiety, anger, and frustration (Alberti & Emmons, 1986; Herman, 1978; Jakubowski & Lange, 1978; Smith, 1992; Weinfeld & Donohue, 1989). The goal of assertive communication is to relate to others in ways that lead to maximum dialogue and an outcome that is positive for each participant. All who participate must know that they are respected and that their ideas have worth. This manner of communicating builds trust.

Assertive behavior can be better understood in contrast to two other ways of communicating: nonassertive (or passive) and aggressive. Nonassertive communication is characterized by inaction. The passive individual does not participate in the communication process, but rather maintains silence or quietly acquiesces. In essence, such passivity negates any opportunity for dialogue that would help clarify, expand, or change the pattern of thought or action under discussion. Those who are passive often feel as if they have nothing of value to say or that they will be punished if they speak up. Passive communicators violate their own rights by allowing others to infringe on them (Alberti & Emmons, 1986; Lee & Crockett, 1994; Weinfeld & Donohue, 1989).

People who communicate aggressively do so in a way that violates the rights of other speakers and listeners. In contrast to assertive communication, communicating aggressively is hostile, competitive, and one-way. Aggressive

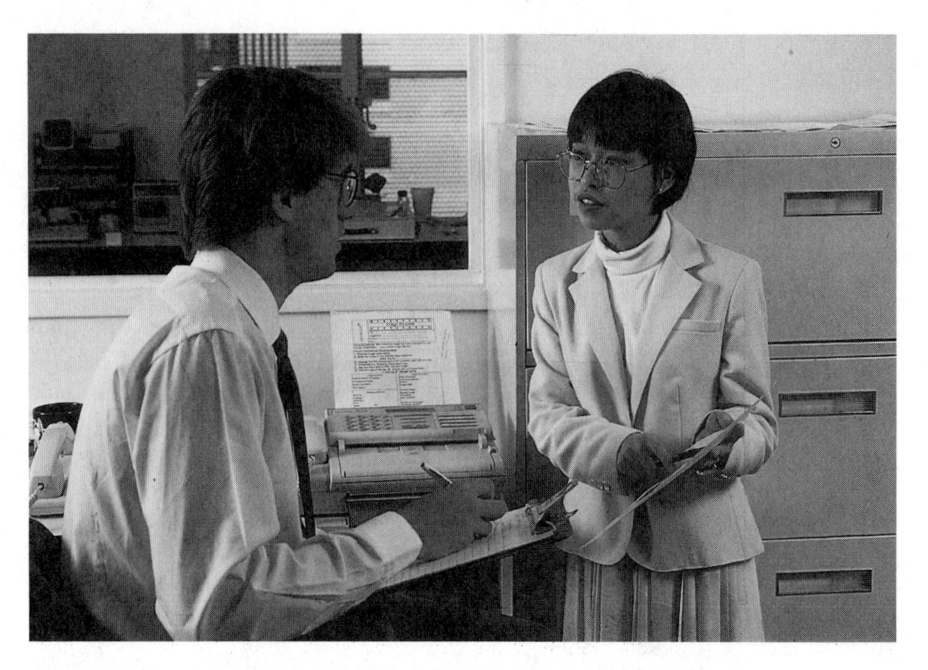

Case managers must be able to communicate assertively with both clients and colleagues.

communicators wish to dominate the interaction and to give orders; they are not seeking dialogue. They want their opinion to be the final word heard and to govern the decision. Such people stand up for their own rights in a way that violates the rights of others (Alberti & Emmons, 1986; Lee & Crockett, 1994; Weinfeld & Donohue, 1989).

Assertive communication works in the case manager's favor, for every facet of case management involves communicating with others. Whether initiating assertive communication or responding to passive or aggressive clients or colleagues, assertive communication is an important skill.

Communicating assertively also has much to do with the principles of case management discussed in Chapter 1. Both empowerment and advocacy benefit from assertive communication. Advocacy is by definition standing up for the rights of those who cannot represent themselves, and when case managers engage in advocacy, they are acting assertively. Judith Slater describes assertiveness in a hospital: "I think that it takes some real assertiveness skills to be willing to advocate. . . . It was kind of scary to get up there and have this doctor yelling at you. But you have to look at the patient and say 'What do they need?' " (personal communication, October 14, 1995).

Empowerment is the activity of instilling the belief that individuals have rights and abilities and can solve problems for themselves. Assertiveness promotes the self-confidence and self-respect that supports empowerment. At times, clients and their families do not want to be empowered, and this creates a very difficult situation for case managers. Clients may be angry when they do not receive what they feel they are entitled to or when they are asked to help themselves. Alan Tiano at Pima Health System describes such an encounter. "Some of our clients believe that they should be getting more. . . . We have one right now . . . a very assertive family who did not want to provide help . . . thought the system should provide all the services" (personal communication, October 7, 1994).

GUIDELINES FOR ASSERTIVE BEHAVIOR

Two guidelines are useful when learning assertive behavior: respect personal feelings and know personal rights. When individuals are not acting assertively, one of these principles has usually been forgotten (Alberti & Emmons, 1986; Weinfeld & Donohue, 1989).

Respect personal feelings When acting assertively, the communicator must respect his or her own feelings and those of others. For example, the case manager must make sure his or her view of a particular situation gets heard. People often just walk away from situations because they do not know how they are feeling, do not trust their feelings, or do not think that they have the right to those feelings. Exploring personal feelings, recognizing them as legitimate, and listening to others' feelings provide the basis for assertive communication.

Know personal rights Each person has a number of rights, which provide a foundation for a sense of personal worth and dignity. Awareness of these rights and acceptance of them make it easier to learn how to be assertive.

How Case Managers Use Assertiveness

Assertiveness helps case managers communicate more effectively with clients and colleagues in five areas (Smith, 1992; Sundel & Sundel, 1980). As you read about each area, you will learn why assertiveness is needed and how to use it, and you will have an opportunity for practice. Ms. Song's work as a case manager will provide examples.

SHARING PERSONAL VIEWS AND PROVIDING FEEDBACK

At times, case managers are asked by clients or colleagues to provide judgments about particular matters. At other times, they themselves feel the need to provide feedback. In both cases, there are ways of communicating that will be seen as helpful and thoughtful rather than critical or hurtful. As a general rule, advice that may be perceived as negative is better received when it is requested instead of volunteered. In case management in particular, sharing personal views or providing feedback is more positive for the listener when there is no demand to take the advice. The client's rights to autonomy, self-determination, and empowerment are important to remember. Opinions and beliefs are just that, meant only to be helpful. They are not law and should not be presented as such.

Providing feedback can help generate alternatives or stimulate a discussion of the issues, which is a positive outcome in itself and can strengthen the helping relationship. Guidelines for expressing personal views and providing feedback are given by Smith (1992) and Sundel and Sundel, (1980):

- Get consent before expressing a personal view.
- Remember that the client or colleague is a unique individual.
- Use effective communication techniques.
- Use a clear, firm voice.
- Look at the client or colleague when speaking.
- Use hand gestures and facial expressions.
- Give specific information to the listener.
- Express why you hold the opinion.
- Explain that the views expressed represent only one perspective.
- Ask for a response from the listener.

One of Ms. Song's clients is Georgia, a 19-year-old mother. She has a 1-year-old daughter and a 2-year-old son. Currently, Georgia is dating a man who is a known drug user. Ms. Song worries about Georgia and her relationship with this young man; one of her main worries is that Georgia is in jeopardy for becoming HIV positive. Ms. Song wants to bring up the subject.

---

1. Why should Ms. Song give Georgia feedback?
2. Why should Ms. Song remain silent on this subject?
3. If Ms. Song were to discuss the matter with Georgia, describe how she might do this.
4. Evaluate your answer to Question 3, using the guidelines presented above.

CONFRONTATION

Confrontation is conveying information or beginning a discussion that involves a clash of ideas or divergent opinions (Ellis & Hartley, 1991). It happens in a direct manner, often face to face. Many individuals avoid confrontation, since by definition it raises difficult issues. Often, people fear damaging their relationships with others, or they do not want to hurt them. Confrontation is more than just dialogue; it is asking for change (Brammer & MacDonald, 1996; Egan, 1982; Ellis & Hartley, 1991; Smith, 1992). Case managers' confrontations with clients or colleagues tend to occur in certain situations. These occasions fall into two broad categories: when destructive thoughts have been expressed, and when rights have been violated.

The following process allows difficult issues to be raised in a way that respects the rights of both the speaker and the listener and promotes honest, fair, and positive outcomes (Ellis & Hartley, 1991; Smith; 1992; Sundel & Sundel, 1980).

- Use a calm voice.
- Focus on the behavior that is troublesome.
- Do not label the client or colleague.
- Explain why the behavior is a problem.
- Listen to the response.
- Paraphrase the response.
- Describe a suggested change in behavior.
- Be specific about the benefits of change.
- Ask for feedback about the suggestions.

One of Ms. Song's clients, Misha, was institutionalized in her early teens for a severe bipolar disorder. Each year, she is readmitted two or three times. She now lives with her father, but he is refusing to house her any longer. Ms. Song believes that Misha needs the stability of her father's home. Ms. Song needs to talk with Misha's father about his decision to ask Misha to find another place to live.

---

1. In this instance, what are Ms. Song's rights?
2. What are Misha's father's rights?
3. What are Misha's rights?
4. In what ways can Ms. Song confront Misha's father?
5. What are the anticipated outcomes?
6. Evaluate your answer to Question 4 using the guidelines presented above.

SAYING "NO" TO UNREASONABLE REQUESTS

It is often difficult for case managers to say "no" to requests from colleagues or clients. Their commitment to help others makes them inclined to say "yes" to any request. However, a request is unreasonable if the action would violate ethical or legal codes. Also, colleagues or clients may ask for action that causes case managers to be frustrated, anxious, or physically or emotionally threatened. To be reasonable, a request must take into consideration the rights of case managers as well as those of clients and colleagues.

Just as there are times to say "no" to requests, it is also significant *how* "no" is said. If it is done in such a way that the rights of all parties are respected, communication can continue after the refusal.

Ms. Song is the envy of her colleagues because of her skill on the computer. She is continually asked to help retrieve files that have been mistakenly deleted, to transmit E-mail messages, to format a document, or to enter client data into the state data system. Just yesterday, she was interrupted three times while she was with clients, twice while she was on the phone, and once in a staff meeting. By the end of the day, she was so frustrated that she responded to the last request with a "no," shut her door, and took two aspirin. This morning, she has decided how she will deal with these interruptions. Her plan is to advise her colleagues that although she is happy to assist with their computer problems, she can only do it from 1:30 to 2:00 each day. She also plans to talk with her supervisor about computer training for the staff.

---

1. How are the requests for computer assistance affecting Ms. Song's work?
2. What do you think she would like to say to her co-workers?
3. Is her plan a good one?
4. Speaking assertively, how might she inform her colleagues of her new policy?

ENCOURAGING OTHERS TO SAY "YES"

Case managers' work often requires asking for support or help. The type of support needed can take many forms: how to do a particular job, advice about difficult clients or colleagues, additional support staff or equipment, or the assumption of greater responsibility on the part of colleagues or families. To ask for support effectively, case managers use a combination of the guidelines for confronting and for saying "no." To encourage others to say "yes," case managers first present their needs so that others can understand them.

- Focus on the specific problem.
- Describe why it is a problem.
- Detail how to make the change being sought.
- List the benefits of the proposed change.

Strategies for respecting the rights of others are also incorporated.

- Give time for consideration of the matter.
- Make sure that requests do not violate ethical or legal principles.
- Present the gains and losses to all involved.
- Create an environment where it is OK to say "no."

Ms. Song's youngest client is a 15-year-old who has a 6-month-old daughter. Sushon and her baby now live with Sushon's parents and her younger brother. Ms. Song, Sushon, and her family developed a plan of care for Sushon that includes child care, parenting classes, special high school classes, and a support group for young mothers. For this plan to work, Sushon must live at home, and her parents and brother must provide some child care. The parents want Sushon to get a job and contribute financially to the rent and groceries. They do not believe that she needs to go to school, parent training, or the support group.

---

1. What rights must be considered here?
2. What does Ms. Song wish to ask the parents?
3. Describe how she will make her request.
4. Evaluate your answer to Question 3 based on the guidelines presented earlier.

USING ASSERTIVENESS WITH AGGRESSIVE CLIENTS AND COLLEAGUES

From time to time, everyone encounters people who communicate in an aggressive way. It is difficult to turn an aggressive encounter into an assertive one. At times, one's reaction is passive because the aggressiveness is so surprising; and sometimes aggressiveness is returned for aggressiveness. Either a passive or an aggressive response leaves the feeling that rights have been violated and that communication is closed. Here are some ways to respond to people who communicate aggressively (Smith, 1992).

- Try to gather more information.
- Provide feedback concerning the aggressive behavior.
- Make clear what part of the behavior is not appropriate.
- Describe how the behavior needs to change.

Ms. Song comes from a family that is quiet and mannerly. It is very much part of her culture to put problems out of her mind. Difficulties are to be suffered, not discussed. She has never been yelled at or seen a fight, except at school. Aggressiveness in clients is difficult for her to understand, and she is not clear how she should respond. Late last week, Jo, a 22-year-old client, and her boyfriend David came in to see Ms. Song. They had a fight in Ms. Song's office. Ms. Song sat stunned while they called each other names. Then David hit Jo in the face and dashed out of the room.

---

1. What is the dilemma for Ms. Song?
2. If she is to handle the situation assertively, what should be her goals?
3. Describe how she might handle the situation.
4. Based on the guidelines stated above, evaluate your response to Question 3.

Chapter Summary

Professional development is important for the case manager. Without the opportunity to grow and develop professionally, case managers are more likely to experience burnout. Several areas of growth can help foster a more positive environment for the case manager: developing a personal philosophy of helping, learning how to manage time effectively, and learning how to act assertively.

Chapter Review

Key Terms

Burnout
Empowerment
Personal philosophy
Assertiveness

Reviewing the Chapter

1. Define *burnout.*
2. How do the nature of clients, the bureaucracy, and personality traits contribute to burnout?
3. What are some signs of burnout?
4. Identify the three components of a philosophy of case management.
5. Review the nine principles of case management and describe their relationship to a personal philosophy of case management.
6. What are the ethics of case management?
7. Discuss some choices you have made in the past that will influence your philosophy of case management.
8. Following the three steps to develop a personal philosophy of case management, write your own statement.
9. Why is time management a critical skill in case management?
10. Describe the different approaches to time management that reflect people's personal styles.
11. After recording how you spend your time for one week, analyze the data according to the various activities performed.
12. List the steps involved in planning how to use time.
13. Name and illustrate the three time management issues related to case management.
14. How do you know when you are not managing time wisely?
15. What are the characteristics of communicating assertively?
16. Distinguish between passive, assertive, and aggressive ways of communicating.
17. Give two guidelines for learning assertive behavior.
18. Illustrate how you might behave assertively in each of five areas that are important for effective case management.

Questions for Discussion

1. Why do you think it is important for case managers to avoid burnout?
2. Given your study of the field of human services, do you believe that most case managers have a personal philosophy? Support your response.
3. Do you think that it is a good idea for case managers to act assertively? Why or why not?
4. What strategy might you employ to manage your own time?

References

Alberti, R., & Emmons, M. L. (1986). *Your perfect right.* San Luis Obispo, CA: Impact.

Arches, J. (1991). Social structure, burnout, and job satisfaction. *Social Work, 36*(3), 193–212.

Brammer, L. M., & MacDonald, G. (1996). *The helping relationship.* Boston: Allyn & Bacon.

Cassell, J., & Mulkey, W. (1985). *Rehabilitation caseload management.* Austin, TX: Pro-Ed.

Clark, J., & Clark, S. (1992). *Prioritize, organize.* Shawnee Mission, KS: National Press.

Cole, B. S., Pearl, L. F., & Welsch, M. J. (1989). Education of social workers for intervention with families of children with special needs. *Child and Adolescent Social Work, 6*(4), 327–337.

Eagan, M. (1993). Resilience at the front lines: Hospital social work with AIDS patients and burnout. *Social Work in Health Care, 18*(2), 109–125.

Egan, G. (1982). *The skilled helper: Model, skills, and methods for effective helping.* Pacific Grove, CA: Brooks/Cole.

Ellis, J. R., & Hartley, C. L. (1991). *Managing and coordinating nursing care.* New York: Lippincott.

Eriksen, K. (1981). *Human services today.* Reston, VA: Reston Publishing.

Fritz, R. (1991). *Think like a manager.* Shawnee Mission, KS: National Press.

Gillies, D. (1994). *Nursing management.* Philadelphia: Saunders.

Grant, A. B. (1994). *The professional nurse.* Springhouse, PA: Springhouse.

Halfon, N., Berkowitz, G., & Klee, L. (1993). Development of an integrated case management program for vulnerable children. *Child Welfare, 72*(4), 379–396.

Hamilton, P. (1992). *Realities of contemporary nursing.* Reading, MA: Addison-Wesley Nursing/Benjamin-Cummings.

Herman, S. J. (1978). *Becoming assertive.* New York: Van Nostrand.

Hiscott, R. D., & Connop, P. J. (1990). The health and well being of mental health professionals. *Canadian Journal of Public Health, 81*(6), 422–426.

Holbrook, T. L. (1983). Social casework practice: Who controls the worker's time? *Social Casework, 64,* 323–330.

Jakubowski, P., & Lange, A. J. (1978). *The assertive opinion.* Champaign, IL: Research Press.

Kerfoot, K. (1994). Help relieve stress. In T. Porter-O'Grady (Ed.), *The nurse manager problem solver* (pp. 255–256). St. Louis, MO: Mosby.

Lachman, V. (1994). Stress management. In T. Porter-O'Grady (Ed.), *The nurse manager problem solver* (p. 255). St. Louis, MO: Mosby.

Laekin, A. (1976). *How to get control of your time and your life.* New York: Signet.

Lee, S., & Crockett, M. S. (1994). Effect of assertiveness training on levels of stress and assertiveness experienced by nurses in Taiwan, Republic of China. *Mental Health Nursing, 15*(4), 419–432.

Maslach, C. (1978). Job burnout: How people cope. *Public Welfare, 36,* 56–58.

Priddy, D. (1990). A social worker's agony: Working with children affected by crack/cocaine. *Journal of the National Association of Social Workers, 35*(3), 197–201.

Ratliff, N. (1988). Stress and burnout in the helping professions. *Social Casework, 69*(3), 147–154.

Rokeach, M. (1973). *The nature of human values.* New York: Free Press.

Savicki, V., & Cooley, E. J. (1994). Burnout in child protective service workers: A longitudinal study. *Journal of Organizational Behavior, 15,* 655–666.

Sheridan, M. S. (1988). Time management in health care social work. *Social Work in Health Care, 13*(3), 91–99.

Smith, S. (1992). *Communications in nursing.* St. Louis: Mosby.

Sundel, S. S., & Sundel, M. (1980). *Be assertive: A practical guide for human service workers.* Beverly Hills, CA: Sage.

Truitt, M. R. (1991). *The supervisor's handbook.* Shawnee Mission, KS: National Press.

Weinfeld, R. S., & Donohue, E. M. (1989). *Communicating like a manager.* Baltimore: Williams and Wilkins.

Williams, R. L., & Long, J. D. (1991). *Manage your life.* Dallas: Houghton Mifflin.

Wilson, H. C., & Kniesl, C. R. (1988). *Psychiatric nursing.* Reading, MA: Addison-Wesley.

Chapter Twelve

Case Management Today

It would help to have computer skills. When a parole officer is hired who doesn't have these skills, they receive training to learn how to use the computer. We talk with our supervisors about a case often on an informal basis. We document something if there is a question about a parolee going inpatient. Suppose he has a great job and loses his job due to being placed in an inpatient program. It may be more detrimental than the inpatient program [is beneficial]. I talk to my supervisor about it and see what he/she recommends. Then I document it. Such and such was discussed on this date and this is what was decided. . . . There is so much paperwork. . . . It has gotten to the point it is almost all paperwork. Every contact you have with a parolee or anybody connected to the case must be documented. At the end of the month, we have our administrative reports we have to do, which

are our travel claims, our statistics, and our caseload management forms. I write a lot of contact notes. I write everything down. I might even go back to something that I have written to jog my memory sometimes on something. One reason I document everything is to cover myself because of all the liability involved. That is one of my reasons for writing so much.

—Angela LaRue, Office of Parole, Knoxville, Tennessee, personal communication, October 18, 1995

The team leader who is going to supervise a particular case manager looks at resumes and brings the applicant in. We have an all-day interview process. If the supervisor is interested in that person, we will bring him or her back. That way other team leaders and members get to meet with the person. . . . We have an orientation process once a person is hired. It is specific to intensive case management and involves all departments because we all work together. . . . The supervision that is provided is real intense. We feel that it is needed because it is a very stressful job. There are a lot of pressures. Lots of things come up that, without supervision, no one would know how to handle. I was real green. I came to this pretty blind, but I think case management is a job that you are either going to like or not like. It is going to have to be something that you really want to do.

—Paula Hudson, Helen Ross McNabb Center, Inc., Knoxville, Tennessee, personal communication, April 27, 1995

Our supervisors don't carry caseloads. We have a formal staff meeting on Fridays when all the staff get together and the supervisors tell us what is happening on the state level. We are developing a structure that includes more one-on-one supervision, in order to go over cases to see what is going on. How do we interact professionally? In what direction are we heading? Are we burning out? What do we need to do? Then the Case Manager II's will do that with the Case Manager I's on their team. . . . Our agency is computerized. We enter huge amounts of data every day on children and families, their movement from place to place, their insurance provider, the services they are getting, the educational situation. . . .

—Thom Prassa, East Tennessee Community Health Agency, Knoxville, Tennessee, personal communication, July 10, 1994

You are learning about case management at a particularly interesting time in its history. Human services is under close scrutiny for cost effectiveness, efficient delivery, and appropriate outcomes. The challenges for a case manager are many: limited resources, complex client problems, changing societal values, and a bureaucracy that is growing and changing at the same time. The chapter-opening quotations illustrate the daily work of case management: writing reports, documenting services, using supervision, and mastering technology. This final chapter offers a look at each of these areas.

The chapter begins with a description of case management today. Throughout the text, you have read what helping professionals from across the nation have to say about case management—their work, their agencies, and their clients. Analyzing complete interviews with these professionals provides a comprehensive view of case management today. What is case management like? What are the frustrations? What are the rewards? What do prospective case managers need to know? As you read about their work, you will see that the content of the chapters of this book has been informed by the voices of those who are actually doing case management.

For each main section of the chapter, you should be able to accomplish the objectives listed.

CASE MANAGERS SPEAK

- Summarize each of the four themes that characterize case management today.
- List the knowledge and skills needed to be an effective case manager.

DOCUMENTATION AND REPORT WRITING

- Make a case for documentation in case management.
- Write guidelines for good documentation and report writing.

USING SUPERVISION

- Discuss ways in which case managers can take initiative in supervisory activities.
- Design a plan for your professional development.

TECHNOLOGY AND CASE MANAGEMENT

- Offer a rationale for the use of technology in case management.
- Illustrate how technology is used in human services.

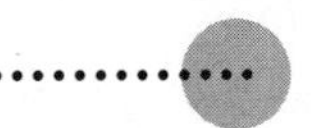

Case Managers Speak

The quotations that appear throughout the text are excerpts from interviews that we have conducted since 1993. When the complete interviews were analyzed, four themes emerged: knowing and learning, handling the "terrible bureaucracy," gathering information, and client problems and goals (McClam & Woodside, 1994a). These themes reflect the experiences of case managers, their struggles to learn the case management process, and their commitment to teach that process to others. Reading about these themes will add to your understanding of the realities of case management today.

Knowing and Learning

Responses related to this theme often focused on how professionals learn to be case managers. A key phrase is "Come with or be prepared to obtain." Some case managers believe that certain knowledge and skills necessary for effective case management cannot be taught; they use phrases such as "their own perception," "instinct," "kind of experience," and "street know-how." They suggest that "caseworkers . . . are born" and already know "what nobody is going to teach you."

However, becoming a good case manager also involves learning. Interviewees discussed the learning that occurs before they come to a job—in classes, field experience, other work experience, or on the job. Case managers talked about the need for on-the-job training, stressing that case management is "a skill that you keep on learning." They also emphasized teaching case management skills to others. One case manager supervisor "had to impose structure on a staff person." While teaching how to conduct needs assessments, another commented, "I don't think this is something people are generally born knowing; you have to learn it."

So there are different opinions about the roles of personal aptitude and training in becoming a good case manager. Clearly, however, a person brings some personal attitudes, characteristics, and ways of doing things to the job—for example, patience, persistence, and creativity. On the other hand, certain aspects of the work must be learned during academic preparation, agency orientation sessions, on-the-job training, supervision, and professional development opportunities (see Chapter 11).

Handling the Bureaucracy

In discussions about the knowledge and skills that case managers need to perform their jobs, a recurring phrase was "the terrible bureaucracy." Problems related to bureaucracy included paperwork, juggling priorities, mandates, and eligibility standards.

"Papers are constantly going in and out." The case manager must have "really advanced organizational skills to set up your work space" and good writing skills. The pressures of paperwork require a well-structured filing program and time to record the activities of the day. Juggling priorities is also a constant struggle. There are internal pressures, regulatory pressures, and pressures from clients. Through it all, case managers must constantly decide "what can wait and what is urgent." Those interviewed emphasized time management skills (see Chapter 11) and the awareness that not everything that needs to be done can be accomplished with the resources available.

Case managers who were interviewed also expressed concern about mandates. They were knowledgeable about relevant mandates from state and federal governments but found that they were not always easy to follow. At times, state regulations may forbid what clients want or need, making for conflict in determining what care to provide. Often, there is so much "going on in the political arena" that case managers must "be assertive" when dealing with "hundreds of outside agencies, government and private."

Eligibility was also a concern. Case managers say that they sometimes spend "too much time doing entitlement," so they must possess keen skills in dealing with eligibility criteria. Eligibility problems raise the question of what can be done for clients who are not able to get services through the particular agency. Case managers provide the "first point of entry," but they feel that their mandate "is to connect them [clients] to the right agencies."

The discussion of eligibility also touched on funding issues. Case managers worry about an emerging structure in which there is a charge for every service—"one reimbursement for everything." Also, the fact that clients can often choose the services they will receive creates competition among agencies. Clients are seen as customers who are free "to go to another agency they find more amicable or friendlier." This creates an atmosphere where everyone is "hungry for money." "For any organization to survive, you've got to know what is out there."

Gathering Information

The third theme that became apparent throughout the interviews was the importance of gathering information. Salient aspects of this process included the interview, the multifaceted nature of information gathering, the presence of "little bits and pieces," and the development of a meaningful relationship.

For many case managers, the interview is the single most important method for gathering information about the client. Whether the interaction is a short intake interview, a more comprehensive psychosocial study, or a medical evaluation, the focus is on understanding the client. Information gathered during interviews is used to screen for eligibility, establish goals, and develop a plan of services. (Important skills for effective interviewing are the ones discussed in Chapter 5.)

The gathering of information is directly linked to making sense of the information; some described it as "detective work." For each case, "Something is going on," and "Our job is to look into every corner" and "to gather every bit of information possible." Eventually, understanding emerges—"things just begin to fall into place." Unfortunately, there are barriers to understanding the information gathered. First, the initial information given by the client cannot always be trusted. It is sometimes necessary to "go beyond the surface of people's problems" and "to pick up something . . . not seen." Also, the case manager must not make his or her own assessment too early in the process. What's more, "Sometimes you miss the forest for the trees"—there may be so much information available that you are overwhelmed.

During information gathering, case managers begin "developing a meaningful relationship with the client." To gather information, they need to know "how to make people feel comfortable" and "become really involved." As the person begins to feel comfortable with them, "suddenly a relationship develops." This relationship is the key to the information-gathering process. As comfort and trust grow, more information is shared. Case managers say that they are there "to give, not to receive" and that they "give as much as they can, though it is not a lot sometimes." The reward is the satisfaction of seeing clients begin to get a handle on their lives. However, case managers express caution about getting too emotionally involved with clients.

Client Problems and Client Goals

The fourth theme that emerged from the interviews was identifying client problems and developing client goals. Case managers distinguished between the presenting problem and the underlying problem. Professionals realize that they must sort through the "initial stuff" to get at the problems that emerge later in the case management process. Their collective experience is that clients express their needs in different ways. Some are "very articulate and very forthcoming with what they need," whereas others "don't know exactly what they need and can't articulate it." At times, the client needs to talk "just to let off steam." Other times, case managers must focus the discussion on needs so as to determine whether the individual is eligible or whether the agency can provide what he or she needs.

Establishing goals requires flexibility and patience. Most case managers find that the best approach is to establish an initial plan and to alter it as circumstances demand. Each plan consists of small steps, encouraging the client to stay realistic and "not tackle too much too soon." Empowering and advocacy also emerged as important concepts in the discussion of goal setting. Empowering is "teaching them those skills so they can do without me," and advocacy is to be undertaken "on behalf of the client and the agency."

In conclusion, the case managers who were interviewed affirm the increasing importance of case management in human service delivery. The challenges and frustrations of working in the human service bureaucracy call for survival skills to help them cope with feelings of stress and pressure, the complications of diversity, and emotional involvement with their clients and their problems. Time management, professional development, organizational knowledge, and the expertise and skills for competent case management all help the case manager cope with the demands of service delivery.

Documentation and Report Writing

Nearly all the chapters in this text make reference to documentation and report writing. These activities begin with assessment and plan development and continue throughout case management to termination. Follow-up studies of helping professionals, interviews with employers, and interviews with service providers all indicate that writing consumes a great deal of their time at work (McClam & Woodside, 1994b; Woodside & McClam, 1994).

Report writing and documentation take many forms. During assessment, planning, and service coordination, case managers add to their case notes, write a plan for services, and prepare justifications for services. They also write case summaries, reports, and updates. They communicate with other professionals—for example, to request a medical or psychological evaluation or arrange a service such as housing or therapy. Many of these interactions occur in writing. How well case managers present the case reflects on their credibility as helping professionals, so good writing skills are essential. This section will summarize why documentation and report writing are important and review guidelines for good writing.

The Importance of Documentation

Accurate, timely, and objective documentation brings numerous benefits to the client, the agency, and the helper (Mitchell, 1991; Wilson, 1980). First, for all involved with the case, documentation provides a record of what has occurred and presents a picture of ongoing progress toward established goals. For example, Roy Roger Johnson (in Chapter 1) had five different counselors, so documentation was a key factor in maintaining continuity of services. Second, documentation is valuable for the purposes of accountability and quality control. Some agencies conduct random reviews of the records of service delivery. For the case manager and the client, accountability is important to confirm that promised services are delivered, that each participant follows through, and that desired outcomes are achieved. Third, clear communication about the case helps avoid or settle litigation in the event of any disputes about service delivery. Remember that records may turn out to be court evidence of the facts of diagnosis, activities, and services.

Agencies require documentation for several reasons. As stated previously, agencies have specific criteria for eligibility for services. Common criteria include the presence of a physical, mental, or emotional disability; income; unemployment; and the expectation that services will improve the client's situation. Of course, the criteria for eligibility depend on the mission of the agency and the services offered. No matter what the criteria are, agencies require that the case file include documentation that establishes eligibility. For example, a medical examination is confirmation of a physical disability, and copies of the previous month's bills serve to document the client's living expenses.

A second reason for documentation is that an agency's accountability depends on maintaining statistical records of client cases. This is particularly true of agencies that must report to state or federal agencies, but it also applies to organizations that report to a board of directors. Records of the number of clients, diagnoses, and services rendered provide essential information about the demand for services, the types of services delivered, and the outcomes of service delivery. Agencies use such information in accounting for what has occurred in the past year as well as in planning for future needs.

Although many case managers bemoan the time and the energy required to document, it has many advantages for them. Effective documentation means organizing a mass of information into a structured presentation of data, observations, interviews, and services. For example, documenting goes hand in hand with effective interviewing. It provides structure to the interviewing process by organizing what is known and what additional information is needed. Documenting is also beneficial to case managers because it provides the means to assess their performance. Supervisors review case records to track the delivery of services to the client. Outcomes of service delivery and utilization of agency resources are part of such a review. Finally, documentation facilitates interdisciplinary communication. Case notes can effectively communicate to co-workers or team members any relevant information about a specific client, services received, and progress made.

All these factors benefit clients, either directly or indirectly. Certainly, clients get more effective service delivery if the case file shows good documentation. Good documentation has direct financial benefits for those clients who are eligible for third-party payments. Written documentation is necessary for reimbursement to be paid, either to the client or to the provider, regardless of whether the provider is an agency or an individual such as a physician.

As you can see, there are many benefits to documentation. In most cases, case managers are those responsible for the various types of documentation, so skills in report writing are essential. The next subsection gives some general guidelines for writing.

Effective Writing Skills

There are two basic principles of documentation and report writing. The first is that the purpose is to inform. A social history, for example, shows the client within the framework of his or her social environment, representing strengths and weaknesses and bringing the client into focus as an individual. A social history that contains no relevant information, or waxes eloquent about your experience working with the client, does not do its job of informing the reader about the social development and situation of the client. Likewise, case notes have their purpose of informing the reader about what is happening with a client, including service delivery, progress, and behavior. Since case notes are usually in chronological order, the reader can also learn what has happened in the past.

The second principle is that the reader's needs must be considered. Most documentation and reports in human services are read by other professionals (although it is important to remember that the client does have access to this information). Knowing who the reader is and what information he or she needs is essential in determining the level of terminology, content, and reading level.

Other general guidelines have to do with the "three P's: purpose, pattern, and process. *Purpose* is the reason you are writing the report. Is it to inform, communicate, document, propose, make comparisons, evaluate, summarize, or record? Knowing your purpose guides you in all aspects of writing the report. *Pattern* is a matter of *when* you write. Usually, the agency determines the appropriate times to write case notes, summaries, evaluations, social histories, correspondence, or the shift log. For example, are case notes written daily, weekly, or sporadically? Is a summary to be prepared after the intake interview, once assessment is completed, at plan development, or after termination? The final "P" is *process: how* documentation occurs—by hand, dictaphone, or word processor. The case manager may have more flexibility to choose the method that best suits his or her working style.

Let's look at case notes first, and then reports. We will present common mistakes and some general guidelines for writing correctly.

CASE NOTES

Often, case notes are too short or too vague. Phrases such as "improve self-esteem," "difficulty getting along with peers," and "client is ambivalent" are unhelpful. Mitchell (1991, pp. 15–16) suggests the following key points in writing case notes:

- Make notes grammatically clear and correct.
- Use precise language; reduce the potential for misinterpretation.

- Use only adjectives that are defined, necessary, and clinically appropriate; when possible, replace an adjective with a verb that describes behavior.
- Avoid cliches like the plague.

In practical terms, look for ways to eliminate words. Case notes are not lengthy or chatty. The phrase "the client and I" can be replaced with "we." It is perfectly acceptable to use abbreviations, as long as you are sure that anyone reading the case notes will understand them. It is also wise to eliminate such words as "seems" or "appears." If these terms *are* used, follow them with "as evidenced by. . . ." At times, the client's actual words make the most accurate case note, eliminating from the record any misinterpretation the case manager might introduce if he or she was dealing with ambiguous phrasing on the client's part. Finally, use correct grammar; your credibility depends on it. Here are some examples of clear and grammatical case notes.

> I asked client if she had a suicide plan. Client said she did not.
>
> The client seemed shy, as evidenced by sitting alone, not talking, and keeping his head down.
>
> Client worked at task for 10 minutes, then asked for a break.
>
> Client says she feels that no progress is being made: "I don't know why I'm here or what you are trying to do with me."

Of course, each agency is likely to have specific guidelines for case notes. The series of case notes presented in Chapter 4 are done in the format dictated by the residential treatment center where the client was living. Whatever the format, good case managers always make entries while the facts are fresh in their minds. They also bear in mind that the client, as well as other staff members, may read the notes, and they can be subpoenaed. A good rule in this connection is to write legibly and in ink.

REPORT WRITING

It is a fact of life that case management includes writing—a lot of writing! Rather than wasting time agonizing about writing the summaries for the files stacked on your desk, or procrastinating about writing those referral letters, adopt approaches to writing tasks that make them easier to complete.

Writing is a skill—one that improves the more you do it. If you don't write, you won't improve. It's that simple. Michael Keene, an expert on technical writing, suggests that writers *attack* the first draft, *write* fast, and *commit* to doing multiple revisions (personal communication, March 8, 1996). He emphasizes getting that first draft down on paper. Don't be distracted by spelling or grammar; that comes with the second draft.

Being aware of common writing problems will help you avoid them. Wilson (1980) lists incorrect spelling, poor organization of thoughts, and misuse of verb tenses as the most frequent offenses. Other major problems are unnecessary duplication of content, missing information, failure to label an interpretation or impression as such, and a writing style that is unclear, vague, and wordy (Tallent, 1993).

How does one write effectively? Mitchell (1991) believes that good listening is the place to start. Do you remember the discussion of listening in Chapter 5? Case managers listen to clients, family members, supervisors, and other professionals, each of whom has pieces to the puzzle. On a practical level, once you are ready to write and are certain that you know your purpose and know the reader, then you may find helpful the following writing principles, adapted from Gunning (1953):

1. Keep your average sentence length short.
2. Vary the length of your sentences.
3. Prefer the simple to the complex and the familiar to the far-fetched. Choose the common everyday word when you can. Don't try to impress the reader with your knowledge, and don't attempt to write about things with which you are not familiar.
4. Use action verbs whenever possible. For example, instead of "There are certain behaviors that limit," write "Certain behaviors limit. . . ."
5. Avoid unnecessary words—use as few words as you can to say what needs to be said.
6. Use descriptive words when possible. General statements are more effective if they are supported by specific facts. Observations of behavior should generally be phrased as descriptive statements.
7. Write the way you talk. The goal in written communication is to express ideas and information as clearly as possible, so that the reader won't have to guess at what you mean.
8. Write to express, not to impress.

Using Supervision

Most case managers have supervisors with the authority to determine their responsibilities and to evaluate them. Professionals often believe that supervision is a one-directional experience, with communication flowing only from the supervisor downward. In fact, supervision can be a two-way process in which case managers take an active role. Good supervision, in addition to ensuring quality care for clients, should contribute to the job satisfaction and professional development of case managers.

Responsibilities of the Supervisor

To learn how to participate fully in the supervisory process, it is important to understand the responsibilities of those who supervise case managers. According to Gillies (1994, p. 343), **supervision** "includes inspecting another's work, evaluating her or his performance, and approving or correcting the performance." In case management, there must be oversight by a professional who can place it within the larger context of the organization and the human service delivery system. Inherent tension is often felt between supervisors and case managers because supervision inevitably involves making a judgment of performance that can jeopardize the worker's job security.

In addition to performance evaluation, supervisors also work to do the following: (1) develop the potential of case managers; (2) set challenging, but not impossible, expectations; (3) motivate case managers to perform well; (4) emphasize the importance of team effort in case management; and (5) communicate the responsibilities of the case manager (Rhinehart, 1969). Speaking of her supervisory responsibilities, Roz Jaffe describes types of meetings that she schedules. "One is the staff meeting to go over program issues that pertain to staff and clients who require in-depth discussion. . . . Care plan review, where different disciplines come together and give their assessment of what has happened to a client. . . . Overview to talk about people at risk. . . . In-service meetings on particular issues, like depression with the elderly or elderly abuse" (personal communication, May 3, 1994). Roz Jaffe uses such meetings to meet many of the goals of supervision. The activities of a supervisor include planning for professional development, giving guidance and direction, providing up-to-date information, conducting performance appraisals, and offering support.

Case managers can take an active role in the supervisory process by helping supervisors perform their responsibilities. Case managers should ask themselves, "What can I do to help this process?" For example, the first supervisory activity listed above is planning for professional development. Instead of waiting for a supervisor to develop a plan, the case manager could assume that responsibility. For example, Ms. Song (see Chapter 11) has initiated the planning process by presenting a tentative plan that she has developed (see Figure 12.1.) This approach to planning encourages feedback and assistance from her supervisor.

For each of the supervisory activities listed next, case managers can take the initiative in the ways suggested (Austin, Kopp, & Smith, 1986; Ellis & Hartley, 1991; Gillies, 1994; Miller, 1988; Pecora & Austin, 1987).

Planning for professional development As case managers carry out their responsibilities, they become more aware of the knowledge, skills, and experiences that they need to improve their performance. With the help of supervisors, they can plan a productive professional development program.

Several of Ms. Song's clients have been involved with treatment for alcohol and drug abuse, and she feels that she needs more information about the latest biomedical research about the treatment of substance abuse. The local hospital is sponsoring a roundtable discussion on this research during Wednesday grand rounds. She arranges to attend these meetings.

Asking for guidance and direction Case management responsibilities are determined by the policies and procedures of the agency and interpreted by those who supervise. It is important to know these policies and procedures. Case managers can let their supervisors know which policies and procedures

PROFESSIONAL DEVELOPMENT PLAN

To: Ms. Lowe

From: Ms. Song

Date: January 1999

Subject: Professional development plan

I have worked on a plan for my professional development. My main focus is the skills and knowledge that I need to better meet the needs of the young mothers I serve. I have set aside one day per month for professional development.

Date	*Activity*	*Goal*	*Cost*
January	attend a workshop on making referrals	increase skill	$50.00
February	visit the new interfaith clinic	expand network	free
March	attend our agency fund-raiser	expand network	free
April	attend state training session	increase knowledge of rules and regulations	$25.00
May			
June			
July			
August	attend federal training session	increase knowledge of rules and regulations	$75.00
September	attend computer workshop	increase skill	free
October	attend conference on assessment	increase skill	$20.00
November			
December			

As you can see, my plan is not complete. Any suggestions or revisions would be appreciated. I am looking forward to your feedback. Thank you.

Figure 12.1 Ms. Song's tentative plan

they are following, which are helpful, and which are troublesome. They can ask their supervisors for help in interpreting policies and procedures. The supervisors may be able to suggest ways to use the policies and procedures to help meet client needs.

Ms. Song also works with the families of her young female clients. Right now the policy is to make a maximum of two visits to the home before beginning the delivery of services, but sometimes two visits are not enough to establish relationships with the young women and their family members. Often, the young mothers want to meet alone with Ms. Song, just to have their own say in private. To meet such a request, Ms. Song must sometimes violate agency policy. If she does make the visit, she cannot record it, and any information she gains is not in the record. Ms. Song asks her supervisor for advice.

Seeking up-to-date information Guidelines and policies in human service delivery are always changing. The identification of the target population, services available, funding sources, agreements with other agencies, goals for programming and services, and new ways to handle a crisis are among the changes that may affect case management. Much of the information about changes comes from supervisors through the chain of command. Case managers need to let their supervisors know how these changes affect how they work with their clients.

The new state block grant will fund the work program for unemployed homeless males through the Department of Human Services. This new funding structure, to begin July 1, will cause a reshuffling of personnel. Ms. Song is not clear about how this new grant will affect her work and that of her colleagues. She asks for a meeting with her supervisor to clarify the situation.

Supporting performance appraisal It is the responsibility of supervisors to do formal evaluations of case managers, but the process can begin in a more informal way if the workers ask supervisors for feedback. One way to get feedback is to submit a self-assessment, like the one shown in Figure 12.2. The case manager should make an appointment to share the assessment with the supervisor and ask for feedback about its accuracy. This meeting is also a good time to ask how to improve identified weaknesses or fill gaps in knowledge and skills. This dialogue can improve communication between supervisor and case manager, and it is less stressful than formal evaluation.

Ms. Song is having real difficulty identifying the underlying problems her clients are facing. In her initial contact with clients, she hears the young women tell their stories and describe their problems. But she is often surprised by the problems that emerge later on. Since Rebecca Lowe, Ms. Song's supervisor, stressed identification of client problems as one area in which Ms. Song would be evaluated, she approaches the supervisor before the formal appraisal process begins, asking for her help.

Asking for support It is important for case managers to ask for support from their supervisor when they need it. Once case managers have used all the alternatives that they know, the next step is to tell their supervisor about the situation and the actions already taken and to ask for feedback.

ASSESSING MY WORK AS A CASE MANAGER

	Needs work	Adequate	Good
Information-related work			
1. Has effective communication with clients			
2. Has effective communication with colleagues			
3. Uses verbal cues			
4. Uses nonverbal cues			
5. Has ability to handle feedback			
6. Has ability to problem-solve			
7. Demonstrates acceptance of diversity			
8. Writes clear, accurate reports			
Self-development			
9. Has ability to lead staff meetings			
10. Takes initiative in an effort to influence events			
11. Is calm in tense situations			
12. Makes timely decisions			
13. Handles ambiguity			
14. Meets quality standards			
15. Manages time			
16. Is appropriately assertive			
17. Follows through, with good attention to detail			
Coordination			
18. Plans and prioritizes effectively			
19. Uncovers relevant information			
20. Uses critical thinking to evaluate information			
21. Builds consensus			
22. Understands outcomes of decisions			
23. Makes timely judgments			
24. Is organized			
25. Maintains paperwork			
26. Wishes to be accountable			

Figure 12.2 Self-assessment

SOURCE: Adapted from *Delivering Human Services*, by Alexis Halley, Michael Austin, and Judy Kopp, pp. 539–541.

Right now, Ms. Song is working through a challenging problem with a doctor at the local residential mental health facility. Two of her clients were once in residential treatment programs and are still on medication. Ms. Song would like to talk to the doctor about how the medication affects their parenting and their ability to learn new vocational skills. According to the doctor, he is unavailable for consultation; he recommends that Ms. Song talk with a colleague—who unfortunately does not know the answers. After trying several approaches to engage the doctor's attention, Ms. Song asks her supervisor for other ways to get this information or involve the doctor in a dialogue about her patients. Her supervisor has had the same trouble herself with this doctor. She is sympathetic and promises to help Ms. Song solve this dilemma.

These are but a few ways case managers can be active participants in their own supervision. They should also observe their supervisors, read their job descriptions, and listen as they talk about their responsibilities. In this way, they can build an understanding that will facilitate the supervisory process.

Obviously, communication is a key element in all of the above suggestions for using supervision. As communication improves, case managers will find better approaches. The case workers at the Third Avenue Service Center involve their supervisors in their most difficult cases. "We work together with the director and the supervisor. We usually come together to combine all of the problems we are facing in a case. . . . It is not only one problem, it is preventive" (Teresa Rodriguez, personal communication, May 5, 1994).

The performance review process can also contribute to professional development. Let's look at the definition of performance review, the issues that are especially relevant for case managers, and ways to use it to enhance professional development.

Performance Review

The **performance review** is "designed to measure the extent to which the work is achieving the requirements of his or her position" (Pecora & Austin, 1987, p. 56). Kadushin, quoted in Pecora and Austin (1987, p. 56), says good performance evaluation must be "based on clearly, specified, realistic, and achievable criteria reflecting agency standards."

The performance review of case managers is conducted for three purposes. First, the review provides information to be used for personnel decisions, such as raises, promotions, transfers, and termination. Second, performance evaluations encourage professional development by assessing the worker's strengths and weaknesses and developing plans to increase skills and knowledge (Ellis & Hartley, 1991; Gillies, 1994; Lewis & Lewis, 1983). Third, performance reviews assume additional purposes in the managed care environment—accountability and cost containment. Whether funds are used to purchase a prearranged package of services, or each individual service is reimbursed, the activities of case managers have financial implications for the services clients can receive. The timing of performance reviews varies from organization to organization, but many beginning case managers undergo a review three or six months after assuming their responsibilities, and most are then reviewed once a year (Ellis & Hartley, 1991; Gillies, 1994; Lewis & Lewis, 1983; Rhinehart, 1969).

DIFFICULTIES OF PERFORMANCE REVIEW

The performance review can be stressful for case managers, for a variety of reasons (Miller, 1988; Pecora & Austin, 1987; Rhinehart, 1969; Smith, 1992). First, case managers have various pressing responsibilities and are often in a position to make autonomous decisions and to carry out plans without supervision. Therefore, supervisors may be unaware of their work and may second-guess the case manager's decision making. The people who work with case managers on a regular basis and are thus in a better position to judge their performance are often not employees of the agency (or work in another unit within the agency).

Another difficult issue regarding performance evaluation is associated with the empowerment of clients. In the interest of this principle, case managers encourage clients to assume as much responsibility as possible. Client responsibility should be seen as a positive outcome of case management—not an opportunity for the case manager to have the clients do most of the work. How clients feel about their level of empowerment should certainly be a consideration in supervisors' evaluation of case managers, though it must be borne in mind that the client is not qualified to give a definitive evaluation. Supervisors reviewing the performance of case managers face a challenge, because of the variety and complexity of their tasks and the difficulty of measuring outcomes fairly. In spite of these difficulties, case managers must be evaluated. Defining tasks and clarifying expected outcomes are important activities to undertake before the performance evaluation.

Compounding the difficulties described above is the evaluation anxiety that many case managers experience. This anxiety can jeopardize both job satisfaction and performance, since the affected workers are worried about how others are judging their activities (Smith, 1992). Sometimes case managers suffering from such anxiety become more focused on themselves than on their tasks, which can lead to self-doubt and self-blame when things do not go right. They also spend a lot of time comparing themselves to others (Dweck & Wortman, 1982; Smith, 1992; Wine, 1982). This is similar to the test anxiety that students experience in school. As you may know from experience, test anxiety does not always lead to positive results.

PREPARING FOR PERFORMANCE EVALUATION

In preparing to give performance reviews, supervisors have definite ideas about case management and the characteristics they would like their case managers to demonstrate (Austin, Kopp, & Smith, 1986; Ellis & Hartley, 1991; Hamilton, 1992; Miller, 1988; Pecora & Austin, 1987; Rhinehart, 1969). Positive attitudes, critical thinking, and stability are generally regarded as evidence of a professional approach to case management. Table 12.1 lists more specific desirable traits under these three categories.

The categories in the table exclude other significant attributes of good case managers—knowledge of roles and responsibilities, understanding of human beings, and expertise in the case management process. These have been

TABLE 12.1 PREFERRED CHARACTERISTICS FOR CASE MANAGERS

Positive attitudes	*Critical thinking*	*Stability*
willingness to learn and improve	making good judgments	good physical health
willingness to follow directions	making good decisions	emotional stability
willingness to cooperate with others	handling a crisis	dependability
initiative		good work habits
liking the job		
acceptance of criticism		

discussed in more detail earlier in the text. The characteristics listed in the table are necessary for good case management work regardless of what responsibilities are assumed, tasks undertaken, or populations served.

Case managers who understand performance review and know what makes for successful case management are ready to guide their own professional development, without waiting for the review. As case managers assume an active role in improving their performance, they can expect positive outcomes. They begin to identify areas in which they wish to improve; as they work on their competencies, their sense of confidence increases; and, of course, they are prepared for the performance review, whenever it may occur. Ms. Song provides some excellent examples of taking the initiative in her professional development.

Ms. Song has decided to focus on eight areas of professional development—two or three from each of the three main categories (positive attitudes, critical thinking, and stability). The following are examples of how she targets areas for improvement and develops specific skills and behaviors that turn troublesome situations into more positive ones.

Positive attitude: initiative

At 5:30 P.M., Ms. Song receives a call from Ms. Gonzales, who is getting ready to go to work and cannot get her car to start. Although Ms. Song has discussed alternative transportation scenarios with Ms. Gonzales, none of them will work today. Ms. Song calls the department's van operator and asks her to pick up Ms. Gonzales on the way to the day care center. It is after 7:00 P.M. by the time Ms. Song has finally finished her work. Often she is able to leave at 5:30, but she works late when the situation demands it.

Positive attitude: willingness to learn and improve

For the past month, Ms. Song has changed her way of working with an interdisciplinary team, and the work seems to be going more smoothly and efficiently. Earlier, almost all of the work had fallen to her; her colleagues had expected her to take notes and write up their reports. Now the team members bring their reports to the meeting. Ms. Song and two other team members developed a standard format for the reports, so information can be entered on the computer and then compiled into one report.

Positive attitude: Willingness to cooperate with others

Ms. Song does not perform her responsibilities in isolation; much of her work is done in meetings with others. Ms. Song is learning that proposals she makes are more likely to be accepted if they are based on the comments of others. She has often been frustrated in meetings, feeling that no one was really listening to her or showing any desire to write a single cohesive plan for the client. Recently she has changed her approach to meetings. She now tries to listen to everyone before voicing her own opinion or making recommendations. When she does speak, she tries to build on others' comments.

Critical thinking: making good judgments

Ms. Song provides the initial assessment of a mother's eligibility for the work program. She gathers data from the applicant and others and uses it to draw conclusions about relevant matters, such as present financial status. She also gathers data for making interpretations and subjective judgments, such as rating her clients' emotional stability, dependability, initiative, and maturity. These ratings require good judgment on her part. Negative outcomes could result if her interpretations are faulty: Applicants might be admitted to the program but not complete it, thereby taking space from applicants who might stand a better chance of success. Funding is contingent on clients completing the program and gaining employment, so it is important to choose the applicants with the highest chance of success. Ms. Song worries about excluding people for whom the program can offer a new chance. She begins working with her team to sharpen her insights regarding these ratings.

Critical thinking: handling a crisis

Few crisis situations develop in the work of Ms. Song and her co-workers. Most of their clients encounter some difficulties while they are in the program, but those are usually handled with traditional problem-solving methods. The staff still undergoes training for crises, and there is a standard policy on how to handle those that do arise. Since crises happen so rarely, Ms. Song wonders if she will be able to react appropriately when one does occur. She has asked co-workers about the history of the agency—in particular, what crises have occurred in the past, how they were handled, and who handled them. Ms. Song is also trying to identify crises that have occurred in her own life. She analyzes her approach to crises and assesses her strengths and weaknesses.

Stability: emotional stability

There are periods when Ms. Song has so much to do that she does not even know where to start. She then works after hours and on weekends to catch up. Recently, she attended a time management seminar conducted especially for case managers in the state human service system. The leader stressed that professionals must determine how much time they are willing to work and then decide how they want to spend that time. Priority setting is not just a skill to be used at work; it should extend into balancing the time spent on personal and professional activities. Ms. Song knows that she must learn to do this if she is to avoid burnout.

Stability: good work habits

Completing tasks on time has been a point of pride for Ms. Song since she was a little girl. This value has served her well in her case management work. She has many deadlines: for contacts with clients and other professionals, reports to submit, visits to make, and meetings to set and attend. She knows that meeting these deadlines contributes to her own job satisfaction. Since the work of others depends on her completing her tasks, Ms. Song also knows how important it is to meet the goals established or to ask for help. Completing work on time, with high quality, enhances her belief in herself and the respect of colleagues.

Stability: dependability

For a long time, Ms. Song did not feel that she was contributing to the agency's goals. She did not really even know what the organization's goals were until she began working on a grant proposal to acquire mental health support for the agency's clients. One of the questions the funding source asked was "How does this program relate to the mission and goals of the organization?" She then studied the mission and goals of the agency and made the case that mental health services for young mothers would help "empower these young women to become self-sufficient." Ms. Song became more committed not only to her young clients, but also to her agency. Several times in planning meetings, she has been able to set aside her own goals in favor of those suggested by others, because she has gained a broader understanding of the mission of the agency and her own place in it.

The importance of this discussion is to expand your awareness of supervisors' expectations for case managers. As discussed earlier, a good way to reduce anxiety about the performance review is to assess performance before the evaluation occurs. Beginning a plan early changes the emphasis of the review, focusing on professional development. It also puts the case manager in a position to reduce work-related stress, since he or she feels in control of the work.

Technology and Case Management

Recent advances in technology are beginning to change how human service work is done. Case managers can expect the computer revolution to continue changing their work in the future. Technological progress has coincided with the economic crises of the 1980s and 1990s. Because resources have been scarce, using technology has been viewed as a way of "doing more with less." Human service agencies are exploring ways to use technology to manage cases more effectively. The goals of "information technology applications [are] to improve management systems and better monitor, evaluate and account for service delivery and indirectly increase productivity" (Penzias, 1993, p. 18).

Grasso and Epstein (1993) point out that not everyone favors the use of technology in human services. Advocates of technology believe that it will allow services to be provided more effectively and efficiently. Those who oppose its use argue that technology creates "a bureaucracy for workers and clients and trivializes professional practice" (p. 374). Some educators believe that helpers fear computers and will resist their use in service delivery. Studies of the acceptance of computers in practice have yielded revealing results. Case managers begin to use computer systems successfully if they expect that it will improve the administration of the agency and facilitate quality service delivery to clients. Satisfaction with the technology is associated with using it in the daily

Human service workers are finding that technology can play a useful role in case management.

routine, belief in the agency's ability to implement the computer system, involvement of the workers in the development and implementation of the system, and belief that the computerization process has been well-planned (Cwikel & Monnickendam, 1993; Finnegan, 1996; Monnickendam & Eaglstein, 1993). The introduction of technology into human service delivery is a complicated effort that will have far-reaching effects (Grasso & Epstein, 1993). Each agency must integrate the technology to meet its own goals, and the particular approach taken is bound to influence the delivery of services.

Technology can facilitate case management in several areas, including documentation, treatment planning, quality assurance, communication, implementation strategies, and professional training. Each of these areas is discussed below, and examples are provided.

Using the computer to improve documentation is one of the more obvious ways in which technology can support the work of case managers. They can use the computer to store information from the client's intake interview, data gathered from other professionals during the assessment phase, the social history, goals and objectives, the treatment plan, and records of client progress.

Once the information has been entered, it is accessible to the case manager and others on a "need to know" basis.

For example, in Alzheimer's respite services, a network is used to link the clinical information databases of seven separate programs. This system was designed to meet two goals: (1) to create a unified database of inquires for services, client demographics, and use of services; and (2) to provide each agency with on-site access to the database so that they can track clients and monitor usage of services (Looman & Deimling, 1993). The database includes information on inquiry/intake, assessment, service tracking, and client satisfaction. (See Figure 12.3.)

Information systems such as the one outlined in Figure 12.3 have both positive and negative aspects. According to the users of this system, the availability of information has created a new working environment in which *quantitative* information is used to write reports, plan outreach, and target client groups. It also helps agencies standardize their procedures and information gathering. Once case managers become used to the system, they realize what computers can and cannot do and how they can be used to support case management work. The downside of using the technology includes the extensive time commitment to develop and learn to use the system. Training is an ongoing effort, as new staff members are hired and programs changed. Data entry is time-consuming, and many professionals are unhappy with the lack of qualitative information (Looman & Deimling, 1993).

Streamlining the recording of information about the client, the treatment plan, and progress made is useful in the documentation process. Technology also has the potential to make this information usable in a more sophisticated way. Databases can be designed to contain information about specific difficulties clients experience, the treatments used, and the progress made. Software can readily synthesize this information, making it easier to establish standards of quality care. A case manager, in planning options to meet a specific client need such as housing or parental education, can access the database and determine what alternatives have been used previously, with what degree of success.

A substance abuse agency developed a computer-assisted program evaluation to meet several needs. It was designed to track services delivered and guide the treatment process. Among the agency's objectives was to make sure clients stayed in the program long enough to overcome their addictions. First, case managers entered intake information, clinical files, and daily logs into the database. They also entered information about client progress and length of stay. A computer program was written to manage all of this information. Then the computer program made recommendations for the treatment for each client, based on past practices of the agency. This information is given to the professional staff, who consult it in developing treatment plans designed to increase retention in the program (Waters, Robertson, & Kerr, 1993).

INQUIRY/INTAKE

Inquiry Form

Demographics (impaired person's area of residence [census tract], diagnosis, age, gender, living arrangement and duration of living arrangement)
Social support (major caregiver's area of residence and relationship to impaired person; identity of other caregivers)
Types of respite assistance requested (3)
Physician information
Source of referral
Probable program type and funding source
Inquiry disposition

Respondent Group

Persons making service inquiries to the different programs

ASSESSMENT

Assessment Form

Demographics, household composition, contact information
Sources of informal assistance
Sources of formal assistance/use of other services
References for respite service
Recipient's health status, conditions, medications, nutritional status, any limitations, and use of assistive devices
Recipient's ADL, IADL, and mobility limitations
Recipient's MSQ test score
Major caregiver's ratings of the care recipient's cognitive, social, and moral functioning
Demographics for major caregiver
Caregiver strain: Major caregiver's self-ratings of their current health and any physical health change, emotional health change, relationship strain, any restrictions, sense of mastery
Discharge information, program type, funding source(s), fee subsidy, assistance duration

Figure 12.3 Components of an information system

SOURCE: Copyright © 1993 The Haworth Press, Binghamton, NY. From *The Maturation of a Multiagency's Computerization Effort for Alzheimer's Respite Services,* by W. J. Looman and G. T. Deimling, pp. 100–101. Reprinted with permission.

Respondent Group
Persons who qualify for respite service and receive in-person assessment from program staff

SERVICE DELIVERY

Service Data Form
Type/amount of service units used

Respondent Group
Persons who receive respite service

FOLLOW-UP

Follow-up Form
Need for additional respite service
Use of other support services
Reported satisfaction with respite service
Perceived benefits of respite service use for self/care recipient
Caregiver's reported satisfaction with type, amount, duration, and provider of respite service

Respondent Group
Persons who have received respite service for 3+ months

Figure 12.3 *(continued)*

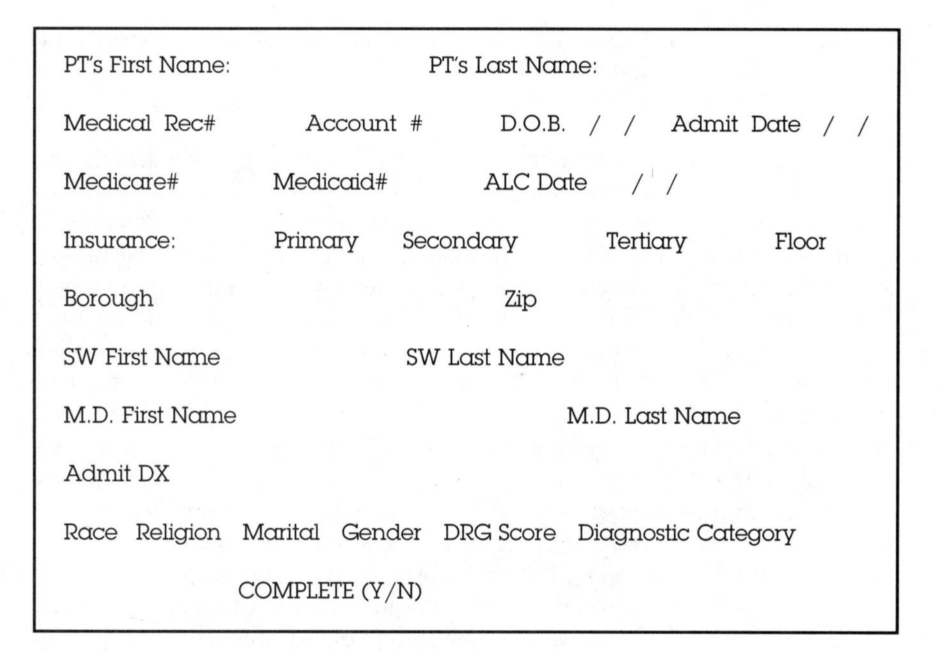

PT's First Name: PT's Last Name:

Medical Rec# Account # D.O.B. / / Admit Date / /

Medicare# Medicaid# ALC Date / /

Insurance: Primary Secondary Tertiary Floor

Borough Zip

SW First Name SW Last Name

M.D. First Name M.D. Last Name

Admit DX

Race Religion Marital Gender DRG Score Diagnostic Category

COMPLETE (Y/N)

Figure 12.4 Sample menu for ALC software

SOURCE: Copyright © 1993 The Haworth Press, Binghamton, NY. From *Technology in People Services: Research, Theory, and Applications*, by C. Auerbach et al., p. 39. Reprinted with permission.

A similar database was built for a case management system for "alternate level of care" (ALC) patients (Auerbach, Cohen, Ambrose, Quitkin, & Rock, 1993). These patients are still hospitalized after the acute stage of their illness. They must be treated very carefully. Discharging them too soon diminishes the quality of their care; on the other hand, if they stay too long, the hospital suffers a financial loss. This program organized the information about ALC patients: their demographic characteristics, any unique problems related to discharge, and any factors having to do with length of stay. A sample menu (Figure 12.4) shows how some basic information about the client is recorded.

An analysis of this database yields valuable information about this client group. For example, the range in the length of ALC stay spans 141 days, with a mean of 8 days and a median of 14 days. Also notable was the different length of stay when broken down by particular discharge plan. For clients discharged to nursing homes, the length of stay in ALC averaged 18.6 days; for clients with an alternative care plan, the ALC averaged 11.3 days. Difficulties associated with discharge also influenced the number of ALC days. For example, patients with family-related discharge problems were ALC for approximately one month. When working with new clients, case managers use these data to develop a plan that takes into consideration the trends uncovered by the software.

The case manager is committed to quality care of the client. The two preceding examples show ways in which technology can be useful in planning and delivering quality services. It is key to establish a database that integrates all the data about a client and makes it available to all helping professionals involved. Such an integrated database can create a paperless process in which all case management records are available in one automated file, on line.

The computer facilitates case management work by simplifying tasks that require repetition, speed, and accuracy (Mowry, 1992). This allows case managers to use the data to "deal with exceptions, to set goals, to establish priorities, and to perceive relationships" (Mowry, 1992, p. 156). Having routine matters handled more efficiently by computers allows more time for the case manager to reflect and work with clients, families, and other professionals.

One of the most exciting contributions that technology has made is the facilitation of communication. Case management is a people-intensive activity, and its success depends on timely communication among case manager, client, family, and other professionals. The new technology is quite adaptable to this purpose. Among the systems being used are electronic mail and teleconferencing. **Electronic mail** allows people to contact others who are on similar computer systems; the message takes only seconds or minutes to travel from sender to recipient.

Teleconferencing also links individuals at different locations in real time; participants can exchange information and ideas. Teleconferencing is used in the human service delivery system to conduct meetings, introduce professional development activities, and share innovative programs.

Many human service organizations are also now using the Internet and the World Wide Web to broaden their audiences. One way that this medium can be used to improve communications is through a **listserv**—a bulletin board or discussion group established by individuals or organizations, available to anyone having a special interest in the topic addressed. For example, case managers working in child welfare might participate in a listserv to share ideas and provide support for colleagues with similar work experience and common concerns.

Another popular medium of exchange is through a **home page** on the World Wide Web. Individuals or organizations establish a web site that provides particular information to readers. One home page of interest is that of the Human Service Alliance. It provides information about the Alliance's mission, goals, and services. Text, pictures, sound, and video can be put into a web site. Usually, the content is a combination of what the individual or organization wants to tell its readers and what the organization thinks that its readers might want to know.

In Bombay, India, the Coordination Committee for Vulnerable Children (CCVC) has established computer links with the state government, the Municipal Corporation, police, and community organizations to coordinate a special program (Scaria, 1993). These organizations work together to provide programming for children in jeopardy. The target population includes street children and others who are considered disadvantaged. A primary task of this project is to collect information about these children and to track them over time. A secondary activity is to coordinate the many agencies who come into contact with these children, helping them cooperate in identification and provision of services.

Many computer-assisted programs have been developed for use as implementation strategies with clients. These include ways of interviewing clients, use of debit cards for clients to buy services and goods, counseling and therapeutic programs, and skill-building educational tools. Many human service professionals worry that "high tech" may get in the way of relationship building and the development of rapport and trust. A saying has evolved from this concern: "high tech–high touch." It means that the personal element must be retained when computers are used as part of an implementation strategy.

The use of technology in implementation is exemplified by the development of teleshopping in the United Kingdom (Cahill, 1993). This service was designed for the inner-city elderly and persons with disabilities, who, for a variety of reasons, could not easily travel to shop in the suburbs, where there are more quality options and lower prices. This service is a partnership between the supermarket chain, the local city council, and a nearby university. Terminals were located in inner-city homes for the aged, day centers, and libraries. Those who could not travel to a terminal could place their orders by phone. Each customer was given a catalog; he or she could order by using a product code and entering the location of a terminal outlet near the inner-city location where the products could be retrieved. This successful program has been expanded to meet more specialized needs of other populations (Cahill, 1993).

OPTEXT Adventure System is a good example of learning software developed with the "high tech–high touch" approach to providing direct services. This computer game has as its target audience children who are preparing for another family placement after an adoption has failed. This game is played by the child and the case manager, with the purpose of experiencing some of the difficulties that occur within families. The program provides multiple scenarios, based on the particular needs of the child. It presents a problem and several alternative solutions, then displays the results of choices made. It is used to illuminate decision making in the family context (Cowan, 1994).

Computer-assisted learning packages are also used to train helping professionals. As early as the 1970s, programs were developed to teach basic skills in communication and interviewing. Later, software was developed to help beginning helpers explore their values and interests. Today's software is more sophisticated, able to provide training and simulations that are more complex and true to life. Examples are poverty policy software, complete with a database on violent crime; a computer simulation of the criminal justice system; counseling simulations with an interactive videodisc approach; videodisc programs for crisis counseling and organizational assessment; a computer-based approach to child protective service training; and interactive videodiscs focusing on interpersonal practice methods (*Computers in Human Services,* 1994).

One such program is Problem Solving in Case Management (PIC): A Computer Assisted Instruction Simulation (Gray, 1994). It presents a simulation of an individual with severe mental illness; the student engages the virtual client in the case management process. It was adapted from teaching models in medical education, rehabilitation counseling, and behavioral counseling education. Based on good teaching practice, the program is self-paced and nonjudgmental and uses the trial-and-error approach. Using pragmatic problem solving, it realistically portrays case management as a process with challenges and barriers.

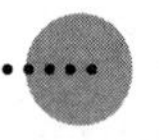

Chapter Summary

This chapter has explored modern-day case management in light of an analysis of interviews with helping professionals who do case management. The data analysis found that the key themes were knowing and learning; dealing with the bureaucracy; gathering information; and client problems and goals. Also discussed were documentation and report writing, using supervision, and the role of technology, all of which have great importance in the case management process.

Chapter Review

◆ *Key Terms*

Bureaucracy
Documentation
Supervision
Performance review
Technology
Electronic mail
Teleconferencing
Listserv
Home page

◆ *Reviewing the Chapter*

1. What are the four key themes in case management today?
2. Distinguish between the "knowing" and the "learning" of case management.
3. Why do case managers use the word *terrible* to describe the bureaucracy?
4. Discuss the interview as a form of gathering information.
5. What do case managers tell us about clients, their problems, and their goals?
6. Discuss the importance of documentation in case management.
7. How does documentation benefit the agency, the client, and the case manager?
8. Name some writing problems that plague case managers.
9. List some general guidelines that will help you with your writing.
10. Describe the roles and activities of supervisors.
11. Illustrate how you might take the initiative with regard to supervisory activities.
12. What are three reasons for conducting performance reviews?
13. What are the arguments for and against technology in human services?
14. How does computerization facilitate case management?
15. Discuss specific ways to use technology in case management.

◆ *Questions for Discussion*

1. What do you think is the most difficult problem in maintaining a positive relationship with your supervisor? Why do you think this difficulty exists?
2. After reading this chapter, what evidence can you give that technology can change how services are delivered?
3. Suppose you had case management responsibilities. What kind of plan would you devise to improve your ability to provide quality documentation?
4. What do you predict will happen to case management as the use of technology becomes more pervasive?

References

Auerbach, C., Cohen, C., Ambrose, D., Quitkin, E., & Rock, B. (1993). The design of a case management system for ALC patients: A preliminary report. In M. Leiderman, C. Guzetta, L. Struminger, & M. Monnickendam (Eds.), *Technology in people services: Research, theory, and applications* (pp. 33–56). New York: Haworth.

Austin, M., Kopp, J., & Smith, P. (1986). *Delivering human services.* New York: Longman.

Cahill, M. (1993). Computer technology and human services in the 90s: Advancing theory and practice: Teleshopping and social services in the United Kingdom. In

M. Leiderman, C. Guzetta, L. Struminger, & M. Monnickendam (Eds.), *Technology in people services: Research, theory, and applications* (pp. 231–245). New York: Haworth.

Computers in Human Services. (1994). [On-line]. Abstracts from: http://www.uta.edu/cussn/vol8.htm

Cowan, L. (1994). OPTEXT Adventure System—Software development in practice—A case history. [On-line]. *Computers in Human Services, 11*(3/4). Abstract from: http://www.uta.edu.cussn/vol11.htm

Cwikel, J., & Monnickendam, M. (1993). Factors in acceptance of advanced information technology among social workers: An exploratory study. [On-line]. *Computers in Human Services, 9*(3/4). Abstract from: http://www.uta.edu.cussn/vol9.htm

Dweck, C. S., & Wortman, C. B. (1982). Learned helplessness, anxiety and achievement motivation. In H. W. Krohne & L. Laux (Eds.), *Achievement, stress, and anxiety* (pp. 93–125). Washington, DC: Hemisphere.

Ellis, J. R., & Hartley, C. L. (1991). *Managing and coordinating nursing care.* New York: Lippincott.

Finnegan, D. J. (1996). Unraveling social workers' ambivalence toward computer technology: An analysis of the attitudes of social work students towards computers and social work practice. [On-line]. *Computers in Human Services, 13*(1). Abstract from: http://www.uta.edu.cussn/vol13.htm

Gillies, D. (1994). *Nursing management.* Philadelphia: Saunders.

Grasso, A. J., & Epstein, I. (1993). Computer technology and the human services: Does it make a difference? In M. Leiderman, C. Guzetta, L. Struminger, & M. Monnickendam (Eds.), *Technology in people services: Research, theory, and applications* (pp. 373–382). New York: Haworth.

Gray, J. I. (1993). Constructing a computer-assisted instructional package to teach case management skills. In M. Leiderman, C. Guzetta, L. Struminger, & M. Monnickendam (Eds.), *Technology in people services: Research, theory, and applications* (pp. 263–277). New York: Haworth.

Gunning, R. (1953). *The technique of clear writing.* New York: McGraw-Hill.

Hamilton, P. (1992). *Realities of contemporary nursing.* Reading, MA: Addison-Wesley Nursing/Benjamin-Cummings.

Lewis, J. A., & Lewis, M. D. (1983). *Management of human service programs.* Pacific Grove, CA: Brooks/Cole.

Looman, W. J., & Deimling, G. T. (1993). The maturation of a multiagency computerization effort for Alzheimer's respite services. In M. Leiderman, C. Guzetta, L. Struminger, & M. Monnickendam (Eds.), *Technology in people services: Research, theory, and applications* (pp. 97–110). New York: Haworth.

McClam, T., & Woodside, M. (1994a). The practitioner's voice: Case management for effective service delivery. *Human Service Education, 14*(1), 39–45.

McClam, T., & Woodside, M. (1994b). *Problem solving in the helping professions.* Pacific Grove, CA: Brooks/Cole.

Miller, P. (1988). *Powerful leadership skills for women.* Shawnee Mission, KS: National Press.

Mitchell, R. W. (1991). *Documentation in counseling records.* Alexandria, VA: American Association for Counseling and Development.

Monnickendam, M., & Eaglstein, A. S. (1993). Computer acceptance by social workers: Some unexpected research findings. In M. Leiderman, C. Guzetta, L. Struminger, & M. Monnickendam (Eds.), *Technology in people services: Research, theory, and applications* (pp. 409–453). New York: Haworth.

Mowry, M. M. (1992). Computerization and quality. In M. Johnson & J. C. McCloskey, (Eds.), *The delivery of quality health care* (pp. 153–171). St. Louis: Mosby.

Pecora, P., & Austin, M. (1987). *Managing human services personnel.* Beverly Hills, CA: Sage.

Penzias, A. A. (1993). Information technology applications, productivity in human services. In M. Leiderman, C. Guzetta, L. Struminger, & M. Monnickendam (Eds.), *Technology in people services: Research, theory, and applications* (pp. 17–31). New York: Haworth.

Rhinehart, E. L. (1969). *Management of nursing care.* New York: Macmillan.

Scaria, J. J. (1993). CCVC: An innovation in successful coordination. In M. Leiderman, C. Guzetta, L. Struminger, & M. Monnickendam (Eds.), *Technology in people services: Research, theory, and applications* (pp. 187–195). New York: Haworth.

Smith, S. (1992). *Communications in nursing.* St. Louis: Mosby.

Sundel, S. S., & Sundel, M. (1980). *Be assertive: A practical guide for human service workers.* Beverly Hills, CA: Sage.

Tallent, N. (1993). *Psychological report writing* (4th ed.). Englewood Cliffs, NJ: Prentice Hall.

Truitt, M. R. (1991). *The supervisor's handbook.* Shawnee Mission, KS: National Press.

Waters, J., Robertson, J. G., & Kerr, D. (1993). Computer-assisted drug prevention and treatment program evaluation. [On-line]. *Computers in Human Services, 9*(1/2). Abstract from: http://www.uta.edu/cussn/vol9.htm

Wilson, S. J. (1980). *Recording: Guidelines for social workers.* New York: Free Press.

Wine, J. E. (1982). Evaluation anxiety: A cognitive-construct. In H. W. Krohne & L. Laux (Eds.), *Achievement, stress, and anxiety* (pp. 207–219). Washington, DC: Hemisphere.

Woodside, M., & McClam, T. (1994). *Introduction to human services.* Pacific Grove, CA: Brooks/Cole.

Glossary

Accountability Being responsible to another person for one's actions and use of resources.

Achievement tests Tests for evaluating an individual's present level of functioning.

Active listening "Hearing" verbal and nonverbal messages, as well as what is not said—the underlying thoughts and feelings of the person communicating.

Adversarial advocacy model Model in which the advocate only supports the needs of the clients, with little concern for the context or for the needs of others involved.

Advocacy Speaking or writing in defense of a person or cause; pleading a case or standing up for people's rights.

Ageism Discrimination or unfavorable opinions based on a person's age.

Applicant An individual who requests services or who is referred for services at a human service agency.

Aptitude tests Tests that give an indication of an individual's potential for learning or acquiring a skill.

Assertiveness Expressing oneself with confidence and conviction while respecting the rights of others.

Assessment A phase of case management that involves evaluating the need or request for services and determining eligibility for services.

Assessment interview An interaction that provides information for the evaluation of an individual.

Attending behavior Ways in which a person communicates interest and attention—for instance, eye contact and attentive body language.

Authority The power to control resources and action.

Autonomy Giving the client the right to make choices.

Broker A role in which the case manager acts as a go-between, linking those who seek services and those who provide them.

Budget A numerical expression of an agency's expected income and planned expenditures for a specified period.

Burnout Emotional exhaustion resulting from the stress of interpersonal contact.

Case documentation Records of interactions and services, in various forms depending on the agency, the services offered, the length of service, and who provides the service.

Case history interview A comprehensive interview that includes open-ended questions and specific questions. It may include family history and a chronology of major life events.

Case manager A helping professional whose primary work is linking clients to services and monitoring the process.

Case review A periodic examination of a client's case.

Caseload management The process case managers use to organize their work with multiple cases. It involves setting goals, establishing priorities, managing time, and handling crises.

Chain of command The flow of authority in an agency or organization, from the position with the most authority to the one with the least.

Client An applicant whose request for services has been approved by an agency.

Client empowerment Developing the client's self-sufficiency to enable him or her to manage life without total dependence on the human service delivery system.

Confidentiality A guarantee to the client that information disclosed during the helping process will be kept in confidence.

Conflict A clash between a professional and another professional, or a client, or the agency.

Content recording An account of what was said by each participant in an interview.

Continuity of care Comprehensive care provided during and after service delivery.

Continuous improvement A movement whose purpose is improving service delivery.

Contracts Agreements that secure funding from governmental agencies or private corporations.

Cost containment The process of planning resource allocation to prevent expenditures from increasing.

Deinstitutionalization Moving clients from self-contained institutions to various community-based settings such as halfway houses.

Departmental team A small number of professionals who have similar job responsibilities and support each other's work.

Documentation Written presentation of data, observations, interviews, and services.

DSM-IV *The Diagnostic and Statistical Manual* (fourth edition), published by the American Psychological Association, which classifies all types of mental disorders.

Duty to warn Helping professionals must violate the confidentiality that has been promised a client in order to warn others that the client is a threat to self or to others.

Effective communication Verbal and nonverbal messages (through greetings, eye contact, and responses) that let the client "know" the interviewer.

Electronic mail A fast way of communicating through computers that are on similar systems.

Employment interview An assessment procedure for hiring and promotion decisions.

Empowerment The belief that all individuals, regardless of their needs, have integrity and worth. Because of this principle, the case manager places the client in a central role in case management.

Equal access to services Nondiscrimination in granting services. Because of the commitment to equal access, the case manager assumes the role of advocate and develops ways of extending services.

Ethnocentrism The belief that one's own group has desirable characteristics and others outside the group are substandard.

Expediter In this capacity, the case manager ensures that services are delivered in an efficient and effective manner.

Fees The cost of the services provided.

Goal A statement describing a broad intent or a desirable condition.

Halo effect A favorable or unfavorable early impression that biases the judgment of the observer or interviewer.

HMO Health maintenance organization—a managed care model that is very structured and controlled and emphasizes positive health promotion.

Home page A World Wide Web site that provides various information.

Implementation The stage of case management when service delivery occurs; the case manager either provides the services or oversees their delivery.

Informal structure The everyday way of doing the business of the organization, determined by personal relationships and influence and varying from traditional organizational structure and formal lines of communication.

Information and retrieval system Knowing what services are locally available to clients and accessing those services.

Intake interview A structured interview, usually occurring when a person applies for services, which is guided by a set of questions in the form of an application.

Intake summary A type of summary recording that is written at some point during the assessment phase.

Integration of services Coordinated effort on the part of many agencies and professions.

Intelligence tests Tests that measure intelligence—in the form that the test creators define the term.

Interagency feedback logs Records that provide feedback to the agencies that deliver services, to help ensure quality information and referral services.

Interdisciplinary team A team that includes professionals from various disciplines, each of which represents a service the client might receive.

Intergroup conflict A struggle between groups.

Interpersonal conflict Conflict between individuals.

Interview A face-to-face meeting between the case manager and the applicant. It may have a number of purposes, including information getting or giving, therapy, resolution of a disagreement, or the consideration of a joint undertaking.

Intragroup conflict Conflict that emerges within an established group.

Intrapersonal conflict Conflict within an individual.

Job description A general description of the work for which the case manager is held responsible.

Listserv A bulletin board or discussion group that is established by individuals or organizations on the World Wide Web.

Managed care An agreement with providers of health care and mental health care to guarantee services.

Maximum performance tests Tests on which examinees are asked to do their best at something.

Medical consultation An appointment with a physician to interpret available medical data, determine any implications for health and employment, and recommend further medical care if needed.

Medical diagnosis An appraisal of an individual's general health status, establishing whether a physical or mental impairment is present.

Mental status examination A structured interview consisting of questions designed to evaluate an individual's current mental status, considering factors such as appearance, behavior, and general intellectual processes.

Mission statement A summary of the guiding principles of an agency.

Mobilizer One who works with other community members to make available new resources or services.

Monitoring services Reviewing the services that are received by the client, the conditions that may have changed since planning, and progress toward the goals and objectives of the plan.

Objective An intended result of service provision rather than the service itself.

Open inquiries Questions that elicit broad answers, allowing the client to express thoughts, feelings, and ideas.

Organization-based case management A model of case management that focuses on ways of configuring services that are comprehensive and meet the needs of clients who have multiple problems.

Organizational chart A symbolic representation of authority, accountability, and information flow within an organization or agency.

Organizational climate The conditions of the work environment that affect how people experience their work.

Performance review A measurement of the extent to which the worker is achieving the requirements of his or her position.

Philosophy A system of thought that reflects beliefs, values, and knowledge, which form the basis for understanding specific phenomena.

Physical examination Inspection, palpation (feeling), percussion (sounding out), and auscultation (listening) by a physician. Typically, this is conducted from the skin inward through various orifices and from the top of the head to the toes.

Plan development A process that includes setting goals, deciding on objectives, and deciding on specific interventions.

Planning A phase of case management in which the case manager and client turn their attention to developing a service plan.

POS (point of service) A managed care option in which clients are encouraged to use providers in the managed care system, but do not lose all their benefits if they choose medical care outside the system.

PPO (preferred provider organization) A kind of plan that falls between the traditional HMO and the standard indemnity health insurance plan.

Pretreatment review The reviewing and approval by a managed care professional of a treatment plan before service delivery.

Privileged communication A legal concept referring to the right of clients not to have their communications with a professional used in court without their consent.

Proactive advocacy The case manager's integration of advocacy into other case management responsibilities.

Process recording A narrative telling of an interaction with another individual.

Psychological evaluation An integral part of the client study process, having as its objective the understanding of the individual by measuring characteristics that pertain to behavior.

Psychological report A document that reports on the evaluation of behavioral characteristics and mental capacity.

Psychological test A device for measuring characteristics that pertain to behavior.

Quality assurance Programs that focus on developing standards of care.

Quality care An emphasis on providing the best services to the client, in terms of both effectiveness and efficiency.

Racism Discrimination or unfavorable opinions based on a person's race.

Record Any written information related to a client's case, including history, observations, examinations, diagnoses, consultations, and financial and social information.

Referral Connecting the client to a resource within the agency or at another agency.

Resource selection Choosing the individual, program, or agency to meet the client's needs.

Responsible advocacy model An advocacy model that uses methods such as negotiation, compromise, and persuasion to support clients' cases.

Responsibility-based case management A model of service delivery in which case management can be performed by the family, a supportive care network, volunteers, or the client.

Role-based case management A model of case management that centers on the roles and responsibilities the case manager is expected to perform.

Second-opinion mandates The requirement that there be consultation with a second professional before a treatment plan can be approved.

Sexism Discrimination or unfavorable opinions based on a person's gender.

Social diagnosis A systematic way in which helping professionals gather information and study the nature of client problems.

Social history The telling of the client's story in his or her own words, with guidance from the helper, reflecting the client's life and individual characteristics.

Social service directory A catalog listing the problems handled and services delivered by other agencies.

Sources of error Potential biases that affect an interview's reliability and validity.

Staff notes Comments written at the time of each visit, contact, or interaction that any helping professional has with a client.

Structured clinical interview An interview consisting of specific questions, asked in a designated order.

Structured interviews Directive and focused interviews, usually guided by a form or a set of questions that elicit specific information.

Supervision Overseeing another's work, evaluating his or her performance, and approving or correcting it.

Teamwork Professionals sharing responsibility for clients.

Teleconferencing A method of communication that links different locations in real time; participants can exchange information and ideas.

Test A measurement device.

Treatment team A group of professionals who meet to review client problems, evaluate information gathered, and make recommendations about priorities, goals, and expected outcomes.

Violence Any physically or verbally assaultive behaviors, including harm to oneself or others or the destruction or damaging of property.

Word root The main part or stem of a word.

Index

Accountability, 305, 355, 33
Achievement tests, 159
Adversarial advocacy model, 231
Advocacy, 28, 35, 44, 66, 231–236, 280, 295–297, 307, 320, 334
  client problems, 233
  guidelines, 234–235
  models, 231
Advocate, 27, 63, 74, 82
African American clients, 117
Aftercare, 42
Ageism, 114–115
Alexander, Mark, 170, 260–261, 307
Applicant, 9, 12–14
Application, 9–10, 16, 89–101
  evaluation of, 98–101
Aptitude tests, 159
Assertiveness, 319–325
  definition, 319
  guidelines, 321–322
Assessment, 24, 35–36, 48, 73, 74–75, 77, 83, 87–108, 141–142, 334
  application, 89–101
  definition, 88
  documentation, 101–108
  eligibility, 88, 92, 100–101
  information, 16–19
  initial contact, 9–16, 25
  stage, 87–108
Assessment interview, 153
Assessor, 70
Attending behavior, 122
Autonomy, 291–295
  client preference, 293–295

Badaines, Leslie, 8, 302
Barriers, 114–115
Baum, Rosalyn, 91, 280
Bourque, Suzy, 29, 33, 123, 279–280, 317
Broker, 47, 64, 69, 70, 71–72, 73, 74, 77, 82, 150, 222–223, 226
Brown, Carolyn, 87
Budgets, 262–265
Bureaucracy, 332–333
Burnout, 303–306
  definition, 303
  personality traits, 305
  prevention, 306

Care coordination, 4, 48
Care coordinator, 39–40, 41
Case assignment, 101
Case file, 171–207
  educational, 197, 207
  medical, 171–181
  psychological, 181–191
  social, 191–197, 198–207
  vocational, 207–214
Case history interview, 154
Caseload management, 4, 316
Case management
  definition, 3, 4–5, 6, 35, 38, 51
  ethics, 308
  goal, 5, 26–30, 307–308
  history, 41–51
  principles, 26–30, 307
  process, 7–24, 35–41, 51
Case management models, 69–82
Case management process, 7–24, 35–41, 51
  assessment, 3, 4, 9–19, 24, 25
  implementation, 22–24
  planning, 20–22, 24, 25
Case management today, 331–334
  bureaucracy, 332–333
  themes, 331–334
Case manager, 39–40
  assertive, 321–325
  attributes, 344–345
  client as, 81–82
  family as, 78
  job titles, 4–5, 6–7, 39–40
  roles, 63–68, 69
  volunteer as, 81
Case notes, 99, 102, 105–108. *See also* Staff notes
Case report, 74

Case review, 24
Casita Maria Settlement House, 7–8, 88, 89–90, 112, 115–116, 120–121, 139, 217, 218, 301, 302
Chain of command, 253
Clarification, 132
Client as care manager, 81–82
Client empowerment. *See* Empowerment
Client participation, 17, 20–21, 24–25, 38–39, 220, 291, 307
  initial contact, 17
  planning, 20–21
Client preferences, 220–221, 293–295
Clients
  African American, 117
  elderly, 118
  Latino, 117
  Native American, 116–117
  violent, 283–285
Client satisfaction, 270
Closed inquiries, 126, 136, 154–155
Code of ethics, 252
Collaborator, 64, 65, 70, 74
Colleague, 64, 65, 74, 77
Communication skills, 118–135, 343
  assertive, 319–325
  confrontation, 322–323
  congruence, 119–121
  listening, 117, 119, 338
  questioning, 119, 123–131
Community organizer, 65
Community resources, 23, 150–153, 218, 220
  referral, 221–226
  selection, 220–221
Comprehensive service center, 75–76
Computers, 348–350
Confidentiality, 98, 116, 285, 288
Conflict, 243–244
  stages, 244–245
Confrontation, 133, 322–323
Consultant, 65, 77
Continuity of care, 26–27, 37–38, 42, 307
Coordination, 70, 73, 77, 78
Coordinator, 63–64
Cost-benefit analysis, 41
Cost containment, 22–23, 56, 70, 72–73
Cost effectiveness, 38
Counselor, 65–66, 70, 72
Cultural sensitivity, 116–118, 119–120, 155

Data gathering, 70, 88, 98–99, 153–165, 333
  interviewing for, 153–156
  questions, 112
  testing, 156–165
"Daughter from California" syndrome, 281–283
Deinstitutionalization, 35–47
Departmental team, 237
Diagnostic and Statistical Manual of Mental Disorders, 185
Documentation, 14, 18, 24–25, 68, 88, 99, 101–108, 142, 220
  benefits, 334–335
  computers, 349–350

East Tennessee Community Health Agency, 7, 26, 34, 170, 237, 249, 250, 307, 309, 330
Educational information, 197–207
Education for All Handicapped Children Act of 1975, 48–50, 56
Effectiveness, 316–317
Efficiency, 316–317
Ehlers, Kim, 30, 62, 236–237, 250, 252, 258, 262–263, 316
Elderly clients, 118
Electronic mail, 354
Eligibility, 9, 17, 18–19, 20, 24, 27, 88, 92, 100–101, 332
Employment interview, 154–155
Empowerment, 25, 29–30, 35, 48, 77, 291, 308, 320–321, 334, 344
Equal access to services, 27
Ethical decision making, 282–283
Ethical issues, 279–299
  autonomy, 291–295
  confidentiality, 285–288
  duty to warn, 288–291
  exceptions, 295–297
  family disagreements, 281–283
  violent clients, 283–285
Ethics
  confidentiality, 308
  self-determination, 308
  self-worth, 308
  values, 308–309
Ethnocentrism, 114–116
Evaluation, 30, 48, 49, 70, 307
  of quality, 35
Evaluator, 66, 70, 74, 77, 82
Expediter, 66–67, 70, 74, 77
Expenditures, 264–265

Families as case managers, 78
Family Counseling Agency, 29, 33, 242, 280
Family disagreements, 281–283
Family service center, 87
Family Support Act of 1988, 48, 50–51
Fee for service, 51–52, 53

Fees, 265–266
Flecha, Sandra, 89–90, 112, 301

Gatekeepers, 78, 81
Gatekeeping, 280
Generalist, 70, 71
Goals, 143–146, 147–148, 334–335
  definition, 144
  writing, 145
Grants and contracts, 205

Halo effect, 155
Health maintenance organization, 52, 54–55, 57
Health Maintenance Organization Act of 1973, 52
Helen Ross McNabb Center, Inc., 2, 3, 27, 91, 254, 280, 330
Henry Street Settlement House, 44
History, 34, 41–51
  legislation, 48–51
  organizations, 42–43
  Red Cross, 44, 46–47
  settlement houses, 43–46
HMO. *See* health maintenance organization
Hudson, Paula, 3, 27, 91, 254, 330
Hull House, 43–45
Human development liaison specialist, 4

Implementation, 22–23, 50, 66, 77, 83, 150, 355–356
  service coordination, 17, 22, 35, 43, 47, 48
  service provision, 22, 39
Improving services, 267–275
  continuous improvement, 270–275
  quality analysis, 268–269
  quality assurance, 269–270
  utilization review, 268–269
Individual Family Service Plan, 78–80
Individualized education program (IEP), 49–50
Individuals with Disabilities Education Act, 48–50
Informal structure, 258–260
Information and referral, 150–153
  setting up, 152–153
Information and referral system, 220–226
Information assessment, 17, 98–100
Information evaluation, 17, 42
Information gathering, 17, 20, 46
Information management, 42, 43–44, 68
Information systems, 350–353
Informing, 133
Initial contact, 9–16, 88, 92
Intake interview, 70, 75, 97, 101, 112
Intake summary, 102, 104–105, 106
Integration of services, 26, 38, 44, 48, 307
Intelligence tests, 159
Interagency coordination, 43
Interagency feedback log, 151
Interdisciplinary team, 76, 238, 283, 285
Intergroup conflict, 244
Interpersonal conflict, 244
Interpretation, 133
Interview, 88, 92–98, 333
  content, 94
  definition, 92, 94
  objectives, 92–93
  outcomes, 93
  process, 94
  reliability, 156
  sources of error, 155–156
  structure, 96–97
  types, 153–156
Interviewer attitudes and characteristics, 113–118
  respect, 114
  self-awareness, 113
Interviewing, 114–136, 335
  barriers, 14–115
  pitfalls, 135–136
  skills, 103, 120–135
Interviewing skills, 103, 120–135
  listening, 117, 121–123
  questioning, 123–131
  responding, 131–135
Intragroup conflict, 244
Intrapersonal conflict, 244

Jaffe, Roz, 2, 3, 26–27, 91–92, 259, 339
Jewish Home and Hospital for the Aged, 2, 3, 26–27, 91–92, 239–240, 339
Job description, 252–253

Lanning, Margaret, 100, 111–112, 169
La Rue, Angela, 62, 70, 266, 329–330
Latino clients, 117
Legislation, 48–51
  Education for All Handicapped Children Act of 1975, 48–50, 56
  Family Support Act of 1988, 48, 50–51
  Individuals with Disabilities Education Act, 48–50
  Older Americans Act of 1965, 48
Listening, 117, 121–123, 338
  active, 119
  attending behavior, 122
  guidelines, 122–123
  S-O-L-E-R, 122
Listserv, 354
Lowe, Cathy, 140, 301

Managed care, 6, 34, 40–41, 51–57, 343
advantages, 56
definition, 52, 53–54
disadvantages, 56–57
history, 51–53
models, 54–57
Massachusetts School of Idiotic and Feebleminded Youth, 42
Medical consultation, 172
Medical diagnosis, 172–173
Medical information, 11–12, 15, 17, 171–181
exams, 157–175
report, 174–177
terminology, 175–181
Mental Measurements Yearbook, 161
Mental status examination, 97, 154
Mikol, Margaret, 1, 3, 30, 99–100, 140
Minimal responses, 131
Mission statement, 251–252
Mitchell Area Adjustment Training Center, 30, 63, 236–237, 250, 262–263, 316
Mobilizer, 225, 226
Monitoring services, 23, 226–229
Morgan, Jana Berry, 2, 3, 254, 258–259, 280
Multi-service center, 75

Native American clients, 116–117
Networking, 3, 21–22, 23, 77

Objectives, 146–150
Older Americans Act of 1965, 48
Open inquiries, 126–130, 154–155
Organizational chart, 254–255
Organizational climate, 260–261
Organizational structure, 251–261
informal, 258–260
mission, 251–252
organizational chart, 254
policies, 252
Organization-based case management, 69, 73–77
Outpatient services, 42

Paraphrase, 131
Parole office, 330
Performance appraisal, 341
Performance review, 343–348
Philosophy, 306–310
Pima Health System, 218, 252, 263, 279, 321
Plan, 73, 74–75
Plan development, 77, 83, 143–150, 334
goals, 143–146, 147–148
objectives, 146–150
Planner, 67, 70, 74, 77, 82
Planning, 20–22, 37, 39, 49, 67, 70, 73, 140–150
continuity, 37–38
data gathering, 153–165
information and referral, 150–153
plan development, 143–150
Point of service, 52, 54, 56, 57
Policies and procedures, 252
POS. *See* Point of service
PPO. *See* Preferred provider organization
Prassa, Thom, 7, 34, 237, 249, 260–261, 309, 330
Preferred provider organization, 52, 54, 55, 57, 269
Pretreatment review, 269
Principles of case management, 26–30, 307
Private giving, 266
Private Industry Council, 28, 29, 88, 94–95, 111, 140–141, 301
Privileged communication, 98
Proactive advocacy, 235
Problem identification, 333–334
Problem solver, 67–68, 70, 74, 77
Process recording, 102–104
Psychological evaluation, 181–191
process, 184–191
referral, 182–184
reports, 185, 191
Psychological report, 185–191
Psychological tests, 157–165
Psychosocial rehabilitative center, 76–77

Quality assurance, 27, 35, 53
Quality care, 27
Questioning, 123–131
advantages, 124–125
closed and open inquiries, 126–130, 136
disadvantages, 125–126
Questions, 112, 119, 153–155
closed, 136

Racism, 114–115
Recordkeeper, 68, 70, 74, 77
Recordkeeping, 42, 43, 44, 101–108
process recording, 102–105
summary recording, 102, 104
Red Cross, 44, 46–47
Referral, 3, 23, 70, 72, 90, 92, 220, 221–226
broker, 222–223
Reflection, 131–132
Reliability, 162
Report writing, 24, 101–108, 334–338
Resource allocation, 262–267
budgets, 262–265

Resource allocation *(continued)*
expenditures, 264–265
revenue sources, 265–267
Resources, 75
Responding, 131–135
Responsibility-based case management, 69–70, 77–82
Responsible advocacy model, 231
Resto, Zulma, 120–121, 302
Revenue, 265–267
Rivera, Marta, 88, 115–116
Rodriquez, Teresa, 305
Role-based case management, 69, 70–72
Roles, 63–68, 69

Second-opinion mandates, 269
Self-interview, 103
Service coordination, 3, 4, 17, 22, 35, 43, 47, 48, 218–231, 305
advantages, 219
broker, 222–223
client participation, 220
documentation, 220
guidelines, 223–226
referral, 220, 222–226
Service integration, 75
Service monitor, 68, 70, 73, 74, 77
Service provision, 22, 39
Settlement houses, 43–46
Sexism, 114–115
Sick Kids Need Involved People of New York, 1, 3, 30, 99–100, 112, 140, 169
Single point of access, 70, 73
Slater, Judith, 61, 217–218, 249, 250, 253, 320
Smith, Linda, 90, 121, 123, 140, 170, 306
Social diagnosis, 46
Social history, 191–197, 198–207, 336
guidelines, 191–192
Social service directory, 151
S-O-L-E-R, 122
Staffing, 24
Staff notes, 102, 105–108, 220, 336–337. *See also* Case notes
guidelines, 336–337
Standards for Educational and Psychological Testing, 161
Structured clinical interview, 153–154
Structured interview, 97
Stueck, Janelle, 28, 29, 88, 94–95, 100, 111, 139–140, 141
Summarizing, 132
Summary recording, 102, 104
Supervision, 260, 338–348
communication, 343
Supervisor responsibilities, 338–343
Supportive care, 78–80
System modifier, 68, 74, 77

Tarasoff court decision, 288–289
Team leaders, 238–240
Teams, 76, 236–242, 283, 285
leaders, 238–240
types, 237–238
Teamwork, 3, 23, 64, 76, 77, 236–242
barriers, 240–242
Technology, 348–356
implementation, 355–356
Teleconferencing, 354
Tennessee School for the Deaf, 8, 90, 121, 123, 140, 170, 232–233, 302, 306
Termination, 25, 73
Testing, 156–165, 184
administration, 163–164
interpretation, 164–165
language, 169
resources, 161–162
selection, 161–163
Tests, 156–165
definition, 157–158
selection, 161–163
types, 159–160
Tests in Print, 161
Third Avenue Family Service Center, 28, 33, 231, 305, 343
Tiano, Alan, 218, 279
Time management, 311–319
assessment, 312–313
guidelines, 314–315
Treatment team, 236–237

Unstructured interview, 97–98
Utilization review, 40

Validity, 162
Vega, Yolanda, 7–8, 139, 217
Violent clients, 283–285
Vocational information, 207–214
Volunteers, 76, 77, 78, 81
as case managers, 81

Washington, Linda, 28
Working with others, 63, 65, 229–246. *See also* Teams; Teamwork
communication, 230
role models, 242–243
support, 242–246
World Wide Web, 354
Writing principles, 338
Writing skills, 336–338

Photo Credits

TO THE OWNER OF THIS BOOK:

We hope that you have found *Generalist Case Management: A Method of Human Service Delivery* useful. So that this book can be improved in a future edition, would you take the time to complete this sheet and return it? Thank you.

School and address: ______________________________

Department: ______________________________

Instructor's name: ______________________________

1. What I like most about this book is: ______________________________

____

____

2. What I like least about this book is: ______________________________

____

____

3. My general reaction to this book is: ______________________________

____

4. The name of the course in which I used this book is: ______________________________

____

5. Were all of the chapters of the book assigned for you to read? ______________

If not, which ones weren't? ______________________________

6. In the space below, or on a separate sheet of paper, please write specific suggestions for improving this book and anything else you'd care to share about your experience in using the book.

____

____

____

____

____

Optional:

Your name: ______________________ Date: ______________________

May Brooks/Cole quote you, either in promotion for *Generalist Case Management: A Method of Human Service Delivery* or in future publishing ventures?

Yes: _________ No: _________

Sincerely,

Marianne Woodside
Tricia McClam

FOLD HERE

NO POSTAGE NECESSARY IF MAILED IN THE UNITED STATES

BUSINESS REPLY MAIL

FIRST CLASS PERMIT NO. 358 PACIFIC GROVE, CA

POSTAGE WILL BE PAID BY ADDRESSEE

ATT: *Marianne Woodside & Tricia McClam*

Brooks/Cole Publishing Company
511 Forest Lodge Road
Pacific Grove, California 93950-9968

FOLD HERE

Brooks/Cole Publishing is dedicated to publishing quality books for the helping professions. If you would like to learn more about our publications, please use this mailer to request our catalogue.

Name: ____

Street Address: ____________________________________

City, State, and Zip: ________________________________

FOLD HERE

BUSINESS REPLY MAIL

FIRST CLASS PERMIT NO. 358 PACIFIC GROVE, CA

POSTAGE WILL BE PAID BY ADDRESSEE

ATT: *Human Services Catalogue*

Brooks/Cole Publishing Company
511 Forest Lodge Road
Pacific Grove, California 93950-9968

FOLD HERE